I0041808

MATHEMATICS

Its Content, Methods, and Meaning

VOLUME ONE

MATHEMATICS

Its Content, Methods, and Meaning

EDITED BY

A. D. Aleksandrov, A. N. Kolmogorov, M. A. Lavrent'ev

TRANSLATED BY

S. H. Gould and T. Bartha

THE M.I.T. PRESS
Massachusetts Institute of Technology
Cambridge, Massachusetts

MIT Press

0262510057

ALEKSANDROV
MATH CONT V1

М А Т Е М А Т И К А
ЕЕ СОДЕРЖАНИЕ, МЕТОДЫ И ЗНАЧЕНИЕ
Издательство Академии Наук СССР
Москва 1956

Sixth Printing, 1989

First MIT Press paperback edition, 1969

Translation aided by grant NSF–G 16422 from the
National Science Foundation

Copyright © 1963 by the American Mathematical Society

All rights reserved. No portion of this book may be reproduced
without the written permission of the publisher.

Library of Congress Card Number: 64–7547
Printed in the United States of America

PREFACE TO
THE RUSSIAN EDITION

Mathematics, which originated in antiquity in the needs of daily life, has developed into an immense system of widely varied disciplines. Like the other sciences, it reflects the laws of the material world around us and serves as a powerful instrument for our knowledge and mastery of nature. But the high level of abstraction peculiar to mathematics means that its newer branches are relatively inaccessible to nonspecialists. This abstract character of mathematics gave birth even in antiquity to idealistic notions about its independence of the material world.

In preparing the present volume, the authors have kept in mind the goal of acquainting a sufficiently wide circle of the Soviet intelligentsia with the various mathematical disciplines, their content and methods, the foundations on which they are based, and the paths along which they have developed.

As a minimum of necessary mathematical knowledge on the part of the reader, we have assumed only secondary-school mathematics, but the volumes differ from one another with respect to the accessibility of the material contained in them. Readers wishing to acquaint themselves for the first time with the elements of higher mathematics may profitably read the first few chapters, but for a complete understanding of the subsequent parts it will be necessary to have made some study of corresponding textbooks. The book as a whole will be understood in a fundamental way only by readers who already have some acquaintance with the applications of mathematical analysis; that is to say, with the differential and integral calculus. For such readers, namely teachers of mathematics and instructors in engineering and the natural sciences, it will be particularly important to read those chapters which introduce the newer branches of mathematics.

v

Naturally it has not been possible, within the limits of one book, to exhaust all the riches of even the most fundamental results of mathematical research; a certain freedom in the choice of material has been inevitable here. But along general lines, the present book will give an idea of the present state of mathematics, its origins, and its probable future development. For this reason the book is also intended to some extent for persons already acquainted with most of the factual material in it. It may perhaps help to remove a certain narrowness of outlook occasionally to be found in some of our younger mathematicians.

The separate chapters of the book are written by various authors, whose names are given in the Contents. But as a whole the book is the result of collaboration. Its general plan, the choice of material, the successive versions of individual chapters, were all submitted to general discussion, and improvements were made on the basis of a lively exchange of opinions. Mathematicians from several cities in the Soviet Union were given an opportunity, in the form of organized discussion, to make many valuable remarks concerning the original version of the text. Their opinions and suggestions were taken into account by the authors.

The authors of some of the chapters also took a direct share in preparing the final version of other chapters: The introductory part of Chapter II was written essentially by B. N. Delone, while D. K. Faddeev played an active role in the preparation of Chapter IV and Chapter XX.

A share in the work was also taken by several persons other than the authors of the individual chapters: §4 of Chapter XIV was written by L. V. Kantorovič, §6 of Chapter VI by O. A. Ladyženskaja, §5 of Chapter 10 by A. G. Postnikov; work was done on the text of Chapter V by O. A. Oleĭnik and on Chapter XI by Ju. V. Prohorov.

Certain sections of Chapters I, II, VII, and XVII were written by V. A. Zalgaller. The editing of the final text was done by V. A. Zalgaller and V. S. Videnskiĭ with the cooperation of T. V. Rogozkinaja and A. P. Leonovaja.

The greater part of the illustrations were prepared by E. P. Sen'kin.

Moscow
1956   EDITORIAL BOARD

FOREWORD BY THE
EDITOR OF THE TRANSLATION

Mathematics, in view of its abstractness, offers greater difficulty to the expositor than any other science. Yet its rapidly increasing role in modern life creates both a need and a desire for good exposition.

In recent years many popular books about mathematics have appeared in the English language, and some of them have enjoyed an immense sale. But for the most part they have contained little serious mathematical instruction, and many of them have neglected the twentieth century, the undisputed "golden age" of mathematics. Although they are admirable in many other ways, they have not yet undertaken the ultimate task of mathematical exposition, namely the large-scale organization of modern mathematics in such a way that the reader is constantly delighted by the obvious economizing of his own time and effort. Anyone who reads through some of the chapters in the present book will realize how well this task has been carried out by the Soviet authors, in the systematic collaboration they have described in their preface.

Such a book, written for "a wide circle of the intelligentsia," must also discuss the general cultural importance of mathematics and its continuous development from the earliest beginnings of history down to the present day. To form an opinion of the book from this point of view the reader need only glance through the first chapter in Part 1 and the introduction to certain other chapters; for example, Analysis, or Analytic Geometry.

In translating the passages on the history and cultural significance of mathematical ideas, the translators have naturally been aware of even greater difficulties than are usually associated with the translation of scientific texts. As organizer of the group, I express my profound gratitude to the other two translators, Tamas Bartha and Kurt Hirsch, for their skillful cooperation.

The present translation, which was originally published by the American Mathematical Society, will now enjoy a more general distribution in its new format. In thus making the book more widely available the Society has been influenced by various expressions of opinion from American mathematicians. For example, ". . . the book will contribute materially to a better understanding by the public of what mathematicians are up to. . . . It will be useful to many mathematicians, physicists and chemists, as well as to laymen. . . . Whether a physicist wishes to know what a Lie algebra is and how it is related to a Lie group, or an undergraduate would like to begin the study of homology, or a crystallographer is interested in Fedorov groups, or an engineer in probability, or any scientist in computing machines, he will find here a connected, lucid account."

In its first edition this translation has been widely read by mathematicians and students of mathematics. We now look forward to its wider usefulness in the general English-speaking world.

August, 1964

S. H. GOULD
Editor of Translations
American Mathematical Society
Providence, Rhode Island

CONTENTS

ix

PART 2

PART 1

CHAPTER I

A GENERAL VIEW
OF MATHEMATICS

An adequate presentation of any science cannot consist of detailed information alone, however extensive. It must also provide a proper view of the essential nature of the science as a whole. The purpose of the present chapter is to give a general picture of the essential nature of mathematics. For this purpose there is no great need to introduce any of the details of recent mathematical theories, since elementary mathematics and the history of the science already provide a sufficient foundation for general conclusions.

§1. The Characteristic Features of Mathematics

1. Abstractions, proofs, applications.   With even a superficial knowledge of mathematics, it is easy to recognize certain characteristic features: its abstractness, its precision, its logical rigor, the indisputable character of its conclusions, and finally, the exceptionally broad range of its applications.

The abstractness of mathematics is easy to see. We operate with abstract numbers without worrying about how to relate them in each case to concrete objects. In school we study the abstract multiplication table, that is, a table for multiplying one abstract number by another, not a number of boys by a number of apples, or a number of apples by the price of an apple.

Similarly in geometry we consider, for example, straight lines and not stretched threads, the concept of a geometric line being obtained by abstraction from all other properties, excepting only extension in one direction. More generally, the concept of a geometric figure is the result of abstraction from all the properties of actual objects except their spatial form and dimensions.

Abstractions of this sort are characteristic for the whole of mathematics. The concept of a whole number and of a geometric figure are only two of the earliest and most elementary of its concepts. They have been followed by a mass of others, too numerous to describe, extending to such abstractions as complex numbers, functions, integrals, differentials, functionals, n-dimensional, and even infinite-dimensional spaces, and so forth. These abstractions, piled up as it were on one another, have reached such a degree of generalization that they apparently lose all connection with daily life and the "ordinary mortal" understands nothing about them beyond the mere fact that "all this is incomprehensible."

In reality, of course, the case is not so at all. Although the concept of n-dimensional space is no doubt extremely abstract, yet it does have a completely real content, which is not very difficult to understand. In the present book it will be our task to emphasize and clarify the concrete content of such abstract concepts as those mentioned earlier, so that the reader may convince himself that they are all connected with actual life, both in their origin and in their applications.

But abstraction is not the exclusive property of mathematics; it is characteristic of every science, even of all mental activity in general. Consequently, the abstractness of mathematical concepts does not in itself give a complete description of the peculiar character of mathematics.

The abstractions of mathematics are distinguished by three features. In the first place, they deal above all else with quantitative relations and spatial forms, abstracting them from all other properties of objects. Second, they occur in a sequence of increasing degrees of abstraction, going very much further in this direction than the abstractions of other sciences. We will illustrate these two features in detail later, using as examples the fundamental notions of number and figure. Finally, and this is obvious, mathematics as such moves almost wholly in the field of abstract concepts and their interrelations. While the natural scientist turns constantly to experiment for proof of his assertions, the mathematician employs only argument and computation.

It is true that mathematicians also make constant use, to assist them in the discovery of their theorems and methods, of models and physical analogues, and they have recourse to various completely concrete examples. These examples serve as the actual source of the theory and as a means of discovering its theorems, but no theorem definitely belongs to mathematics until it has been rigorously proved by a logical argument. If a geometer, reporting a newly discovered theorem, were to demonstrate it by means of models and to confine himself to such a demonstration, no mathematician would admit that the theorem had been proved. The demand for a proof of a theorem is well known in high school geometry, but it pervades the whole of mathematics. We could measure the angles at the base of a thousand isosceles triangles with extreme accuracy, but such a procedure would never provide us with a mathematical proof of the theorem that the angles at the base of an isosceles triangle are equal. Mathematics demands that this result be deduced from the fundamental concepts of geometry, which at the present time, in view of the fact that geometry is nowadays developed on a rigorous basis, are precisely formulated in the axioms. And so it is in every case. To prove a theorem means for the mathematician to deduce it by a logical argument from the fundamental properties of the concepts occurring in that theorem. In this way, not only the concepts but also the methods of mathematics are abstract and theoretical.

The results of mathematics are distinguished by a high degree of logical rigor, and a mathematical argument is conducted with such scrupulousness as to make it incontestable and completely convincing to anyone who understands it. The scrupulousness and cogency of mathematical proofs are already well known in a high school course. Mathematical truths are in fact the prototype of the completely incontestable. Not for nothing do people say "as clear as two and two are four." Here the relation "two and two are four" is introduced as the very image of the irrefutable and incontestable.

But the rigor of mathematics is not absolute; it is in a process of continual development; the principles of mathematics have not congealed once and for all but have a life of their own and may even be the subject of scientific quarrels.

In the final analysis the vitality of mathematics arises from the fact that its concepts and results, for all their abstractness, originate, as we shall see, in the actual world and find widely varied application in the other sciences, in engineering, and in all the practical affairs of daily life; to realize this is the most important prerequisite for understanding mathematics.

The exceptional breadth of its applications is another characteristic feature of mathematics.

In the first place we make constant use, almost every hour, in industry and in private and social life, of the most varied concepts and results of mathematics, without thinking about them at all; for example, we use arithmetic to compute our expenses or geometry to calculate the floor area of an apartment. Of course, the rules here are very simple, but we should remember that in some period of antiquity they represented the most advanced mathematical achievements of the age.

Second, modern technology would be impossible without mathematics. There is probably not a single technical process which can be carried through without more or less complicated calculations; and mathematics plays a very important role in the development of new branches of technology.

Finally, it is true that every science, to a greater or lesser degree, makes essential use of mathematics. The "exact sciences," mechanics, astronomy, physics, and to a great extent chemistry, express their laws, as every schoolboy knows, by means of formulas and make extensive use of mathematical apparatus in developing their theories. The progress of these sciences would have been completely impossible without mathematics. For this reason the requirements of mechanics, astronomy, and physics have always exercised a direct and decisive influence on the development of mathematics.

In other sciences mathematics plays a smaller role, but here too it finds important applications. Of course, in the study of such complicated phenomena as occur in biology and sociology, the mathematical method cannot play the same role as, let us say, in physics. In all cases, but especially where the phenomena are most complicated, we must bear in mind, if we are not to lose our way in meaningless play with formulas, that the application of mathematics is significant only if the concrete phenomena have already been made the subject of a profound theory. In one way or another, mathematics is applied in almost every science, from mechanics to political economy.

Let us recall some particularly brilliant applications of mathematics in the exact sciences and in technology.

The planet Neptune, one of the most distant in the Solar System, was discovered in the year 1846 on the basis of mathematical calculations. By analyzing certain irregularities in the motion of Uranus, the astronomers Adams and Leverrier came to the conclusion that these irregularities were caused by the gravitational attraction of another planet. Leverrier calculated on the basis of the laws of mechanics exactly where this planet must be, and an observer to whom he communicated his results caught sight of it in his telescope in the exact position indicated by Leverrier. This discovery was a triumph not only for mechanics and astronomy, and in particular for the system of Copernicus, but also for the powers of mathematical calculation.

Another example, no less impressive, was the discovery of electromagnetic waves. The English physicist Maxwell, by generalizing the laws of electromagnetic phenomena as established by experiment, was able to express these laws in the form of equations. From these equations he deduced, by purely mathematical methods, that electromagnetic waves could exist and that they must be propagated with the speed of light. On the basis of this result, he proposed the electromagnetic theory of light, which was later developed and deepened in every direction. Moreover, Maxwell's results led to the search for electromagnetic waves of purely electrical origin, arising for example from an oscillating charge. These waves were actually discovered by Hertz. Shortly afterwards, A. S. Popov, by discovering means for exciting, transmitting, and receiving electromagnetic oscillations made them available for a wide range of applications and thereby laid the foundations for the whole technology of radio. In the discovery of radio, now the common possession of everyone, an important role was played by the results of a purely mathematical deduction.

So from observation, as for example of the deflection of a magnetic needle by an electric current, science proceeds to generalization, to a theory of the phenomena, and to formulation of laws and to mathematical expression of them. From these laws come new deductions, and finally, the theory is embodied in practice, which in its turn provides powerful new impulses for the development of the theory.

It is particularly remarkable that even the most abstract constructions of mathematics, arising within that science itself, without any immediate motivation from the natural sciences or from technology, nevertheless have fruitful applications. For example, imaginary numbers first came to light in algebra, and for a long time their significance in the actual world remained uncomprehended, a circumstance indicated by their very name. But when about 1800 a geometrical interpretation (see Chapter IV, §2) was given to them, imaginary numbers became firmly established in mathematics, giving rise to the extensive theory of functions of a complex variable, i.e., of a variable of the form $x + y \sqrt{-1}$. This theory of "imaginary" functions of an "imaginary" variable proved itself to be far from imaginary, but rather a very practical means of solving technological problems. Thus, the fundamental results of N. E. Jukovski concerning the lift on the wing of an airplane are proved by means of this theory. The same theory is useful, for example, in the solution of problems concerning the oozing of water under a dam, problems whose importance is obvious during the present period of construction of huge hydroelectric stations.

Another example, equally impressive, is provided by non-Euclidean geometry,* which arose from the efforts, extending for 2000 years from the time of Euclid, to prove the parallel axiom, a problem of purely

---

* Here we merely point out this example without further explanation, for which the reader may turn to Chapter XVII.

mathematical interest. N. I. Lobačevskiĭ himself, the founder of the new geometry, was careful to label his geometry "imaginary," since he could not see any meaning for it in the actual world, although he was confident that such a meaning would eventually be found. The results of his geometry appeared to the majority of mathematicians to be not only "imaginary" but even unimaginable and absurd. Nevertheless, his ideas laid the foundation for a new development of geometry, namely the creation of theories of various non-Euclidean spaces; and these ideas subsequently became the basis of the general theory of relativity, in which the mathematical apparatus consists of a form of non-Euclidean geometry of four-dimensional space. Thus the abstract constructions of mathematics, which at the very least seemed incomprehensible, proved themselves a powerful instrument for the development of one of the most important theories of physics. Similarly, in the present-day theory of atomic phenomena, in the so-called quantum mechanics, essential use is made of many extremely abstract mathematical concepts and theories, as for example the concept of infinite-dimensional space.

There is no need to give any further examples, since we have already shown with sufficient emphasis that mathematics finds widespread application in everyday life and in technology and science; in the exact sciences and in the great problems of technology, applications are found even for those theories which arise within mathematics itself. This is one of the characteristic peculiarities of mathematics, along with its abstractness and the rigor and conclusiveness of its results.

2. The essential nature of mathematics. In discussing these special features of mathematics we have been far from explaining its essence; rather we have merely pointed out its external marks. Our task now is to explain the essential nature of these characteristic features. For this purpose it will be necessary to answer, at the very least, the following questions:

What do these abstract mathematical concepts reflect? In other words, what is the actual subject matter of mathematics?

Why do the abstract results of mathematics appear so convincing, and its initial concepts so obvious? In other words, on what foundation do the methods of mathematics rest?

Why, in spite of all its abstractness, does mathematics find such wide application and does not turn out to be merely idle play with abstractions? In other words, how is the significance of mathematics to be explained?

Finally, what forces lead to the further development of mathematics, allowing it to unite abstractness with breadth of application? What is the basis for its continuing growth?

In answering these questions we will form a general picture of the content of mathematics, of its methods, and of its significance and its development; that is, we will understand its essence.

Idealists and metaphysicists not only fall into confusion in their attempts to answer these basic questions but they go so far as to distort mathematics completely, turning it literally inside out. Thus, seeing the extreme abstractness and cogency of mathematical results, the idealist imagines that mathematics issues from pure thought.

In reality, mathematics offers not the slightest support for idealism or metaphysics. We will convince ourselves of this as we attempt, in general outline, to answer the listed questions about the essence of mathematics. For a preliminary clarification of these questions, it is sufficient to examine the foundations of arithmetic and elementary geometry, to which we now turn.

§2. Arithmetic

1. The concept of a whole number. The concept of number (for the time being, we speak only of whole positive numbers), though it is so familiar to us today, was worked out very slowly. This can be seen from the way in which counting has been done by various races who until recent times have remained at a relatively primitive level of social life. Among some of them, there were no names for numbers higher than two or three; among others, counting went further but ended after a few numbers, after which they simply said "many" or "countless." A stock of clearly distinguished names for numbers was only gradually accumulated among the various peoples.

At first these peoples had no concept of what a number is, although they could in their own fashion make judgments about the size of one or another collection of objects met with in their daily life. We must conclude that a number was directly perceived by them as an inseparable property of a collection of objects, a property which they did not, however, clearly distinguish. We are so accustomed to counting that we can hardly imagine this state of affairs, but it is possible to understand it.*

At the next higher level a number already appears as a property of a

---

* In fact, every collection of objects, whether it be a flock of sheep or a pile of firewood, exists and is immediately perceived in all its concreteness and complexity. The distinguishing in it of separate properties and relationships is the result of conscious analysis. Primitive thought does not yet make this analysis, but considers the object only as a whole. Similarly, a man who has not studied music perceives a musical composition without distinguishing in it the details of melody, tonality, and so forth, while at the same time a musician easily analyzes even a complicated symphony.

collection of objects, but it is not yet distinguished from the collection as an "abstract number," as a number in general, not connected with concrete objects. This is obvious from the names of numbers among certain peoples, as "hand" for five and "wholeman" for twenty. Here five is to be understood not abstractly but simply in the sense of "as many as the fingers on a hand," twenty is "as many as the fingers and toes on a man" and so forth. In a completely analogous way, certain peoples had no concept of "black," "hard," or "circular." In order to say that an object is black, they compared it with a crow for example, and to say that there were five objects, they directly compared these objects with a hand. In this way it also came about that various names for numbers were used for various kinds of objects; some numbers for counting people, others for counting boats, and so forth, up to as many as ten different kinds of numbers. Here we do not have abstract numbers, but merely a sort of "appellation," referring only to a definite kind of objects. Among other peoples there were in general no separate names for numbers, as for example, no word for "three," although they could say "three men" or "in three places," and so forth.

Similarly among ourselves, we quite readily say that this or that object is black but much more rarely speak about "blackness" in itself, which is a more abstract concept.*

The number of objects in a given collection is a property of the collection, but the number itself, as such, the "abstract number," is a property abstracted from the concrete collection and considered simply in itself, like "blackness" or "hardness." Just as blackness is the property common to all objects of the color of coal, so the number "five" is the common property of all collections containing as many objects as there are fingers on a hand. In this case the equality of the two numbers is established by simple comparison: We take an object from the collection, bend one finger over, and count in this way up to the end of the collection. More generally, by pairing off the objects of two collections, it is possible, without making any use of numbers at all, to establish whether or not the collections contain the same number of objects. For example, if guests are taking their places at the table they can easily, without any counting, make it clear to the hostess that she has forgotten one setting, since one guest will be without a setting.

------

* In the formation of concepts about properties of objects, such as color or the numerosity of a collection, it is possible to distinguish three steps, which we must not, of course, try to separate too sharply from one another. At the first step the property is defined by direct comparison of objects: like a crow, as many as on a hand. At the second, an adjective appears: a black stone or (the numerical adjective being quite analogous) five trees. At the third step the property is abstracted from the objects and may appear "as such"; for example "blackness," or the abstract number "five."

In this way it is possible to give the following definition of a number: Each separate number like "two," "five," and so forth, is that property of collections of objects which is common to all collections whose objects can be put into one-to-one correspondence with one another and which is different for those collections for which such a correspondence is impossible. In order to discover this property and to distinguish it clearly, that is, in order to form the concept of a definite number and to give it a name "six," "ten," and so forth, it was necessary to compare many collections of objects. For countless generations people repeated the same operation millions of times and in that way discovered numbers and the relations among them.

2. Relations among the whole numbers. Operations with numbers arose in their turn as a reflection of relations among concrete objects. This is observable even in the names of numbers. For example, among certain American Indians the number "twenty-six" is pronounced as "above two tens I place a six," which is clearly a reflection of a concrete method of counting objects. Addition of numbers corresponds to placing together or uniting two or more collections, and it is equally easy to see the concrete meaning of subtraction, multiplication, and division. Multiplication in particular arose to a great extent, it seems clear, from the habit of counting off equal collections: that is, by twos, by threes, and so forth.

In the process of counting, men not only discovered and assimilated the relations among the separate numbers, as for example that two and three are five, but also they gradually established certain general laws. By practical experience, it was discovered that a sum does not depend on the order of the summands and that the result of counting a given set of objects does not depend on the order in which the counting takes place, a fact which is reflected in the essential identity of the "ordinal" and "cardinal" numbers: first, second, third, and one, two, three. In this way the numbers appeared not as separate and independent, but as interrelated with one another.

Some numbers are expressed in terms of others in their very names and in the way they are written. Thus, "twenty" denotes "two (times) ten"; in French, eighty is "four-twenties" (quatre-vingt), ninety is "four-twenties-ten"; and the Roman numerals VIII, IX denote that $8 = 5 + 3$, $9 = 10 - 1$.

In general, there arose not just the separate numbers but a system of numbers with mutual relations and rules.

The subject matter of arithmetic is exactly this, the system of numbers

with its mutual relations and rules.* The separate abstract number by itself
does not have tangible properties, and in general there is very little to be
said about it. If we ask ourselves, for example, about the properties of
the number six, we note that $6 = 5 + 1$, $6 = 3 \cdot 2$, that 6 is a factor of 30
and so forth. But here the number 6 is always connected with other numbers; in fact, the properties of a given number consist precisely of its
relations with other numbers.† Consequently, it is clear that every arithmetical operation determines a connection or relation among numbers.
Thus the subject matter of arithmetic is relations among numbers. But
these relations are the abstract images of actual quantitative relations
among collections of objects; so we may say that arithmetic is the science
of actual quantitative relations considered abstractly, that is, purely as
relations. Arithmetic, as we see, did not arise from pure thought, as the
idealists represent, but is the reflection of definite properties of real things;
it arose from the long practical experience of many generations.

3. Symbols for the numbers.   As social life became more extensive
and complicated, it posed broader problems. Not only was it necessary
to take note of the number of objects in a set and to tell others about it,
a necessity which had already led to formulation of the concept of number
and to names for the numbers, but it became essential to learn to count
increasingly larger collections, of animals in a herd, of objects for exchange,
of days before a fixed date, and so forth, and to communicate the results
of the count to others. This situation absolutely demanded improvement
in the names and also in the symbols for numbers.

The introduction of symbols for the numbers, which apparently occured
as soon as writing began, played a great role in the development of
arithmetic. Moreover, it was the first step toward mathematical signs and
formulas in general. The second step, consisting of the introduction of
signs for arithmetical operations and of a literal designation for the
unknown (x), was taken considerably later.

The concept of number, like every other abstract concept, has no
immediate image; it cannot be exhibited but can only be conceived in the

---

* The word "arithmetic," meaning the "art of calculation," is derived from the
Greek adjective "arithmetic" formed from the noun "arithmos," meaning "number."
The adjective modifies a noun "techne" (art, technique), which is here understood.

† This is understandable from the most general considerations. An arbitrary
abstraction, removed from its concrete basis (just as a number is abstracted from a
concrete collection of objects), has no sense "in itself"; it exists only in its relations
with other concepts. These relations are already implicit in any statement about the
abstraction, in the most incomplete definition of it. Without them the abstraction
lacks content and significance, i.e., it simply does not exist. The content of the concept
of an abstract number lies in the rules, in the mutual relations of the system of numbers.

mind. But thought is formulated in language, so that without a name there can be no concept. The symbol is also a name, except that it is not oral but written and presents itself to the mind in the form of a visible image. For example, if I say "seven," what do you picture to yourself? Probably not a set of seven objects of one kind or another, but rather the symbol "7," which forms a sort of tangible framework for the abstract number "seven." Moreover, a number 18273 is considerably harder to pronounce than to write and cannot be pictured with any accuracy in the form of a set of objects. In this way it came about, though only after some lapse of time, that the symbols gave rise to the conception of numbers so large that they could never have been discovered by direct observation or by enumeration. With the appearance of government, it was necessary to collect taxes, to assemble and outfit an army, and so forth, all of which required operations with very large numbers.

Thus the importance of symbols for the numbers consists, in the first place, in their providing a simple embodiment of the concept of an abstract number.* This is the role of mathematical designations in general: They provide an embodiment of abstract mathematical concepts. Thus $+$ denotes addition, x denotes an unknown number, a an arbitrary given number, and so forth. In the second place the symbols for numbers provide a particularly simple means of carrying out operations on them. Everyone knows how much easier it is "to calculate on paper" than "in one's head." Mathematical signs and formulas have this advantage in general: They allow us to replace a part of our arguments with calculations, with something that is almost mechanical. Moreover, if a calculation is written down, it already possesses a definite authenticity; everything is visible, everything can be checked, and everything is defined by exact rules. As examples one might mention addition by individual columns or any algebraic transformation such as "taking over to the other side of the equation with change of sign." From what has been said, it is clear that without suitable symbols for the numbers arithmetic could not have made much progress. Even more is it true that contemporary mathematics would be impossible without its special signs and formulas.

It is obvious that the extremely convenient method of writing numbers that is in use today could not have been worked out all at once. From ancient times there appeared among various peoples, from the very

---

* It is worth remarking that the concept of number, which was worked out with such difficulty in a long period of time, is mastered nowadays by a child with relative ease. Why? The first reason is, of course, that the child hears and sees adults constantly making use of numbers, and they even teach him to do the same. But a second reason, and this is the one to which we wish to draw special attention, is that the child already has at hand words and symbols for the numbers. He first learns these external symbols for number and only later masters the meaning of them.

Table

Symbols for the Numbers

	Slavic		Chinese			
	Cyrillic	Glagolitic	Ancient	Commercial	Scientific	Greek
0				○	○	
1	ã	✛	一	|	ĺ	ᾱ
2	ɓ̃	℡	二	||	||	β̄
3	Γ̃	℧	三	|||	|||	γ̄
4	Ã	℀	四	✗	||||	δ
5	Є̃	Ꙉ	五	୪	|||||	ε̄
6	Ѕ̃	Э	六	↓	T	s̄
7	Ꙃ̃	⯔	七	⅃	Π	ζ
8	Н̃	⯒	八	⅄	Ⅲ	η̄
9	Ѳ̃	⯑	九	⤲	Ⅲⅰ	θ
10	Ι̃	Ꙃ	十	✛	ⅠО	ι
20	К̃	8	二十	⅏	ⅠⅠО	κ̄
30	Λ̃	Ϻ	三十	⅏	ⅠⅠⅠО	λ
100	Ρ̃	ь	百	⅋	ⅠОО	ρ̄
1000	͵Δ̃	ⱚ	千	千	ⅠООО	͵ᾱ

1

AMONG VARIOUS PEOPLES

	Arabic	Georgian	Egyptian		Roman	Mayan
			Hieroglyphic	Hieratic		
0						━
1	‏ا‎	ჳ	I	ı	I	•
2	‏ب‎	ჲ	II	u	II	••
3	‏ج‎	ჳ	III	uı	III	•••
4	‏د‎	ჟ	IIII	ɯɣ	IV	••••
5	‏ه‎	ჯ	﹗﹗	1	V	━
6	‏و‎	ჳ	﹗﹗﹗	㇄	VI	▬•▬
7	‏ز‎	ჳ	﹗﹗﹗﹗	◠	VII	▬••▬
8	‏ح‎	ჳ	﹗﹗﹗﹗	═	VIII	▬•••▬
9	‏ط‎	ი	﹗﹗﹗﹗﹗	⅃	IX	▬••••▬
10	‏ى‎	ი	∩	∧	X	═
20	‏ك‎	ჳ	∩∩	⅄	XX	
30	‏ل‎	ო	∩∩∩	Ⅹ	XXX	
100	‏م‎	ჩ	ℓ	◜	C	
1000	‏غ‎	ჳ	⅃	⅃	M	

beginnings of their culture, various symbols for the numbers, which were very unlike our contemporary ones not only in their general appearance but also in the principles on which they were chosen. For example, the decimal system was not used everywhere, and among the ancient Babylonians there was a system that was partly decimal and partly sexagesimal. Table 1 gives some of the symbols for numbers among various peoples. In particular, we see that the ancient Greeks, and later also the Russians, made use of letters to designate numbers. Our contemporary "Arabic" symbols and, more generally, our method of forming the numbers, were brought from India to Europe by the Arabs in the 10th century and became firmly rooted there in the course of the next few centuries.

The first peculiarity of our system is that it is a decimal system. But this is not a matter of great importance, since it would have been quite possible to use, for example, a duodecimal system by introducing special symbols for ten and eleven. The most important peculiarity of our system of designating numbers is that it is "positional"; that is, that one and the same number has a different significance depending upon its position. For example, in 372 the number 3 denotes the number of hundreds and 7 the numbers of tens. This method of writing is not only concise and simple but makes calculations very easy. The Roman numerals were in this respect much less convenient, the same number 372 being written in the form CCCLXXII; it is a very laborious task to multiply together two large numbers written in Roman numerals.

Positional writing of numbers demands that in one way or another we take note of any category of numbers that has been omitted, since if we do not do this, we will confuse, for example, thirty-one with three-hundred-and-one. In the position of the omitted category we must place a zero, thereby distinguishing 301 and 31. In a rudimentary form, zero already appears in the late Babylonian cuneiform writings, but its systematic introduction was an achievement of the Indians:* It allowed them to proceed to a completely positional system of writing just as we have it today.

But in this way zero also became a number and entered into the system of numbers. By itself zero is nothing; in the Sanskrit language of ancient India, it is called exactly that: "empty" (çūṅga); but in connection with other numbers, zero acquires content, and well-known properties; for example, an arbitrary number plus zero is the same number, or when an arbitrary number is multiplied by zero it becomes zero.

---

* The first Indian manuscript in which zero appears comes from the end of the 9th century; in it the number 270 is written exactly as we would write it today. But it is probable that zero was introduced in India still earlier, in the 6th century.

4. The theory of numbers as a branch of pure mathematics. Let us return to the arithmetic of the ancients. The oldest texts that have been preserved from Babylon and Egypt go back to the second millennium B.C. These and later texts contain various arithmetical problems with their solutions, among them certain ones that today belong to algebra, such as the solution of quadratic and even cubic equations or progressions; all this being presented, of course, in the form of concrete problems and numerical examples. Among the Babylonians we also find certain tables of squares, cubes, and reciprocals. It is to be supposed that they were already beginning to form mathematical interests which were not immediately connected with practical problems.

In any case arithmetic was well developed in ancient Babylon and Egypt. However, it was not yet a mathematical theory of numbers but rather a collection of solutions for various problems and of rules of calculation. It is exactly in this way that arithmetic is taught up to the present time in our elementary schools and is understood by everyone who is not especially interested in mathematics. This is perfectly legitimate, but arithmetic in this form is still not a mathematical theory. There are no general theorems about numbers.

The transition to theoretical arithmetic proceeded gradually.

As was pointed out, the existence of symbols allows us to operate with numbers so large that it is impossible to visualize them as collections of objects or to arrive at them by the process of counting in succession from the number one. Among primitive tribes special numbers were worked out up to 3, 10, 100 and so forth, but after these came the indefinite "many." In contrast to this situation the use of symbols for numbers enabled the Chinese, the Babylonians, and the Egyptians to proceed to tens of thousands and even to millions. It was at this stage that the possibility was noticed of indefinitely extending the series of numbers, although we do not know how soon this possibility was clearly perceived. Even Archimedes (287–212 B.C.) in his remarkable essay "The Sand Reckoner" took the trouble to describe a method for naming a number greater than the number of grains of sand sufficient to fill up the "sphere of the fixed stars." So the possibility of naming and writing such a number still required at his time a detailed explanation.

By the 3rd century B.C., the Greeks had clearly recognized two important ideas: first, that the sequence of numbers could be indefinitely extended and second, that it was not only possible to operate with arbitrarily given numbers but to discuss numbers in general, to formulate and prove general theorems about them. This idea represents the generalization of an immense amount of earlier experience with concrete numbers, from which arose the rules and methods for *general* reasoning about numbers. A transition took place to a higher level of abstraction: from separate given (though abstract) numbers to number in general, to any possible number.

From the simple process of counting objects one by one, we pass to the unbounded process of formation of numbers by adding one to the number already formed. The sequence of numbers is regarded as being indefinitely continuable, and with it there enters into mathematics the notion of infinity. Of course, we cannot in fact, by the process of adding one, proceed arbitrarily far along the sequence of numbers: Who could reach as far as a million-million, which is almost forty times the number of seconds in a thousand years? But that is not the point; the process of adding ones, the process of forming arbitrary large collections of objects is in principle unlimited, so that the possibility exists of continuing the sequence of numbers beyond all limits. The fact that in actual practice counting is limited is not relevant; an abstraction is made from it. It is with this indefinitely prolonged sequence that general theorems about numbers have to deal.

General theorems about any property of an arbitrary number already contain in implicit form infinitely many assertions about the properties of separate numbers and are therefore qualitatively richer than any particular assertions that could be verified for specific numbers. It is for this reason that general theorems must be proved by general arguments proceeding from the fundamental rule for the formation of the sequence of numbers. Here we perceive a profound peculiarity of mathematics: Mathematics takes as its subject not only given quantitative relationships but all possible quantitative relationships and therefore infinity.

In the famous "Elements" of Euclid, written in the 3rd century B.C., we already find general theorems about whole numbers, in particular, the theorem that there exist arbitrarily large prime numbers.*

Thus arithmetic is transformed into the theory of numbers. It is already removed from particular concrete problems to the region of abstract concepts and arguments. It has become a part of "pure" mathematics. More precisely, this was the moment of the birth of pure mathematics itself with the characteristic features discussed in our first section. We must, of course, take note of the fact that pure mathematics was born simultaneously from arithmetic and geometry and that there were already to be found in the general rules of arithmetic some of the rudiments of algebra, a subject which was separated from arithmetic at a later stage. But we will discuss this later.

It remains now to summarize our conclusions up to this point, since we

---

* We recall that a prime number is defined as a positive integer greater than unity which is divisible without remainder only by the number itself and by unity.

have now traced out, though in very hurried fashion, the process whereby theoretical arithmetic arose from the concept of number.

5. The essential nature of arithmetic. Since the birth of theoretical arithmetic is part of the birth of mathematics, we may reasonably expect that our conclusions about arithmetic will throw light on our earlier questions concerning mathematics in general. Let us recall these questions, particularly in their application to arithmetic.

1. How did the abstract concepts of arithmetic arise and what do they reflect in the actual world?

This question is answered by the earlier remarks about the birth of arithmetic. Its concepts correspond to the quantitative relations of collections of objects. These concepts arose by way of abstraction, as a result of the analysis and generalization of an immense amount of practical experience. They arose gradually; first came numbers connected with concrete objects, then abstract numbers, and finally the concept of number in general, of any possible number. Each of these concepts was made possible by a combination of practical experience and preceding abstract concepts. This, by the way, is one of the fundamental laws of formation of mathematical concepts: They are brought into being by a series of successive abstractions and generalizations, each resting on a combination of experience with preceding abstract concepts. The history of the concepts of arithmetic shows how mistaken is the idealistic view that they arose from "pure thought," from "innate intuition," from "contemplation of a priori forms," or the like.

2. Why are the conclusions of arithmetic so convincing and unalterable?

History answers this question too for us. We see that the conclusions of arithmetic have been worked out slowly and gradually; they reflect experience accumulated in the course of unimaginably many generations and have in this way fixed themselves firmly in the mind of man. They have also fixed themselves in language: in the names for the numbers, in their symbols, in the constant repetition of the same operations with numbers, in their constant application to daily life. It is in this way that they have gained clarity and certainty. The methods of logical reasoning also have the same source. What is essential here is not only the fact that they can be repeated at will but their soundness and perspicuity, which they possess in common with the relations among things in the actual world, relations which are reflected in the concepts of arithmetic and in the rules for logical deduction.

This is the reason why the results of arithmetic are so convincing; its conclusions flow logically from its basic concepts, and both of them, the methods of logic and the concepts of arithmetic, were worked out and firmly fixed in our consciousness by three thousand years of practical experience, on the basis of objective uniformities in the world around us.

3. Why does arithmetic have such wide application in spite of the abstractness of its concepts?

The answer is simple. The concepts and conclusions of arithmetic, which generalize an enormous amount of experience, reflect in abstract form those relationships in the actual world that are met with constantly and everywhere. It is possible to count the objects in a room, the stars, people, atoms, and so forth. Arithmetic considers certain of their general properties, in abstraction from everything particular and concrete, and it is precisely because it considers only these general properties that its conclusions are applicable to so many cases. The possibility of wide application is guaranteed by the very abstractness of arithmetic, although it is important here that this abstraction is not an empty one but is derived from long practical experience. The same is true for all mathematics, and for any abstract concept or theory. The possibilities for application of a theory depend on the breadth of the original material which it generalizes.

At the same time every abstract concept, in particular the concept of number, is limited in its significance as a result of its very abstractness. In the first place, when applied to any concrete object it reflects only one aspect of the object and therefore gives only an incomplete picture of it. How often it happens, for example, that the mere numerical facts say very little about the essence of the matter. In the second place, abstract concepts cannot be applied everywhere without certain limiting conditions; it is impossible to apply arithmetic to concrete problems without first convincing ourselves that their application makes some sense in the particular case. If we speak of addition, for example, and merely unite the objects in thought, then naturally no progress has been made with the objects themselves. But if we apply addition to the actual uniting of the objects, if we in fact put the objects together, for example by throwing them into a pile or setting them on a table, in this case there takes place not merely abstract addition but also an actual process. This process does not consist merely of the arithmetical addition, and in general it may even be impossible to carry it out. For example, the object thrown into a pile may break; wild animals, if placed together, may tear one another apart; the materials put together may enter into a chemical reaction: a liter of water and a liter of alcohol when poured together produced not 2, but 1.9 liters of the mixture as a result of partial solution of the liquids; and so forth.

If other examples are needed they are easy to produce.

To put it briefly, truth is concrete; and it is particularly important to remember this fact with respect to mathematics, exactly because of its abstractness.

4. Finally, the last question we raised had to do with the forces that led to the development of mathematics.

For arithmetic the answer to this question also is clear from its history. We saw how people in the actual world learned to count and to work out the concept of number, and how practical life, by posing more difficult problems, necessitated symbols for the numbers. In a word, the forces that led to the development of arithmetic were the practical needs of social life. These practical needs and the abstract thought arising from them exercise on each other a constant interaction. The abstract concepts provide in themselves a valuable tool for practical life and are constantly improved by their very application. Abstraction from all nonessentials uncovers the kernel of the matter and guarantees success in those cases where a decisive role is played by the properties and relations picked out and preserved by the abstraction; namely, in the case of arithmetic, by the quantitative relations.

Moreover, abstract reflection often goes farther than the immediate demands of a practical problem. Thus the concept of such large numbers as a million or a billion arose on the basis of practical calculations but arose earlier than the practical need to make use of them. There are many such examples in the history of science; it is enough to recall the imaginary numbers mentioned earlier. This is just a particular case of a phenomenon known to everyone, namely the interaction of experience and abstract thought, of practice and theory.

§3. Geometry

1. The concept of a geometric figure. The history of the origin of geometry is essentially similar to that of arithmetic. The earliest geometric concepts and information also go back to prehistoric times and also result from practical activity.

Early man took over geometric forms from nature. The circle and the crescent of the moon, the smooth surface of a lake, the straightness of a ray of light or of a well-proportioned tree existed long before man himself and presented themselves constantly to his observation. But in nature itself our eyes seldom meet with really straight lines, with precise triangles or squares, and it is clear that the chief reason why men gradually worked out a conception of these figures is that their observation of nature was an active one, in the sense that, to meet their practical needs, they manufactured objects more and more regular in shape. They built dwellings, cut stones, enclosed plots of land, stretched bowstrings in their bows, modeled their clay pottery, brought it to perfection and correspondingly formed the notion that a pot is *curved,* but a stretched bowstring is *straight.* In short, they first gave form to their material and only then recognized form as that which is impressed on material and can therefore be considered in itself, as an abstraction from material. By recognizing the form of bodies, man was able to improve his handiwork and thereby to work out still more precisely the abstract notion of form. Thus practical activity served as a basis for the abstract concepts of geometry. It was necessary to manufacture thousands of objects with straight edges, to stretch thousands of threads, to draw upon the ground a large number of straight lines, before men could form a clear notion of the straight line in general, as that quality which is common to all these particular cases. Nowadays we learn early in life to draw a straight line, since we are surrounded by objects with straight edges that are the result of manufacture, and it is only for this reason that in our childhood we already form a clear notion of the straight line. In exactly the same way the notion of geometric magnitudes, of length, area, and volume, arose from practical activity. People measured lengths, determined distances, estimated by eye the area of surfaces and the volumes of bodies, all for their practical purposes. It was in this way that the simplest general laws were discovered, the first geometric relations: for example, that the area of a rectangle is equal to the product of the lengths of its sides. It is useful for a farmer to be aware of such a relation, in order that he may estimate the area he has sowed and consequently the harvest he may expect.

So we see that geometry took its rise from practical activity and from the problems of daily life. On this question the ancient Greek scholar, Eudemus of Rhodes, wrote as follows: "Geometry was discovered by the Egyptians as a result of their measurement of land. This measurement was necessary for them because of the inundations of the Nile, which constantly washed away their boundaries.* There is nothing remarkable in the fact that this science, like the others, arose from the practical needs of men. All knowledge that arises from imperfect circumstances tends to perfect itself. It arises from sense impressions but gradually becomes an object of our contemplation and finally enters the realm of the intellect."

Of course, the measurement of land was not the only problem that led the ancients toward geometry. From the fragmentary texts that have survived, it is possible to form some idea of various problems of the ancient Egyptians and Babylonians and of their methods for solving them. One of the oldest Egyptian texts goes back to 1700 B.C. This is a manual

---

* What is meant here is the boundaries between shares of land. Let us note, parenthetically, that *geometry* means land-measurement (in ancient Greek "ge" is land, and "metron" is measure).

of instruction for "secretaries" (royal officers), written by a certain Ahmes. It contains a collection of problems on calculating the capacity of containers and warehouses, the area of shares of land, the dimensions of earthworks, and so forth..

The Egyptians and Babylonians were able to determine the simplest areas and volumes, they knew with considerable exactness the ratio of the circumference to the diameter of a circle, and perhaps they were even able to calculate the surface area of a sphere; in a word, they already possessed a considerable store of geometrical knowledge. But so far as we can tell, they were still not in possession of geometry as a theoretical science with theorems and proofs. Like the arithmetic of the time, geometry was basically a collection of rules deduced from experience. Moreover, geometry was in general not distinguished from arithmetic. Geometric problems were at the same time problems for calculation in arithmetic.

In the 7th century B.C., geometry passed from Egypt to Greece, where it was further developed by the great materialist philosophers, Thales, Democritus, and others. A considerable contribution to geometry was also made by the successors of Pythagoras, the founders of an idealistic religiophilosophical school.

The development of geometry took the direction of compiling new facts and clarifying their relations with one another. These relations were gradually transformed into logical deductions of certain propositions of geometry from certain others. This had two results: first, the concept of a geometrical theorem and its proof; and second, the clarification of those fundamental propositions from which the others may be deduced, namely, the axioms.

In this way geometry gradually developed into a mathematical theory.

It is well known that systematic expositions of geometry appeared in Greece as far back as the 5th century B.C., but they have not been preserved, for the obvious reason that they were all supplanted by the "Elements" of Euclid (3rd century B.C.). In this work, geometry was presented as such a well-formed system that nothing essential was added to its foundations until the time of N. I. Lobačevskiĭ, more than two thousand years later. The well-known school text of Kiselev, like school books over the whole world, represented in its older editions, nothing but a popular reworking of Euclid. Very few other books in the world have had such a long life as the "Elements" of Euclid, this perfect creation of Greek genius. Of course, mathematics continued to advance, and our understanding of the foundations of geometry has been considerably deepened; nevertheless the "Elements" of Euclid became, and to a great extent remains, the model of a book on pure mathematics. Bringing together the accomplishments of his predecessors, Euclid presented the mathematics of his time as an independent theoretical science; that is, he presented it essentially as it is understood today.

2. The essential nature of geometry.   The history of geometry leads to the same conclusions as that of arithmetic. We see that geometry arose from practical life and that its transformation to a mathematical theory required an immense period of time.

Geometry operates with "geometric bodies" and figures; it studies their mutual relations from the point of view of magnitude and position. But a geometric body is nothing other than an actual body considered solely from the point of view of its spatial form,* in abstraction from all its other properties such as density, color, or weight. A geometric figure is a still more general concept, since in this case it is possible to abstract from spatial extension also; thus a surface has only two dimensions, a line, only one dimension, and a point, none at all. A point is the abstract concept of the end of a line, of a position defined to the limit of precision so that it no longer has any parts. It is in this way that all these concepts are defined by Euclid.

Thus geometry has as its object the spatial forms and relations of actual bodies, removed from their other properties and considered from the purely abstract point of view. It is just this high level of abstraction that distinguishes geometry from the other sciences that also investigate the spatial forms and relations of bodies. In astronomy for example, the mutual positions of bodies are studied, but they are the actual bodies of the sky; in geodesy it is the form of the earth that is studied, in crystallography, the form of crystals, and so forth. In all these other sciences, the form and the position of concrete bodies are studied in their dependence on other properties of the bodies.

This abstraction necessarily leads to the purely theoretical method of geometry; it is no longer possible to set up experiments with breadthless straight lines, with "pure forms." The only possibility is to make use of logical argument, deriving some conclusions from others. A geometrical theorem must be proved by reasoning, otherwise it does not belong to geometry; it does not deal with "pure forms."

The self-evidence of the basic concepts of geometry, the methods of reasoning and the certainty of their conclusions, all have the same source as in arithmetic. The properties of geometric concepts, like the concepts themselves, have been abstracted from the world around us. It was necessary for people to draw innumerable straight lines before they could take it as an axiom that through every two points it is possible to draw a

---

* By form we mean also dimensions.

straight line; they had to move various bodies about and apply them to one another on countless occasions before they could generalize their experience to the notion of superposition of geometric figures and make use of this notion for the proof of theorems, as is done in the well-known theorems about congruence of triangles.

Finally, we must emphasize the generality of geometry. The volume of a sphere is equal to $4/3\pi R^3$ quite independently of whether we are speaking of a spherical vessel, of a steel sphere, of a star, or of a drop of water. Geometry can abstract what is common to all bodies, because every actual body does have more or less definite form, dimensions, and position with respect to other bodies. So it is no cause for wonder that geometry finds application almost as widely as arithmetic. Workmen measuring the dimensions of a building or reading a blueprint, an artillery man determining the distance to his target, a farmer measuring the area of his field, an engineer estimating the volume of earthworks, all these people make use of the elements of geometry. The pilot, the astronomer, the surveyor, the engineer, the physicist, all have need of the precise conclusions of geometry.

A clear example of the abstract-geometrical solution of an important problem in physics is provided by the investigations of the well-known crystallographer and geometer, E. S. Fedorov. The problem he set himself of finding all the possible forms of symmetry for crystals is one of the most fundamental in theoretical crystallography. To solve this problem, Fedorov made an abstraction from all the physical properties of a crystal, considering it only as a regular system of geometric bodies "in place of a system of concrete atoms." Thus the problem became one of finding all the forms of symmetry which could possibly exist in a system of geometric bodies. This purely geometrical problem was completely solved by Fedorov, who found all the possible forms of symmetry, 230 in number. His solution proved to be an important contribution to geometry and was the source of many geometric investigations.

In this example, as in the whole history of geometry, we detect the prime moving force in the development of geometry. It is the mutual influence of practical life and abstract thought. The problem of discovering possible symmetries originated in physical observation of crystals but was transformed into an abstract problem and so gave rise to a new mathematical theory, the theory of regular systems, or of the so-called Fedorov groups.* Subsequently this theory not only found brilliant confirmation in the practical observation of crystals but also served as a general guide in the development of crystallography, giving rise to new investigations, both in experimental physics and in pure mathematics.

---

* Compare Chapter XX.

§4. Arithmetic and Geometry

1. The origin of fractions in the interrelation of arithmetic and geometry.
Up to now we have considered arithmetic and geometry apart from each
other. Their mutual relation, and consequently the more general interrelation of all mathematical theories, has so far escaped our attention.
Nevertheless this relation has exceptionally great significance. The interaction of mathematical theories leads to advances in mathematics itself
and also uncovers a rich treasure of mutual relations in the actual world
reflected by the these theories.

Arithmetic and geometry are not only applied to each other but they
also serve thereby as sources for further general ideas, methods, and
theories. In the final analysis, arithmetic and geometry are the two roots
from which has grown the whole of mathematics. Their mutual influence
goes back to the time when both of them had just come into being. Even
the simple measurement of a line represents a union of geometry and
arithmetic. To measure the length of an object we *apply* to it a certain
unit of length and *calculate* how many times it is possible to do this; the
first operation (application) is geometric, the second (calculation) is
arithmetical. Everyone who counts off his steps along a road is already
uniting these two operations.

In general, the measurement of any magnitude combines calculation
with some specific operation which is characteristic of this sort of
magnitude. It is sufficient to mention measurement of a liquid in a gradu-
ated container or measurement of an interval of time by counting the
number of strokes of a pendulum.

But in the process of measurement it turns out, generally speaking, that
the chosen unit is not contained in the measured magnitude an integral
number of times, so that a simple calculation of the number of units is not
sufficient. It becomes necessary to divide up the unit of measurement in
order to express the magnitude more accurately by parts of the unit;
that is, no longer by whole numbers but by fractions. It was in this way
that fractions actually arose, as is shown by an analysis of historical and
other data. They arose from the division and comparison of continuous
magnitudes; in other words, from measurement. The first magnitudes to
be measured were geometric, namely lengths, areas of fields, and volumes
of liquids or friable materials, so that in the earliest appearance of fractions
we see the mutual action of arithmetic and geometry. This interaction
leads to the appearance of an important new concept, namely of fractions,
as an extension of the concept of number from whole numbers to fractional
numbers (or as the mathematicians say, to rational numbers, expressing
the ratio of whole numbers). Fractions did not arise, and could not arise,

from the division of whole numbers, since only whole objects are counted by whole numbers. Three men, three arrows, and so forth, all these make sense, but two-thirds of a man and even two-thirds of an arrow are senseless concepts; even three separate thirds of an arrow will not kill a deer, for this it is necessary to have a *whole* arrow.

2. Incommensurable magnitudes. In the development of the concept of number, arising from the mutual action of arithmetic and geometry, the appearance of fractions was only the first step. The next was the discovery of incommensurable intervals. Let us recall that intervals are called incommensurable if no interval exists which can be applied to each of them a whole number of times or, in other words, if their ratio cannot be expressed by an ordinary fraction; that is, by a ratio of whole numbers.

At first people simply did not think about the question whether every interval can be expressed by a fraction. If in dividing up or measuring an interval they came upon very small parts, they merely discarded them; in practice, it made no sense to speak of infinite precision of measurement. Democritus even advanced the notion that geometrical figures consist of atoms of a particular kind. This notion, which to our view seems quite strange, proved very fruitful in the determination of areas and volumes. An area was calculated as the sum of rows consisting of atoms, and a volume as the sum of atomic layers. It was in this way, for example, that Democritus found the volume of a cone. A reader who understands the integral calculus will note that this method already forms the prototype of the determination of areas and volumes by the methods of the integral calculus. Moreover, in returning in thought to the times of Democritus, one must attempt to free oneself of the customary notions of today, which have become firmly fixed in our minds by the development of mathematics. At the time of Democritus, geometrical figures were not yet separated from actual ones to the same extent as is now the case. Since Democritus considered actual bodies as consisting of atoms, he naturally also regarded geometrical figures in the same light.

But the notion that intervals consist of atoms comes into contradiction with the theorem of Pythagoras, since it follows from this theorem that incommensurable intervals exist. For example, the diagonal of a square is incommensurable with its side; in other words, the ratio of the two cannot be expressed as the ratio of whole numbers.

We shall prove that the side and the diagonal of a square are in fact incommensurable. If a is the side and b is the diagonal of a square, then according to the theorem of Pythagoras $b^2 = a^2 + a^2 = 2a^2$ and therefore

$$\left(\frac{b}{a}\right)^2 = 2.$$

But there is no fraction such that its square is equal to 2. In fact, if we suppose that there is, let p and q be whole numbers for which

$$\left(\frac{p}{q}\right)^2 = 2,$$

where we may assume that p and q have no common factor, since otherwise we could simplify the fraction. But if $(p/q)^2 = 2$, then $p^2 = 2q^2$, and therefore p^2 is divisible by 2. In this case p^2 is also divisible by 4, since it is the square of an even number. So $p^2 = 4q_1$; that is, $2q^2 = 4q_1$, and $q^2 = 2q_1$. From this it follows that q must also be divisible by 2. But this contradicts the supposition that p and q have no common factor. This contradiction proves that the ratio b/a cannot be expressed by a rational number. The diagonal and the side of a square are incommensurable.

This discovery made a great impression on the Greek scientists. Nowadays, when we are accustomed to irrational numbers and calculate freely with square roots, the existence of incommensurable intervals does not disturb us. But in the 5th century B.C., the discovery of such intervals had a completely different aspect for the Greeks. Since they did not have the concept of an irrational number and never wrote a symbol like $\sqrt{2}$, the previous result indicated that the ratio of the diagonal and the side of the square was not represented by any number at all.

In the existence of incommensurable intervals the Greeks discovered a profound paradox inherent in the concept of continuity, one of the expressions of the dialectical contradiction comprised in continuity and motion. Many important Greek philosophers considered this contradiction; particularly well-known among them, because of his paradoxes, is Zeno the Eleatic.

The Greeks founded a theory of ratios of intervals, or of magnitudes in general, which takes into consideration the existence of incommensurable intervals;* it is expounded in the "Elements" of Euclid, and in simplified form is explained today in high school courses in geometry. But to recognize that the ratio of one interval to another (if the second interval is taken as the unit of length, this ratio is simply the length of the first interval) may also be considered as a number, whereby the very concept of number is generalized, to this idea the Greeks were not able to rise: The concept of an irrational number simply did not originate among them.† This step was taken at a later period by the mathematicians

---

* This theory is ascribed to the Greek scientist Eudoxus, who lived in the 4th century B.C.

† As a result of the fact that the theory of the measurement of magnitudes did not become part of arithmetic but passed over into geometry, mathematics among the

of the East; and in general, a mathematically rigorous definition of a real number, not depending immediately on geometry, was given only recently: in the seventies of the last century.* The passage of such an immense period of time after the founding of the theory of ratios shows how difficult it is to discover abstract concepts and give them exact formulation.

3. The real number. In describing the concept of a real number, Newton in his "General Arithmetic" wrote: "by number we mean not so much a collection of units as an abstract ratio of a certain quantity to another quantity taken as the unit." This number (ratio) may be integral, rational, or if the given magnitude is incommensurable with the unit, irrational.

A real number in its original sense is therefore nothing but the ratio of one magnitude to another taken as a unit; in particular cases this is a ratio of intervals, but it may also be a ratio of areas, weights, and so forth.

Consequently, a real number is a ratio of magnitudes in general, considered in abstraction from their concrete nature.

Just as *abstract* whole numbers are of mathematical interest only in their relations with one another, so *abstract* real numbers have content and become an object of mathematical attention only in relation with one another in the system of real numbers.

In the theory of real numbers, just as in arithmetic, it is first necessary to define operations on numbers: addition, subtraction, multiplication, division, and also the relations expressed by such words as "greater than" or "less than." These operations and relations reflect actual connections among the various magnitudes; for example, addition reflects the placing together of intervals. A beginning on operations with abstract real numbers was made in the Middle Ages by the mathematicians of the East. Later came the gradual discovery of the most important property of the system

---

Greeks was engulfed by geometry. Such questions, for example, as the solution of quadratic equations, which today we treat in an algebraic way, they stated and solved geometrically. The "Elements" of Euclid contain a considerable number of such questions, which obviously represented for contemporary mathematicians a summary of the foundations not only of geometry in our sense but of mathematics in general. This domination by geometry continued up to the time when Descartes, on the contrary, subjected geometry to algebra. Traces of the long domination by geometry are preserved, for example, in such names as "square" and "cube" for the second and third powers: "*a* cubed" is a cube with side *a*.

* We are speaking here not of a descriptive definition, but of a definition which serves as the immediate basis for proofs of theorems about the properties of real numbers. It is natural that such definitions should arise at a later period, when the development of mathematics, and in particular of the infinitesimal analysis, required a suitable definition of the real number represented by "the variable *x*." This definition was given in various forms in the seventies of the last century by the German mathematicians Weierstrass, Dedekind, and Cantor.

of real numbers, its continuity. The system of real numbers is the abstract image of all the possible values of a continuously varying magnitude.

In this way, as in the similar case of whole numbers, the arithmetic of real numbers deals with the actual quantitative relations of continuous magnitudes, which it studies in their general form, in complete abstraction from all concrete properties. It is precisely because real numbers deal with what is common to all continuous magnitudes that they have such wide application: The values of various magnitudes, a length, a weight, the strength of an electric current, energy and so forth, are expressed by numbers, and the interdependence or relations among these entities are mirrored as relations among their numerical values.

To show how the general concept of real numbers can serve as the basis of a mathematical theory, we must give their mathematical definition in a formal way. This may be done by various methods, but perhaps the most natural is to proceed from the very process of measurement of magnitudes which actually did lead in practical life to this generalization of the concept of number. We will speak about the length of intervals, but the reader will readily perceive that we could argue in exactly the same way about any other magnitudes which permit indefinite subdivision.

Let us suppose that we wish to measure the interval AB by means of the interval CD taken as a unit (figure 1).

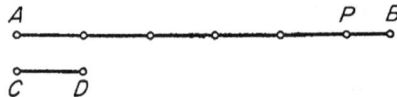

FIG. 1.

We apply the interval CD to AB, beginning for example with the point A, as long as CD goes into AB. Suppose this is n_0 times. If there still remains from the interval AB a remainder PB, then we divide the interval CD into ten parts and measure the remainder with these tenths. Suppose that n_1 of the tenths go into the remainder. If after this there is still a remainder, we divide our measure into ten parts again; that is, we divide CD into a hundred parts, and repeat the same operation, and so forth. Either the process of measurement comes to an end, or it continues. In either case we reach the result that in the interval AB the whole interval CD is contained n_0 times, the tenths are contained n_1 times, the hundredths n_2 times and so forth. In a word, we derive the ratio of AB to CD with increasing accuracy: up to tenths, to hundredths, and so forth. So the ratio itself is represented by a decimal fraction with n_0 units, n_1 tenths and so forth

$$\frac{AB}{CD} = n_0 \cdot n_1 n_2 n_3 \cdots$$

This decimal fraction may be infinite, corresponding to the possibility of indefinite increase in the precision of measurement.

Thus the ratio of two intervals, or of two magnitudes in general, is always representable by a decimal fraction, finite or infinite. But in the decimal fraction there is no longer any trace of the concrete magnitude itself; it represents exactly the abstract ratio, the real number. Thus a real number may be formally defined if we wish, as a finite or infinite decimal fraction.*

Our definition will be complete if we say what we mean by the operations of addition and so forth for decimal fractions. This is done in such a way that the operations defined on decimal fractions correspond to the operations on the magnitudes themselves. Thus, when intervals are put together their lengths are added; that is, the length of the interval $AB + BC$ is equal to the sum of the length AB and BC. In defining the operations on real numbers, there is a difficulty that these numbers are represented in general by *infinite* decimal fractions, while the well-known rules for these operations refer to finite decimal fractions. A rigorous definition of the operations for infinite decimals may be made in the following way. Suppose, for example, that we must add the two numbers a and b. We take the corresponding decimal fractions up to a given decimal place, say the millionth, and add them. We thus obtain the sum $a + b$ with corresponding accuracy, up to two millionths, since the errors in a and b may be added together. So we are able to define the sum of two numbers *with an arbitrary degree of accuracy*, and in that sense their sum is completely defined, although at each stage of the calculation it is known only with a a certain accuracy. But this corresponds to the essential nature of the case, since each of the magnitudes a and b is also measured only with a certain accuracy, and the exact value of each of the corresponding infinite fractions is obtained as the result of an indefinitely extended increase in accuracy. The relations "greater than" and "less than" may then be defined by means of addition: a is greater than b if there exists a magnitude c such that $a = b + c$, where we are speaking, of course, of positive numbers.

The continuity of the sequence of real numbers finds expression in the fact that if the numbers $a_1, a_2, \cdots$ increase and $b_1, b_2, \cdots$ diminish but

---

* Fractions with the periodic digit nine are not considered here, they are identical with the corresponding fraction without nines according to the well-known rule, which is clear from the example: $0.139999 \cdots = 0.140000 \cdots$.

always remain greater than the a_i, then between the one series of numbers and the other there is always a number c. This may be visualized on a straight line if its points are put into correspondence with the numbers (figure 2) according to the well-known rule.

FIG. 2.

Here it is clearly seen that the presence of the number c and of the point corresponding to it signify the absence of a break in the series of numbers, which is what is meant by their continuity.

4. The conflict of opposites: concrete and abstract.   Already in the example of the interaction of arithmetic and geometry we can see that the development of mathematics is a process of conflict among the many contrasting elements: the concrete and the abstract, the particular and the general, the formal and the material, the finite and the infinite, the discrete and the continuous, and so forth. Let us try, for example, to trace the contrast between concrete and abstract in the formation of the concept of a real number. As we have seen, the real number reflects an infinitely improvable process of measurement or, in slightly different terms, an absolutely accurate determination of a magnitude. This corresponds to the fact that in geometry we consider ideally precise forms and dimensions of bodies, abstracting altogether from the mobility of concrete objects and from a certain indefiniteness in their actual forms and dimensions; for example, the interval measured (figure 1) was a completely ideal one.

But ideally precise geometric forms and absolutely precise values for magnitudes represent abstractions. No concrete object has absolutely precise form nor can any concrete magnitude be measured with absolute accuracy, since it does not even *have* an absolutely accurate value. The length of a line segment, for example, has no sense if one tries to make it precise beyond the limits of atomic dimensions. In every case when one passes beyond well-known limits of quantitative accuracy, there appears a qualitative change in the magnitude, and in general it loses its original meaning. For example, the pressure of a gas cannot be made precise beyond the limits of the impact of a single molecule; electric charge ceases to be continuous when one tries to make it precise beyond the charge on an electron and so forth. In view of the absence in nature of objects of ideally precise form, the assertion that the ratio of the diagonal of a square to the side is equal to the $\sqrt{2}$ not only cannot be deduced with absolute accuracy from immediate measurement but does not even have any absolutely accurate meaning for an actual concrete square.

The conclusion that the diagonal and the side of a square are incommensurable comes, as we have seen, from the theorem of Pythagoras. This is a *theoretical* conclusion based on a development of the data of experience; it is a result of the application of logic to the original premises of geometry, which are taken from experience.

In this way the concept of incommensurable intervals, and all the more of real numbers, is not a simple immediate reflection of the facts of experience but goes beyond them. This is quite understandable. The real number does not reflect any given concrete magnitude but rather magnitude in general, in abstraction from all concreteness; in other words, it reflects what is *common* to particular concrete magnitudes. What is common to all of them consists in particular in this, that the value of the magnitude can be determined more and more precisely; and if we abstract from concrete magnitudes, then the limit of this possible increase in precision, which depends on the concrete nature of the magnitude, becomes indefinite and disappears.

In this way a *mathematical* theory of magnitudes, since it considers magnitudes in abstraction from their individual nature, must inevitably consider the possibility of unlimited accuracy for the value of the magnitude and *must* thereby lead to the concept of a real number. At the same time, since it reflects only what is common to various magnitudes, mathematics takes no account of the peculiarity of each individual magnitude.

Since mathematics selects only general properties for consideration, it operates with its clearly defined abstractions quite independently of the actual limits of their applicability, as must happen precisely because these limits are different in different particular cases. These limits depend on the concrete properties of the phenomena under consideration and on the qualitative changes that take place in them. So in making an application of mathematics, it is necessary to verify the actual applicability of the theory in question. To consider matter as continuous and to describe its properties by continuous magnitudes is permissible only if we may abstract from its atomic structure, and this is possible only under well-known conditions.

Nevertheless, the real numbers represent a trustworthy and powerful instrument for the mathematical investigation of actual continuous magnitudes and processes. Their theory is based on practice, on an immense field of applications in physics, technology, and chemistry. Consequently, practice shows that the concept of the real number correctly reflects the general properties of magnitudes. But this correctness is not without limits; it is not possible to consider the theory of real numbers as something absolute, allowing an unlimited abstract development in complete separation from reality. The very concept of the real number is continuing to develop and is in fact still far from being complete.

5. The conflict of opposites: discrete and continuous. The role of another of the mentioned contrasts, the contrast between the discrete and the continuous, may also be illustrated by the development of the concept of number. We have already seen that fractions arose from the division of continuous magnitudes.

On this theme of division there is a humorous question which is extraordinarily instructive. Grandmother has bought three potatoes and must divide them equally between two grandsons. How is she to do it? The answer is: make mashed potatoes.

The joke reveals the very essence of the matter. Separate objects are indivisible in the sense that, when divided, the object almost always ceases to be what it was before, as is clear from the example of "thirds of a man" or "thirds of an arrow." On the other hand, continuous and homogenous magnitudes or objects may easily be divided and put together again without losing their essential character. Mashed potatoes offer an excellent example of a homogeneous object, which in itself is not separated into parts but may nevertheless be divided in practice into as small parts as desired. Lengths, areas, and volumes have the same property. Although they are continuous in their very essence and are not actually divided into parts, nevertheless they offer the possibility of being divided without limit.

Here we encounter two contrasting kinds of objects: on the one hand, the indivisible, separate, discrete objects; and on the other, the objects which are completely divisible and yet are not divided into parts but are continuous. Of course, these contrasting characteristics are always united, since there are no absolutely indivisible and no completely continuous objects. Yet these aspects of the objects have an actual existence, and it often happens that one aspect is decisive in one case and the other in another.

In abstracting forms from their content, mathematics by this very act sharply divides these forms into two classes, the discrete and the continuous.

The mathematical model of a separate object is the unit, and the mathematical model of a collection of discrete objects is a sum of units, which is, so to speak, the image of pure discreteness, purified of all other qualities. On the other hand, the fundamental, original mathematical model of continuity is the geometric figure; in the simplest case, the straight line.

We have before us therefore two contrasts, discreteness and continuity, and their abstract mathematical images: the whole number and the geometric extension. Measurement consists of the union of these contrasts:

The continuous is measured by separate units. But the inseparable units are not enough; we must introduce fractional parts of the original unit. In this way the fractional numbers arise and the concept of numbers develops precisely as a result of the union of the mentioned contrasts.

Then, on a more abstract level, appeared the concept of incommensurable intervals, and, as a result, the real number as an abstract image of unlimited increase in accuracy in the determination of a magnitude. This concept was not formed immediately, and the long path of its development led through many a conflict between these same two contrasting elements, the discrete and the continuous.

In the first place, Democritus represented figures as consisting of atoms and in this way reduced the continuous to the discrete. But the discovery of incommensurable intervals led to the abandonment of such a representation. After this discovery continuous magnitudes were no longer thought of as consisting of separate elements, atoms or points, and they were not represented by numbers, since numbers other than the whole numbers and the fractions were not known at that time.

The contrast between the continuous and the discrete appeared in mathematics again with renewed force in the 17th century, when the foundations of the differential and integral calculus were being laid. Here it was the infinitesimal that was under discussion. In some accounts the infinitesimal was thought of as a real, "actually" infinitesimal, "indivisible" particle of the continuous magnitude, like the atoms of Democritus, except that now the number of these particles was considered to be infinitely great. Calculation of areas and volumes, or in other words integration, was thought of as summation of an infinite number of these infinitely small particles.

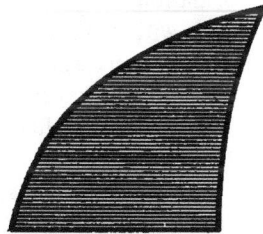

FIG. 3.

An area, for example, was understood as "the sum of the lines from which it is formed" (figure 3). Consequently, the continuous was again reduced to the discrete, but now in a more complicated way, on a higher level. But this point of view also proved unsatisfactory, and, as a counterweight to it, there appeared, on the basis of Newton's work, the notion of *continuous* variables, of the infinitesimal as a *continuous* variable decreasing without limit. This conception finally carried the day at the beginning of the 19th century, when the rigorous theory of limits was founded. An interval was now thought of as consisting not of points or "indivisibles," but as an extension, as a continuous medium, where it was only possible to fix separate points, separate values of a continuous magnitude. Mathematicians then spoke of "extension." In the union of the discrete and the continuous, it was again the continuous that dominated.

But the development of analysis demanded further precision in the theory of variable magnitudes and above all in the general definition of a real number as an arbitrary possible value of a variable magnitude. In the seventies of the last century there arose a theory of real numbers which represents an interval as a set of points, and correspondingly the range of variation of a variable as a set of real numbers. The continuous again consisted of separate discrete points and the properties of continuity were again expressed in the structure of the set of points that formed it. This conception led to immense progress in mathematics and became dominant. But again profound difficulties were discovered in it, and these led to attempts to return on a new level to the notion of pure continuity. Other attempts were made to change the concept of an interval as a set of points. New points of view appeared for the concepts of number, variable, and function. The development of the theory is continuing, and we must await its further progress.

6. Further results of the interaction of arithmetic and geometry. The interaction of geometry and arithmetic played a role elsewhere than in the formation of the concept of a real number. The same interaction of geometry with arithmetic, or more accurately with algebra, also showed itself in the formation of negative and complex numbers, that is of numbers of the form $a + b \sqrt{-1}$. Negative numbers are represented by points of the straight line to the left of the point representing zero. It was exactly this geometric representation which gave imaginary numbers a firm place in mathematics; up to that time they had not been understood. New concepts of magnitude appeared: for example, vectors, which are represented by directed line segments; and tensors, which are still more general magnitudes; in these again algebra is united with geometry.

The union of various mathematical theories has always played a great and sometimes decisive role in the development of mathematics. We shall see this further on in the rise of analytic geometry, differential and integral calculus, the theory of functions of a complex variable, the recent so-called functional analysis, and other theories. Even in the theory of numbers itself, that is in the study of whole numbers, methods are applied with great success which depend on continuity (namely on the infinitesimal analysis) and on geometry. These methods have given rise to extensive chapters in the theory of numbers, the "analytic theory of numbers," and the "geometry of numbers."

From a certain well-known point of view, it is possible to regard the foundations of mathematics as the union of concepts arising from geometry and arithmetic; that is to say, of the general concepts of continuity and of algebraic operations (as generalizations of arithmetic operations). But we will not be able to speak here of these difficult theories. The aim of the present chapter has been to give an impression of the general interaction of concepts, of the union and the conflict between contrasting ideas in mathematics, as illustrated by the interaction of arithmetic and geometry in the development of the concept of number.

§5. The Age of Elementary Mathematics

1. The four periods of mathematics. The development of mathematics cannot be reduced to the simple accumulation of new theorems but includes essential qualitative changes. These qualitative changes take place, however, not in a process of destruction or abolition of already existing theories but in their being deepened and generalized, so as to form more general theories, for which the way has been prepared by preceding developments.

From the most general point of view, we may distinguish in the history of mathematics four fundamental, qualitatively distinct periods. Of course, it is not possible to draw exact boundary lines between these stages, since the essential traits of each period appeared more or less gradually, but the distinctions among the stages and the passages from one to another are completely clear.

The first stage (or period) is the period of the rise of mathematics as an independent and purely theoretical science. It begins in the most ancient times and extends to the 5th century B.C., or perhaps earlier, when the Greeks laid the foundations of "pure" mathematics with its logical connection between theorems and proofs (in that century there appeared, in particular, systematic expositions of geometry like the "Elements" of Hippocrates of Chios). This first stage was the period of the formation of arithmetic and geometry, in the form considered earlier. At this time mathematics consisted of a collection of separate rules deduced from experience and immediately connected with practical life. These rules did not yet form a logically unified system, since the theoretical character of mathematics with its logical proof of theorems was formed very slowly, as material for it was accumulated. Arithmetic and geometry were not separated but were closely interwoven with each other.

The second period may be characterized as the period of elementary mathematics, of the mathematics of constant magnitudes; its simple fundamental results now form the content of a high school course. This period extended for almost 2000 years and ended in the 17th century with the rise of "higher" mathematics. It is with this period that we will be concerned in greater detail in the present section. The following sections will be devoted to the third and fourth periods, namely to the founding and development of analysis and to the period of contemporary mathematics.

2. Mathematics in Greece. The period of elementary mathematics may in its turn be divided into two parts, distinguished by their basic content: the period of the development of geometry (up to the 2nd century A.D.) and the period of the predominance of algebra (from the 2nd to the 17th century). With respect to historical conditions it is divided into three parts, which may be called "Greek," "Eastern," and "European Renaissance." The Greek period coincides in time with the general flowering of Greek culture, beginning with the 7th century B.C., reaching its culmination in the 3rd century B.C. at the time of the great geometers of antiquity, Euclid, Archimedes, and Apollonius, and ending in the 6th century A.D. Mathematics, and especially geometry, enjoyed a wonderful development in Greece. We know the names and the results of numerous Greek mathematicians, although only a few genuine works have come down to us. It is to be remarked that Rome gave nothing to mathematics though it reached its zenith in the 1st century A.D. at a time when the science of Greece, which had been conquered by Rome, was still flourishing.

The Greeks not only developed and systematized elementary geometry to the extent to which it is given in the "Elements" of Euclid and is now taught in our secondary schools, but achieved considerably higher results. They studied the conic sections: ellipse, hyperbola, parabola; they proved certain theorems relating to the elements of what is called projective geometry; guided by the needs of astronomy, they worked out spherical geometry (in the 1st century A.D.) and also the elements of trigonometry, and calculated the first tables of sines (Hipparchus, 2nd century B.C. and Claudius Ptolemy, 2nd century A.D.);* they determined the areas and volumes of a number

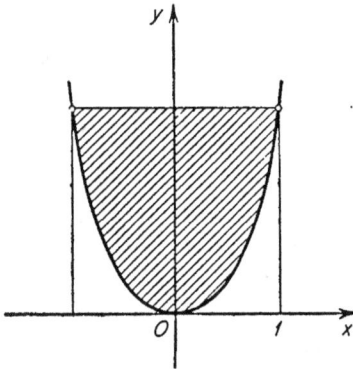

FIG. 4.

---

* Ptolemy is widely known as the author of a system in which the Earth is considered as the center of the universe and the motion of the heavenly bodies is described as proceeding around it. This system was supplanted by the Copernician system.

of complicated figures; for example, Archimedes found the area of the segment of a parabola by proving that it is 2/3 of the area of the rectangle containing it (figure 4). The Greeks were also acquainted with the theorem that of all bodies with a given surface area the sphere has the greatest volume, but their proof has not been preserved and was probably not complete. Such a proof is quite difficult and was first discovered in the 19th century, by means of the integral calculus.

In arithmetic and in the elements of algebra, the Greeks also made considerable progress. As was mentioned earlier, they laid the foundation for the theory of numbers. Here belong, for example, their investigations on prime numbers (the theorem of Euclid on the existence of an infinite number of prime numbers and the "sieve" of Eratosthenes for finding prime numbers) and the solution of equations in whole numbers (Diophantus about 246–330 A.D.).

We have already said that the Greeks discovered irrational magnitudes but considered them geometrically, as line segments. So the problems that today we deal with algebraically were treated geometrically by the Greeks. It was in this way that they solved quadratic equations and transformed irrational expressions. For example, the equation that we today write in the form $x^2 + ax = b^2$, they stated as follows: Find a segment x such that if to the square constructed on it we add a rectangle constructed on the same segment and on the given segment a, we obtain a rectangle equal in area to a given square. This dominance of geometry lasted a long time after the Greeks. They were also acquainted with (geometric) methods for extracting square roots and cube roots and with the properties of arithmetic and geometric progressions.

In this way the Greeks were already in possession of much of the material of contemporary elementary algebra but not, however, of the following essential elements: negative numbers and zero, irrational numbers abstracted entirely from geometry, and finally a well-developed system of literal symbols. It is true that Diophantus made use of literal symbols for the unknown quantity and its powers and also of special symbols for addition, subtraction, and equality, but his algebraic equations were still written with concrete numerical coefficients.

In geometry the Greeks attained what we now call "higher" mathematics. Archimedes made use of integral calculus for the calculation of areas and volumes and Apollonius used analytic geometry in his investigations on conic sections. Apollonius actually gives the equations of these curves*

---

* He gives the "equations" of conic sections referred to a vertex. For example "the equation" of the parabola $y^2 = 2px$ is formulated thus: The square on the side y is equal in area to the rectangle with sides $2p$ and x. Of course, in place of the symbols p, x, y he uses the corresponding line segments.

but expresses them in geometric language. In these equations there does not yet appear the general notion of an arbitrary constant or of a variable magnitude; and the necessary means of expressing such concepts, namely the literal symbols of algebra, appear only at a later age; they alone could convert such investigations into a source of new theories, which would be truly a part of higher mathematics. The founders of these new theories were guided, a thousand years later, by the legacy of the Greek scientists; in fact, the "Geometry" of Descartes (1637), which laid the foundation for analytic geometry, begins with a selection of problems left by the Greeks.

Such is the general rule. The old theories, by giving rise to new and profound problems, outgrow themselves, as it were, and demand for further progress new forms and new ideas. But these forms and ideas may demand new historical conditions for their birth. In ancient society the conditions necessary for the passage to higher mathematics did not and could not exist; they came on the scene with the development of the natural sciences in modern times, a development which in its turn was conditioned in the 16th and 17th centuries by the new demands of technology and of manufacturing and was connected in this way with the birth and development of capitalism.

The Greeks practically exhausted the possibilities of elementary mathematics, which is the explanation of the fact that the brilliant progress of geometry dried up at the beginning of our era and was replaced by trigonometry and algebra in the works of Ptolemy, Diophantus, and others. In fact, one may consider the works of Diophantus as the beginning of the period in which algebra played the leading role. But the society of the ancients, already verging to its decline, was no longer able to advance science in this new direction.

It should be noted that, a few centuries earlier, arithmetic had already reached a high level in China. The Chinese scientists of the 2nd and 1st centuries B.C. described the rules for arithmetical solution of a system of three equations of the first degree. It is here for the first time in history that negative coefficients are made use of and the rules for operating with negative quantities are formulated. But the solutions themselves were sought only in the form of positive numbers, just as later in the works of Diophantus. These Chinese books also include a method for the extraction of square roots and cube roots.

3. The Middle East. With the end of Greek science a period of scientific stagnation began in Europe, the center of mathematical development being shifted to India, Central Asia, and the Arabic countries.*

---

* To give some orientation in the dates we list here the times of some of the out-

For a period of about a thousand years, from the 5th to the 15th century, mathematics developed chiefly in connection with the demands of computation, particularly in astronomy, since the mathematicians of the East were for the most part also astronomers. It is true that they added nothing of importance to Greek geometry; in this field they only preserved for later times the results of the Greeks. But the Indian, Arabic, and Central Asian mathematicians achieved immense successes in the fields of arithmetic and algebra.*

As has been mentioned in §2, the Indians invented our present system of numeration. They also introduced negative numbers, comparing the contrast between positive and negative numbers with the contrast between property and debt or between the two directions on a straight line. Finally, they began to operate with irrational magnitudes exactly as with rational, without representing them geometrically, in contrast to the Greeks. They also had special symbols for the algebraic operations, including extraction of roots. For the very reason that the Indian and Central Asian scholars were no longer embarrassed by the difference between the irrational and rational magnitudes, they were able to overcome the "dominance" of geometry, which was characteristic of Greek mathematics, and to open up paths for the development of contemporary algebra, free of the heavy geometric framework into which it had been forced by the Greeks.

The great poet and mathematician, Omar Khayyam (about 1048–1122), and also the Azerbaijanian mathematician, Nasireddin Tusi (1201–1274), clearly showed that every ratio of magnitudes, whether commensurable or incommensurable, may be called a number; in their works we find the same general definition of number, both rational and irrational, as was introduced above in Newton's formulation, in §4. The magnitude of these achievements becomes particularly clear when we recall that complete recognition of negative and irrational numbers was attained by European mathematicians only very slowly, even after the beginning of the Renaissance of mathematics in Europe. For example, the celebrated French mathematician Viète (1540–1603), to whom algebra owes a great deal, avoided negative numbers, and in England protests against them lasted even into the 18th century. These numbers were considered absurd, since they were less than zero, that is "less than nothing at all." Nowadays they

standing mathematicians of the East. From India: Aryabhata, born about 476 A.D.; Brahmagupta, about 598–660; Bhaskara, 12th century; from Kharizm: Al-Kharizmi, 9th century; Al-Biruni, 973–1048; from Azerbaijan: Nasireddin Tusi, 1201–74; from Samarkand: Gyaseddin Jamschid, 15th century.

* One should keep in mind that it is wrong to associate the development of mathematics in this period chiefly with the Arabs. The term "Arabic" mathematics came into use chiefly because most of the scholars of the East wrote in the Arabic language, which had been spread abroad by the Arab conquests.

have become familiar, if only in the form of negative temperature; everyone reads the newspapers and understands what is mean by "the temperature in Moscow is —8°."

The word "algebra" comes from the name of a treatise of the mathematician and astronomer Mahommed ibn Musa al-Kharizmi (Mahommed, son of Musa, native of Kharizm), who lived in the 9th century. His treatise on algebra was called Al-jebr w'al-muqabala, which means "transposition and removal." By transposition (al-jebr) is understood the transfer of negative terms to the other side of an equation, and by removal (al-muqabala), cancellation of equal terms on both sides.

The Arabic word "al-jebr" became in Latin transcription "algebra" and the word al-muqabala was discarded, which accounts for the modern term "algebra."*

The origin of this term corresponds very well to the actual content of the science itself. Algebra is basically the doctrine of arithmetical operations considered formally from a general point of view, with abstraction from the concrete numbers. Its problems bring to the fore the formal rules for transformation of expressions and solution of equations. Al-Kharizmi placed on the title page of his book the actual names of two most general formal rules, expressing in this way the true spirit of algebra.

Subsequently, Omar Khayyam defined algebra as the science of solving equations. This definition retained its significance up to the end of the 19th century, when algebra, along with the theory of equations, struck out in new directions, essentially changing its character but not changing its spirit of generality as the science of formal operations.

The mathematicians of Central Asia found methods for calculation, both exact and approximate, of the roots of a number of equations; they discovered the general formula for the "binomial of Newton," although they expressed it in words; they greatly advanced and systematized the science of trigonometry, and calculated very accurate tables of sines. These tables were computed, for astronomical purposes, by the mathematician Gyaseddin (about 1427) who was working with the famous Uzbek astronomer Ulug Begh; Gyaseddin also invented decimal fractions 150 years before they were reinvented in Europe.

To sum up, in the course of the Middle Ages in India and in Central Asia the present decimal system of numeration (including fractions) was almost completely built up, as were also elementary algebra and trigonometry. During the same period the achievements of Chinese science began to make their way into the neighboring countries; about the 6th

---

* It is to be noted also that the mathematical term "algorithm," denoting a method or set of rules for computation, comes from the name of the same al-Kharizmi.

century B.C. the Chinese already had methods for the solution of the simplest indeterminate equations, for approximate calculations in geometry, and for the first steps in approximate solution of equations of the third degree. Essentially the only parts of our present high school course in algebra that were not known before the 16th century were logarithms and imaginary numbers. However, there did not yet exist a system of literal symbols: The content of algebra had outdistanced its form. Yet the form was indispensable: The abstraction from concrete numbers and the formulation of general rules demanded a corresponding method of expression; it was essential to have some way of denoting *arbitrary* numbers and operations on them. The algebraic symbolism is the necessary form corresponding to the content of algebra. Just as in remote antiquity it had been necessary, in order to operate with whole numbers, that symbols should be invented for them, so now, to operate with arbitrary numbers and to give general rules for their use, it was necessary to work out corresponding symbols. This task, begun at the time of the Greeks, was not brought to completion until the 17th century, when the present system of symbols was finally set up in the works of Descartes and others.

4. Renaissance Europe. At the time of the Renaissance the Europeans became acquainted with Greek mathematics by way of the Arabic translations. The books of Euclid, Ptolemy, and Al-Kharizmi were translated in the 12th century from Arabic into Latin, the common scientific language of Western Europe, and at the same time, the earlier system of calculation, as derived from the Greeks and Romans, was gradually replaced by the present-day Indian method, which was borrowed by the Europeans from the Arabs.

It was only in the 16th century that European science finally surpassed the achievements of its predecessors. Thus the Italians, Tartaglia and Ferrari, solved the general cubic equation, and later, the general equation of the fourth degree (see Chapter IV). Let us note that although these results are not taught in school, they belong, with respect to the methods employed in them, to elementary algebra. To higher algebra we must however refer the general theory of equations.

During the same period imaginary numbers began for the first time to be used; at first this was done in a purely formal manner, without logical foundation, which came considerably later at the beginning of the 19th century. Our present-day algebraic symbols were also worked out; in particular, literal symbols were used by Viète in 1591 not only for unknown quantities but also for given ones.

Many mathematicians took a share in this development of algebra. At the same time decimal fractions appeared in Europe; they were invented by the Dutch scholar Stevin, who wrote about them in 1585.

Finally, Napier in Great Britain invented logarithms as an aid in astronomical calculations and wrote about them in 1614; Briggs calculated the first decimal tables of logarithms, which were published in 1624.*

At the same time there appeared in Europe the "theory of combinations" and the general formula for the" binomial of Newton";† the progressions being already known, and in this way the structure of elementary algebra was completed. Therewith came to an end, at the beginning of the 17th century, the whole period of the mathematics of constant magnitudes, of elementary mathematics as it is now taught, with a few additions, in our schools. Arithmetic, elementary geometry, trigonometry, and elementary algebra were now essentially complete. There followed a transition to higher mathematics, to the mathematics of variable magnitudes.

It is not to be thought, however, that the development of elementary mathematics ceased at this time; for example, new results were discovered and are being constantly discovered today in elementary geometry. Furthermore, it is precisely because of the subsequent development of higher mathematics that we now understand more clearly the essential nature of elementary mathematics itself. But the leading role in mathematics was now taken over by the concepts of variable magnitude, function, and limit. The problems, that led from elementary mathematics to higher mathematics are nowadays clarified and solved by the concepts and methods of higher mathematics (occasionally they are not solvable at all by elementary methods), and there are other problems which may be stated in terms of elementary mathematics but which serve even today as a source of more general results and even of entire theories. Examples are provided by the earlier mentioned theory of regular systems of figures or by problems of the theory of numbers which are elementary in their formulation but far from elementary in the methods by which they are solved. For further details the reader may consult Chapter X.

---

* It is interesting to note that Napier did not define logarithms as they are defined nowadays, when we say that in the formula $x = a^y$ the number y is the logarithm of x to the base a. This definition of logarithms appeared later. Napier's definition was related to the concepts of a variable magnitude and an infinitesimal and amounted to saying that the logarithm of x is a function $y = f(x)$ whose rate of growth is inversely proportional to x; that is, $y' = c/x$ (see Chapter II). In this way the basis of the definition was essentially a differential equation, defining the logarithm, although differentials had not yet been invented.

† The formula bears the name of Newton not because he was the first to discover it but because he generalized it from integral exponents to arbitrary fractional and irrational exponents.

§6. Mathematics of Variable Magnitudes

1. Variable and function. In the 16th century the investigation of motion was the central problem of physics. The physical sciences were led to this problem, and to the study of various others involving interdependence of variable magnitudes, by the demands of practical life and by the whole development of science itself.

As a reflection of the general properties of change, there arose in mathematics the concepts of a variable magnitude and a function, and it was this cardinal extension of the subject matter of mathematics that determined the transition to a new stage, to the mathematics of variable magnitudes.

The law of motion of a body in a given trajectory, for example along a straight line, is defined by the manner in which the distance covered by the body increases with time.

Thus Galileo (1564–1642) discovered the law of falling bodies by establishing that the distance fallen increases proportionally to the square of the time. This fact is expressed in the well-known formula

$$s = \frac{gt^2}{2},\qquad(1)$$

where g is approximately equal to 9.81 m/sec².

In general, the law of motion expresses the distance covered in the time t. Here the time t and the distance s are respectively the "independent" and the "dependent" variable, and the fact that to each time t there corresponds a definite distance s is what is meant by saying that the distance s is a function of the time t.

The mathematical concepts of variable and function are the abstract generalization of concrete variables (such as time, distance, velocity, angle of rotation, and area of surface traced out) and of the interdependences among them (the distance depends on the time and so forth). Just as the concept of a real number is the abstract image of the actual value of an arbitrary magnitude, so a "variable" is the abstract image of a varying magnitude, which assumes various values during the process under consideration. A mathematical variable x is "something" or, more accurately, "anything" that may take on various numerical values. This is the meaning of a variable in general; in particular, we may understand by it the time, the distance, or any other variable magnitude.

In exactly the same way, a function is the abstract image of the dependence of one magnitude on another. The assertion that y is a function of x means in mathematics only that to each possible value of x there corresponds a definite value of y. This correspondence between the values of y and the values of x is called a function. For example, according to the law of falling bodies, the distance covered corresponds to the time of fall by formula (1). The distance is a function of the time. Let us look at some other examples.

The energy of a falling body is expressed by its mass and its velocity according to the formula

$$E = \frac{mv^2}{2}.$$ (2)

For a given body the energy is a function of the velocity v.

By a familiar law the quantity of heat generated in a conductor in unit time by the passage of an electric current is expressed by the formula

$$Q = \frac{RI^2}{2},$$ (3)

where I is the magnitude of the current and R is the resistance of the conductor. For a given resistance there corresponds to every current I a definite amount of heat Q, generated in unit time. That is, Q is a function of I.

The area of a right-angled triangle S with a given acute angle α and corresponding side x (see figure 5) is expressed by the formula

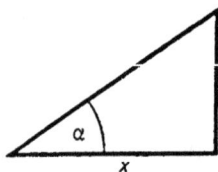

$$S = \tfrac{1}{2} x^2 \tan \alpha.$$ (4)

Fig. 5

For a given angle α the area is a function of the side x.

All these formulas (1)–(4) may be united in the one

$$y = \tfrac{1}{2} ax^2.$$ (5)

This general formula represents a transition from the concrete variable magnitudes t, s, E, Q, v and so forth to the general variables x and y, and from the concrete dependences (1), (2), (3), (4) to their general form (5). Mechanics and the theory of electricity have to do with concrete formulas (1), (2), (3), interrelating concrete magnitudes, but the mathematical theory of functions deals with the general formula (5), without associating this formula with any concrete magnitudes.

The next degree of abstraction from the concrete consists in our examining not a given dependence of y on x, like $y = \tfrac{1}{2} ax^2$, $y = \sin x$, $y = \log x$ and so forth, but the general dependence of y on x expressed in the abstract formula

$$y = f(x).$$

This formula states that the magnitude y is in general some function of x; that is, to each value assumed by x there corresponds, in some fashion or another, a definite value y. The subject matter of mathematics thus consists not only of certain given functions ($y = \frac{1}{2} ax^2$, $y = \sin x$, and so forth), but of *arbitrary* (more accurately, more or less arbitrary) functions. These degrees of abstraction, first from concrete magnitudes and then from concrete functions, are analogous to the degrees of abstraction observed in the formation of the concept of a whole number: First, abstraction from concrete collections of objects led to the concept of whole numbers (1, 3, 12, and so forth), and then a further abstraction led to the concept of an arbitrary whole number in general. This generalization is the result of a profound interraction between analysis and synthesis: analysis of separate interrelations and synthesis, in the form of new concepts, of their common features.

The branch of mathematics devoted to the study of functions is called analysis, or often, infinitesimal analysis, since one of the most important elements in the study of functions is the concept of the infinitesimal (the meaning of this concept and its significance are explained in Chapter II).

Since a function is the abstract image of a dependence of one magnitude on another, we may say that analysis takes as its subject matter dependences between variable magnitudes, not between one concrete magnitude and another but between variables in general, in abstraction from their content. An abstraction of this sort guarantees great breadth of application, since one formula or one theorem contains an infinite number of possible concrete cases. An example of this is given already by our simple formulas (1)–(5). So the complete analogy of analysis with arithmetic and algebra becomes evident. They all originate in definite practical problems and give a general abstract expression to concrete relationships in the actual world.

2. Analytic geometry and analysis. Thus the new period of mathematics, beginning in the 17th century, may be defined as the period of the birth and development of analysis. (This is the third of the three important periods mentioned earlier.) It is to be understood, of course, that no theory arises as a result of the mere formation of new concepts, that analysis could not result from the mere existence of the concepts of variable and function. For the founding of a theory, and all the more of a complete branch of science like mathematical analysis, it is necessary that the new concepts become active, so to speak, that among them there be discovered new relationships, and that they permit the solution of new problems.

But more than that, new concepts can originate and develop, and become more general and precise, only on the basis of the very problems they enable us to solve, only through those theorems of which they form a part. The concepts of variable and function did not arise in complete form in the mind of Galileo, Descartes, Newton, or anybody else. They occurred to many mathematicians (for example Napier in connection with logarithms) and gradually assumed a more or less clear, but still by no means final, form with Newton and Leibnitz, being made still more precise and general in the subsequent development of analysis. Their present-day definition was laid down only in the 19th century, but even it is not *absolutely* rigorous or *altogether* final. The development of the concept of a function is continuing even at the present time.

Mathematical analysis was based on material furnished by the new science of mechanics, and on problems of geometry and algebra. The first definite step toward the mathematics of variable magnitudes was the appearance in 1637 of the "geometry" of Descartes, where the foundations were laid for the so-called analytic geometry. The basic ideas of Descartes are as follows.

Suppose we are given, for example, the equation

$$x^2 + y^2 = a^2. \tag{6}$$

In algebra x and y were understood as unknowns, and since the given equation does not allow us to determine them, it did not present any essential interest for algebra. But Descartes did not consider x and y as unknowns, to be found from the equation, but as *variables*; so that the given equation expresses the interdependence of two variables. Such an equation may be written in general form, by taking all its terms to the left-hand side, thus:

$$F(x, y) = 0.$$

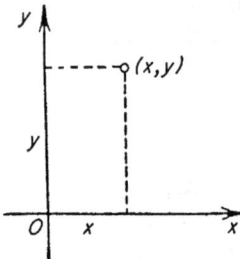

Fig. 6.

Further, Descartes introduced into the plane the coordinates x, y which are now called Cartesian (figure 6). In this way, to each pair of values x and y there corresponds a point, and conversely to each point there corresponds a pair of coordinates x, y. Consequently, the equation $F(x, y) = 0$ determines the geometric locus of those points on the plane whose coordinates satisfy the equation. In general, this will be a curve. For example, equation (6) determines the circumference of a circle of radius a with center at the origin. In fact, as is obvious from figure 7, by the theorem of Pythagoras, $x^2 + y^2$ is the square of the distance from the origin O to the point M with coordinates x and y. So equation (6) represents the geometric locus of those points whose distance from the origin is equal to a, which is the circumference of a circle.

Conversely, a geometric locus of points, given by a geometric condition, may also be given by an equation expressing the same condition in the language of algebra by means of coordinates. For example, the geometric condition defining the circumference of a circle, namely that it is a geometric locus of points equidistant from a given point, may be expressed in algebraic language by equation (6).

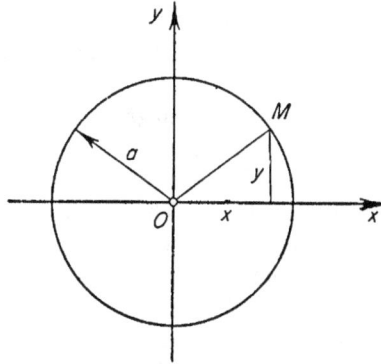

FIG. 7.

Thus the general problem and the general method of analytic geometry are as follows: We represent a given equation in two variables by a curve on the plane, and from the algebraic properties of the equation we investigate the geometric properties of the corresponding curve; and conversely, from the geometric properties of the curve we find the equation, and then from the algebraic properties of the equation we investigate the geometric properties of the curve. In this way geometric problems may be reduced to algebraic, and so finally to computation.

The content of analytic geometry will be discussed in detail in Chapter III. We now wish to direct attention to the fact that, as is evident from our short explanation, it originated in a union of geometry, algebra, and the general idea of a variable magnitude. The main geometric content of the early beginnings of analytic geometry was the theory of conic sections, ellipse, hyperbola, and parabola. This theory, as we have pointed out, was developed by the ancient Greeks; the results of Apollonius already contained in geometric form the equations of the conic sections. The union of this geometric content with algebraic form, developed after the time of the Greeks, and with the general idea of a variable magnitude, arising from the study of motion, produced analytic geometry.

Among the Greeks the conic sections were a subject of purely mathematical interest, but by the time of Descartes they were of practical importance for astronomy, mechanics, and technology. Kepler (1571–1630) discovered that the planets move around the sun in ellipses, and Galileo established the fact that a body thrown in the air, whether it is a stone or a cannonball, moves along a parabola (to the first approximation, if we may neglect air resistance). As a result, the calculation of various magnitudes referring to the conic sections became an urgent necessity, and it was the method of Descartes that solved this problem. So the way was prepared for his method by the preceding development of mathematics, and the method itself was brought into existence by the insistent demands of science and technology.

3. Differential and integral calculus. The next decisive step in the mathematics of variable magnitudes was taken by Newton and Leibnitz during the second half of the 17th century, in the founding of the differential and integral calculus. This was the actual beginning of analysis, since the subject matter of this calculus is the properties of functions themselves, as distinct from the subject matter of analytic geometry, which is geometric figures. In fact Newton and Leibnitz only brought to completion an immense amount of preparatory work, shared by many mathematicians and going back to the methods for determining areas and volumes worked out by the ancient Greeks.

Here we shall not explain the fundamental concepts of differential and integral calculus and of the theories of analysis that followed them, since this will be done in the special chapters devoted to these theories. We wish only to draw attention to the sources of the calculus, which were mainly the new problems of mechanics and the old problems of geometry, the latter consisting of drawing a tangent to a given curve and of determining areas and volumes. These geometric problems had already been studied by the ancients (it is sufficient to mention Archimedes), and also by Kepler, Cavalieri, and others at the beginning of the 17th century. But the decisive event was the discovery of the remarkable relation between these two types of problems and the formulation of a general method for solving them; this was the achievement of Newton and Leibnitz.

This relation, allowing us to connect the problems of mechanics with those of geometry, was discovered because of the possibility, arising from the method of coordinates, of making a graphical representation of the dependence of one variable on another, or in other words of a function. With the help of this graphical representation, it is easy for us to formulate the earlier mentioned relation, between the problems of mechanics and geometry, which was the source of the differential and integral calculus, and consequently to describe the general content of these two types of calculus.

The differential calculus is basically a method for finding the velocity of motion when we know the distance covered at any given time. This problem is solved by "differentiation." It turns out that the problem is completely equivalent to that of drawing a tangent to the curve representing the dependence of distance on time. The velocity at the moment t

is equal to the slope of the tangent to the curve at the point corresponding to *t* (figure 8).

The integral calculus is basically a method of finding the distance covered when the velocity is known, or more generally of finding the total result of the action of a variable magnitude. This problem is obviously

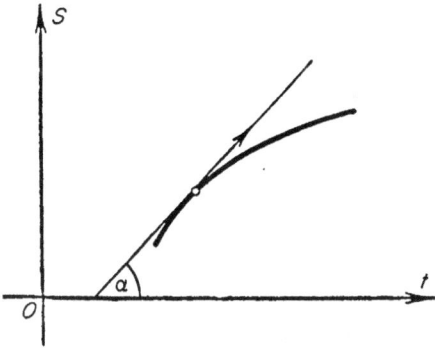

FIG. 8.

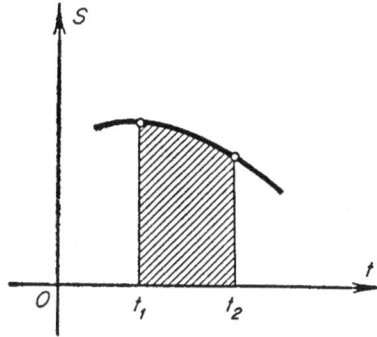

FIG. 9.

the converse of the problem of the differential calculus (the problem of finding the velocity); it is solved by "integration." It turns out that the problem of integration is completely equivalent to that of finding the area under the curve representing the dependence of the velocity on time. The distance covered in the interval of time from the moment t_1 to the moment t_2 is equal to the area under the curve between the straight lines corresponding on the graph to the values t_1 and t_2 (figure 9).

By abstracting from the mechanical formulation of the problems of the calculus and by dealing with functions rather than with dependence of distance or velocity on time, we obtain the general concept of the problems of differential and integral calculus in abstract form.

Fundamental to the calculus, as to the whole subsequent development of analysis, is the concept of a limit, which was formulated somewhat later than the other fundamental concepts of variable and function. In the early days of analysis the role later played by the limit was taken by the somewhat nebulous concept of an infinitesimal. The methods for actual calculation of velocity, given the distance covered (namely, differentiation), and of distance, given the velocity (integration), were founded on a union of algebra with the concept of limit. Analysis originated in the application of these concepts and methods to the aforementioned problems of mechanics and geometry (and also to certain other problems; for example, problems of maxima and minima). The science of analysis was in turn absolutely necessary for the development of mechanics, in the formulation of whose laws its concepts had already appeared in latent form. For example, the second law of Newton, as formulated by Newton himself, states that "the change in momentum is proportional to the acting force" (more precisely: The rate of change of momentum is proportional to the force). Consequently, if we wish to make any use of this law, we must be able to define the rate of change of a variable, that is, to differentiate. (If we state the law in the form that the acceleration is proportional to the force, the problem remains the same, because acceleration is proportional to rate of change of momentum.) Also, it is perfectly clear that in order to state the law governing a motion when the force is variable (in other words, the motion proceeds with a variable acceleration), we must be able to solve the inverse problem of finding a magnitude given its rate of change; in other words, we must be able to integrate. So one might say that Newton was simply *compelled* to invent differentiation and integration in order to develop the science of mechanics.

4. Other branches of analysis. Along with the differential and integral calculus, other branches of analysis arose: The theory of series (see Chapter II, §14), the theory of differential equations (Chapters V and VI), and the application of analysis to geometry, which later became a special branch of geometry, called differential geometry and dealing with the general theory of curves and surfaces (Chapter VII). All these theories were brought to life by the problems of mechanics, physics, and technology.

The theory of differential equations, the most important branch of analysis, has to do with equations in which the unknown is no longer a magnitude but a function, or in other words a law governing the dependence of one magnitude on another or on several others. It is easy to understand how such equations arose. In mechanics we seek to determine the whole law of motion of a body under given conditions and not just one value of the velocity or of the distance covered. In the mechanics of fluids it is necessary to find the distribution of velocity over the whole mass of fluid in motion, or in other words to find the dependence of the velocity on all three space coordinates and on time. Analogously, in the theory of electricity and magnetism we must find the tension in the field throughout all space; that is, the dependence of this tension on the same three space coordinates, and similarly in other cases.

Problems of this sort arose continually in the various branches of mechanics, including hydrodynamics and the theory of elasticity, in acoustics, in the theory of electricity and magnetism, and in the theory of heat. From the very moment of its birth, analysis remained in close contact with mechanics and with physics in general, its most important achievements being invariably connected with the solution of problems posed by the exact sciences. Beginning with Newton, the greatest analysts, D. Bernoulli (1700–1782), L. Euler (1707–1783), J. Lagrange (1736–1813), H. Poincaré (1854–1912), M. V. Ostrogradskiĭ (1801–1861) and A. M. Lyapunov (1857–1918), as well as many others who laid new foundations in analysis, started as a rule from the urgent problems of contemporary physics.

In this way new theories arose: In direct connection with mechanics, Euler and Lagrange founded a new branch of analysis, called the calculus of variations (see Chapter VIII), and at the end of the 19th century Poincaré and Lyapunov, starting again from the problems of mechanics, founded the so-called qualitative theory of differential equations (see Chapter V, §7).

In the 19th century analysis was enriched by an important new branch, the theory of functions of a complex variable (see Chapter IX). The rudiments of it are to be found in the works of Euler and certain other mathematicians, but its transformation into a well-formed theory took place in the middle of the 19th century and was carried out to a great extent by the French mathematician Cauchy (1789–1857). This theory rapidly underwent an imposing development with numerous significant results that allowed mathematicians to penetrate more deeply into many of the laws of analysis and found important applications in problems of mathematics itself, and of physics and technology.

Analysis developed rapidly; not only did it form the center and the most important part of mathematics but it also penetrated into the older regions: algebra, geometry, and even the theory of numbers. Algebra began to be thought of as basically the doctrine of functions expressed in the form of polynomials of one or several variables.* Analytic and differential geometry began to dominate the field of geometry. As far back as Euler, methods of analysis were introduced into the theory of numbers and formed in this way the beginning of the so-called analytic theory of numbers, which contains some of the most profound achievements of the science of whole numbers.

Through the influence of analysis, with its concepts of variable, function, and limit, the whole of mathematics was penetrated by the idea of motion and change, and therefore of dialectic. In exactly the same way, basically through analysis, mathematics was affected by the exact sciences and

---

* Polynomials are functions of the form $y = a_0x^n + a_1x^{n-1} + \cdots + a_n$. The fundamental problem of the algebra of the period, namely the solution of the equation $a_0x^n + a_1x^{n-1} + \cdots + a_n = 0$, simply means the search for values of x for which the function $y = a_0x^n + a_1x^{n-1} + \cdots + a_n$ is equal to zero. The very existence of a solution, of a root of the equation, which is called the fundamental theorem of algebra, is proved by means of analysis (see Chapter IV, §3).

technology and in turn played a role in their development, since it was the means of giving exact expression to their laws and of solving their problems. Just as among the Greeks mathematics was basically geometry, one may say that after Newton it was basically analysis. Of course, analysis did not completely absorb the whole of mathematics; in geometry, in the theory of numbers, and in algebra the problems and methods characteristic of these sciences were everywhere continued. Thus in the 17th century there arose, along with analytic geometry, another branch of geometry, namely projective geometry, in which purely geometric methods played a dominant role. It originated chiefly in problems of the representation of objects on a plane (projection), and as a result it is particularly useful in descriptive geometry.

At the same time there was developed an important new branch of mathematics, the theory of probability, which takes as its subject matter the uniformities observable in large masses of phenomena, such as a long series of rifle shots or tosses of a coin. In the succeeding period it acquired a special importance in physics and technology and its development was conditioned by the problems which came to it from those branches of science. The characteristic feature of this theory is that it deals with the laws of "random events," providing mathematical methods for investigation of the irregularities that necessarily appear in random events. The basic features of the theory of probability will be explained in Chapter XI.

5. Applications of analysis. Analysis in all its branches provided physics and technology with powerful methods for the solution of problems of many different kinds. We have already mentioned the earliest of these: to find the rate of change of a magnitude when we know how the magnitude itself depends on time; to find the area of curvilinear figures and the volumes of solids; and to find the total result of some process or another or the total action of a variable magnitude. Thus, the integral calculus allows us to determine the work done by an expanding gas as the pressure changes according to a well-known law; the same integral calculus allows us to compute, for example, the tension of an electric field with an arbitrarily given system of charges, basing our work on the law of Coulomb which determines the tension of a field resulting from a point charge, and so forth.

Further, analysis provided a method for finding the maximum and the minimum values of a magnitude under given conditions. Thus, with the help of analysis it is easy to determine the shape of a cylindrical cistern which for a given volume will have the smallest surface and consequently will require the smallest outlay of material. It turns out that the cistern will have this property if its height is equal to the diameter of its base (figure 10). Analysis allows us to determine the shape of the curve along which a body must roll in order to fall in the shortest time from one given point to another (this curve is the so-called cycloid; figure 11).

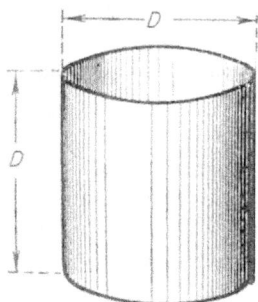

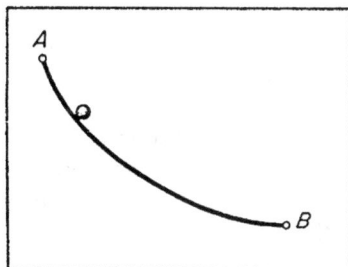

FIG. 10.   FIG. 11.

For the solution of these and other problems the reader may turn to Chapters II and VIII.

Analysis, or more precisely the theory of differential equations, allows us not merely to find separate values for variable magnitudes but also to determine unknown functions; that is, to find laws of dependence of certain variables on others. Thus we have the possibility, on the basis of the general laws of electricity, of computing how the current varies with time in a circuit with arbitrary resistance, capacitance, and self-induction. We can determine laws for the distribution of velocities throughout the whole mass of a fluid under given conditions. We can deduce general laws for the vibration of strings and membranes, and for the propagation of vibrations in various media; here we are referring to sound waves, electromagnetic waves, or elastic vibrations propagated through the Earth by earthquakes or explosions. Parenthetically, we may remark that new methods are thereby provided for searching for useful minerals and for carrying out investigations far below the surface of the Earth. Individual problems of this sort will be found in Chapters V and VI.

Finally, analysis not only provides us with methods for solving special problems; it also gives us general methods for mathematical formulation of the quantitative laws of the exact sciences. As was mentioned, earlier, the general laws of mechanics could not be formulated mathematically without recourse to the concepts of analysis, and without such a formulation we would not be able to solve the problems of mechanics. In exactly the same way the general laws for heat conduction, diffusion through porous materials, propagation of vibrations, the course of chemical reactions, the basic laws of electromagnetism, and many other laws simply could not be given a mathematical formulation without the concepts of analysis. It is only as the result of such a formulation that these laws can be applied to the most varied concrete cases, providing a basis for exact mathematical conclusions in the special problems of heat conduction, vibrations, chemical solution, electromagnetic fields, and other problems of mechanics, astronomy, and all the numerous branches of physics, chemistry, heat engineering, power, machine construction, electrical engineering, and so forth.

6. Critical examination of the foundations of analysis. Just as in the history of geometry among Greeks the rigorous and systematic presentation given by Euclid brought to completion a long previous development, so in the development of analysis there arose the necessity of placing it upon a firmer basis than had been provided by the first creators of its powerful methods: Newton, Euler, Lagrange, and others. As the analysis founded by them grew more extensive, it began, on the one hand, to deal with more profound and difficult problems, and on the other, to require from its very extent a more systematic and carefully reasoned basis. The growth of the theory necessitated a systematization and critical analysis of its foundations. To put a theory on a firm foundation requires examination of its entire development and should by no means be considered as a starting point for the theory itself, since without the theory we would simply have no idea of what it is that we need to provide with a foundation. By the way, certain contemporary formalists forget this fact when they consider it advisable to found and develop a theory starting from axioms that have not been selected on the basis of any analysis of the actual material which they are supposed to summarize. But the axioms themselves require a justification of their content; they only sum up other material and provide a foundation for the logical construction of a theory.*

The necessary period of criticism, systematization, and laying of foundations occurred in analysis at the beginning of the last century. Through the efforts of a number of eminent scientists this important and difficult work was brought to a successful completion. In particular, precise definitions were given for the basic concepts of real number, variable, function, limit, and continuity.

However, as we have already had occasion to mention, none of these definitions may be considered as absolutely rigorous or final. The development of these concepts is continuing. Euclid and all the mathematicians in the course of 2,000 years after him no doubt considered his "Elements"

---

* This double role of the axioms is sometimes lost from view even in works of a methodological character, which thereby attribute to the construction of axioms a significance which does not at all belong to it, namely that of the total construction of a theory.

as the practical limit of logical rigor. But to a contemporary view the Euclidean foundations of geometry seem quite superficial. This historical example shows that we ought not to flatter ourselves with any idea of "absolute" or "final" rigor in contemporary mathematics. In a science that is not yet dead and mummified, there is not and cannot be anything perfect. But we can say with confidence that the foundations of analysis as they exist at present correspond in a quite satisfactory way to the contemporary problems of science and the contemporary conception of logical precision; and second, that the continued deepening of these concepts and the discussions that are now taking place about them give us no cause, and will not give us cause, simply to reject them; these discussions will lead us to a new, more precise, and more profound understanding, the results of which it is still difficult to estimate.

Although the establishment of the basic principles of a theory forms a summary of its development, it does not represent the end of the theory; on the contrary, it is conducive to further development. This is exactly what happened in analysis. In connection with the deepening of its foundations there arose a new mathematical theory, created by the German mathematician Cantor in the seventies of the last century, namely the general theory of infinite sets of arbitrary abstract objects, whether numbers, points, functions or any other "elements". On the basis of these ideas there grew up a new chapter in analysis, the so-called theory of functions of a real variable, whose concepts, along with those of the foundations of analysis and the theory of sets, are explained in Chapter XV. At the same time the general ideas of the theory of sets penetrated every branch of mathematics. But this "set-theoretical point of view" is inseparably connected with a new stage in the development of mathematics, which we will now consider briefly.

§7. Contemporary Mathematics

1. The more advanced character of present-day mathematics. To the four stages of the develoment of mathematics mentioned in §5 there naturally correspond stages in our mathematical education, the material learned at each stage of our study consisting, to a fair degree of approximation, of the basic content of the corresponding period in the history of mathematics.

The basic results of arithmetic and geometry, obtained in the first period of the development of mathematics, form the subject of primary education and are known to us all. For example, when we determine the quantity of material necessary to carry out a certain job, let us say to cover a floor, we are already making use of these first results of mathematics. The most important achievements of the second period, the period of elementary mathematics, are taught in the high schools. The basic results of the third period, the foundations of analysis, the theory of differential equations, higher algebra, and so forth, form the mathematical instruction of an engineer; they are studied in all the schools of higher education, except those devoted purely to the humanities. In this way the basic ideas and results of the mathematics of that period are widely known, use being made of them to some extent by almost every engineer and scientist.

On the other hand, the ideas and results of the present-day period of mathematics are studied almost exclusively in graduate departments of mathematics and physics. Beside mathematical specialists, they are used by researchers in the fields of mechanics and physics, and in a number of the newer branches of technology. Of course, this does not at all mean that they have no practical application, but since they represent the most recent results of science, they are naturally more complicated. Consequently, as we now pass to a general description of the latest stage in the development of mathematics we can no longer consider that everything which we mention briefly will be altogether clear. We will try to present in a few lines the most general character of the new branches of mathematics; their content will be explained in greater detail in the corresponding chapters of the book.

If the present section seems overly difficult it may be passed over at first reading and taken up again after study of the special chapters.

2. Geometry. The beginning of the present-day development of mathematics is characterized by profound changes in all its basic fields: algebra, geometry, and analysis. This change may perhaps be followed most clearly in the field of geometry. In the year 1826 Lobačevskiĭ, and almost simultaneously with him the Hungarian mathematician Janos Bolyai, developed the new non-Euclidean geometry. The ideas of Lobačevskiĭ were far from being immediately understood by all mathematicians. They were too bold and unexpected. But from this moment there began a fundamental new development of geometry; the very conception of what is meant by geometry was changed. Its subject matter and the range of its applications were rapidly extended. The most important step, after Lobačesvskiĭ, in this direction was taken in 1854 by the celebrated German mathematician, Riemann. He clearly formulated the general idea that an unlimited number of "spaces" could be investigated by geometry, and at the same time he indicated their possible significance in the real world. In the new development of geometry two features were characteristic.

In the first place the earlier geometry studied only the *spatial* forms and relations of the material world, and then only to the extent in which they appear in the framework of Euclidean geometry, but now the subject matter of geometry began to include also many *other* forms and relations of the actual world, provided only they were similar to the spatial ones and therefore allowed the use of geometric methods. The term "space" thereby took on in mathematics a new meaning, broader and at the same time more special. Simultaneously, the methods of geometry became much richer and more varied. In their turn they provide us with more complete means for learning about the physical world around us, the world from which geometry in its original form was abstracted.

In the second place, even in Euclidean geometry important progress was made: In it were studied the properties of incomparably more complicated figures, even including arbitrary sets of points. Also a fundamentally new attitude appeared toward the properties of the figures under investigation. Separate groups of properties were distinguished, which could be investigated in abstraction from others, and this very abstraction *within* geometry gave rise to many characteristic branches of the subject, which essentially became independent "geometries." The development of geometry in all these directions is being continued and more and more new "spaces" and their "geometries" are being studied: the space of Lobačesvskiĭ, projective space, Euclidean and other spaces of various dimensions, in particular four-dimensional space, Riemann spaces, Finsler spaces, topological spaces, and so forth. These theories find important application in mathematics itself, outside of geometry, and also in physics and mechanics; particularly noteworthy are their applications in the theory of relativity of contemporary physics, which is a theory of space, time, and gravitation. From what has been said it is clear that we are dealing here with a qualitative change in geometry.

The ideas of contemporary geometry and some of the elements of the theory of various spaces investigated in it will be explained in Chapters XVII and XVIII.

3. Algebra. Algebra too underwent a qualitative change. In the first half of the 19th century new theories arose, which led to changes in its character, and to an extension of its subject matter and its range of application.

In its original form, as pointed out in §5, algebra dealt with mathematical operations on numbers considered from a formal point of view, in abstraction from given concrete numbers. This abstraction found expression in the fact that in algebra magnitudes are denoted by letters, on which calculations are carried out according to well-known formal rules.

Contemporary algebra retains this basis but widens it in a very extensive way. It now considers "magnitudes" of a much more general nature than numbers, and studies operations on these "magnitudes" which are to some extent analogous in their formal properties to the ordinary operations of arithmetic: addition, subtraction, multiplication, and divison. A very simple example is offered by vector magnitudes, which may be added by the well-known parallelogram rule. But the generalization carried out in contemporary algebra is such that even the very term "magnitude" often loses its meaning and one speaks more generally of "elements" on which it is possible to perform operations similar to the usual algebraic ones. For example, two motions carried out one after the other are evidently equivalent to a certain single motion, which is the sum of the two; two algebraic transformations of a formula may be equivalent to a single transformation that produces the same result, and so forth; and so it is possible to speak of a characteristic "addition" of motions or transformations. All this and much else is studied in a general abstract form in contemporary algebra.

The new algebraic theories in this direction arose in the first half of the 19th century in the investigations of a number of mathematicians, among whom we should particularly mention the French mathematician Galois (1811–1832). The concepts, methods, and results of contemporary algebra find important applications in analysis, geometry, physics, and crystallography. In particular, the theory mentioned at the end of §3 concerning the symmetry of crystals, which was developed by E. S. Fedorov, is based on a union of geometry with one of the new algebraic theories, the so-called theory of groups.

As we see, we are dealing here with a fundamental, qualitative generalization of the subject matter of algebra with a change in the very concept of what algebra is. The ideas of contemporary algebra and the basic elements of some of its theories will be explained in Chapter XX and XVI.

4. Analysis.   Analysis in all its branches also made profound progress. In the first place, as was already mentioned in the preceding section, its foundations were made more precise; in particular, its basic concepts were given exact and general definitions: such concepts as function, limit, integral and finally, the basic concept of a variable magnitude (a rigorous definition was given for the real number). A beginning of the process of putting analysis on a more precise foundation was made by the Czech mathematician Bolzano (1781–1848), the French mathematician Cauchy (1789–1857), and a number of others. This greater precision was gained at the same time as the new developments in algebra and geometry were being made; it was brought to completion in its present well-known form in the eighties of the 19th century by the German mathematicians Weierstrass, Dedekind, and Cantor. As was mentioned at the end of §6, Cantor also laid the foundation for the theory of transfinite sets, which plays such a large role in the development of the newer ideas in mathematics.

The increase in precision in the concepts of variable and function in connection with the theory of sets laid the foundation for a further development of analysis. A transition was made to the study of more general functions; and in the same direction the apparatus of analysis, namely the integral and differential calculus, was also generalized. Thus, on the threshold of the present century, there arose the new branch of analysis already mentioned in §6, the so-called theory of functions of a real variable. The development of this theory is chiefly connected with the French mathematicians, Borel and Lebesgue and others, and with N. N. Luzin (1883–1950) and his school. In general, the newer branches of analysis are called modern analysis in contradistinction to the earlier so-called classical analysis.

Other new theories arose in analysis. Thus a special branch was formed by the theory of approximation of functions, which studies questions of the best approximate representation of general functions by various "simple" functions, above all by polynomials, that is by functions of the form

$$a_0 x^n + a_1 x^{n-1} + \cdots + a_{n-1} x + a_n .$$

The theory of approximation of functions has great importance, if only for the reason that it lays down general foundations for the practical calculation of functions, for the approximate replacement of complicated functions by simpler ones. The rudiments of this theory go back to the very beginnings of analysis. Its modern direction was given to it by the great Russian mathematician P. L. Čebyšev (1821–1894). This direction was later developed into the so-called constructive theory of functions, chiefly in the works of Soviet mathematicians, particularly S. N. Bernšteĭn (born 1880), to whom belong the most important results in this field. Chapter XII deals with approximation of functions.

We spoke earlier about the development of the theory of functions of a complex variable. We must still mention the so-called qualitative theory of differential equations, originating in the works of Poincaré (1854–1912) and A. M. Lyapunov (1857–1918), about which some ideas will be given in Chapter V, and also the theory of integral equations. These theories have great practical importance in mechanics, physics, and technology. Thus, the qualitative theory of differential equations provides solutions of problems concerning stability of motion, and the action of mechanisms or of vibrating electric systems and the like. Stability of a process means in the most general sense that if small changes are made in the initial data or in the conditions of the motion, then the motion itself during the whole of its course will change only slightly. The technical significance of questions of this sort hardly needs to be emphasized.

5. Functional analysis. On the ground prepared by the development of analysis and mathematical physics, along with the new ideas of geometry and algebra, there has grown up an extensive new division of mathematics, the so-called functional analysis, which plays an exceptionally important role in modern mathematics. Many mathematicians shared in creating it; let us mention, for example, the greatest German mathematician of recent times, Hilbert (1862–1943), the Hungarian mathematician Riesz (1880–1956) and the Polish mathematician Banach (1892–1945). The separate Chapter XIX is devoted to functional analysis.

The essence of this new branch of mathematics consists briefly in the following. In classical analysis the variable is a magnitude, or "number," but in functional analysis the function itself is regarded as the variable. The properties of the given function are determined here not in themselves but in relation to other functions. What is under study is not a separate function but a whole collection of functions characterized by one property or another; for example, the collection of all continuous functions. Such a collection of functions forms the so-called functional space. This procedure corresponds, for example, to the fact that we may consider the collection of all curves on a surface or of all possible motions of a given mechanical system, thereby defining the properties of the separate curves or motions in their relation to other curves or motions.

The transition from the investigation of separate functions to a *variable* function is similar to the transition from unknown numbers x, y to variables x, y; that is, it is similar to the idea of Descartes mentioned in a preceding paragraph. On the basis of this idea Descartes produced his well-known union of algebra and geometry, of an equation and a curve, which is one of the most important elements in the rise of analysis. Similarly, the union of the concept of a variable function with the ideas of contemporary algebra and geometry produced the new functional analysis. Just as analysis was necessary for the development of the mechanics of the time, so functional analysis provided new methods for the solution of present-day problems of mathematical physics and produced the mathematical apparatus for the new quantum mechanics of the atom. History repeats itself as usual, but in a new way, on a higher plane. As we have said, functional analysis unites the basic ideas and methods of analysis, of modern algebra, and of geometry and in its turn exercises an influence on the development of these branches of mathematics. The problems arising in classical analysis now find new, more general solutions, often almost at a single step, by means of functional analysis. Here, as at a focus, are gathered together, in a very productive way, the most general and abstract ideas of modern mathematics.

From this short sketch, from this mere enumeration of the new directions of analysis (the theory of functions of a real variable, theory of approximation of functions, qualitative theory of differential equations, theory of integral equations, and functional analysis) it may be seen that we are dealing here in fact with an essentially new stage in the development of analysis.

6. Computational mathematics and mathematical logic. At all periods the technical level of the means of computation has had an essential influence on mathematical methods. But the equipment for carrying out calculations which has been at our disposal up until most recent times has been very limited. The simplest devices, such as the abacus, tables of logarithms and the logarithmic sliderule, the calculating machine, and finally more complicated calculators and the automatic calculating machine, these were the basic implements for computation existing up to the forties of the 20th century. These implements made it possible to carry out more or less quickly the separate operations of addition, multiplication, and so forth. But to carry through to final numerical result the practical problems that arise nowadays requires a colossal number of such operations, following one another in a complicated program that sometimes depends on results obtained during the course of the calculation. The solution of such problems proved to be practically impossible or completely valueless on account of the length of the process of solution. But in the last ten years a radical change has taken place in the whole science of computation. Modern calculating machines, constructed on new principles, allow us to make computations with exceptionally great speed and at the same time to carry out complicated chains of calculations automatically, according to extremely flexible programs arranged in advance. Some of the questions connected with the construction and significance of modern calculating machines will be discussed in Chapter XIV.

The new techniques not only enable us to carry out investigations that were formerly quite impracticable but also lead us to change our estimate of the value of many well-known mathematical results. For example, they have given a special stimulus to the development of approximative methods; that is, methods which allow us, by a chain of elementary operations, to reach a desired numerical result with sufficiently great accuracy. The mathematical methods themselves must now be estimated from the point of view of their suitability for corresponding machines.

In close connection with the development of calculating techniques is the subject of mathematical logic. It was developed primarily as a result of intrinsic difficulties arising in mathematics itself, its subject matter being the analysis of mathematical proof. It is itself a branch of mathematics, and includes those branches of general logic that can be objectively formulated and developed by the mathematical method.

Although on the one hand mathematical logic thus goes back to the very sources and foundations of mathematics, it is closely connected, on the other hand, with the most modern questions of computational technique. Naturally, for example, a proof that leads to the setting up of a definite preassigned process, permitting us to approach a desired result with an arbitrary degree of accuracy, is essentially different from more abstract proofs on the existence of the given result.

There also arises here a characteristic range of questions concerning the degree of generality possible in problems that can be dealt with by a method which is completely defined in advance at every step. Profound results have been reached along these lines in mathematical logic, results that are extremely important from a general epistemological point of view.

It would not be an exaggeration to say that with the development of the new computational techniques and the achievements of mathematical logic a new period has begun in modern mathematics, characterized by the fact that its subject matter is not only the study of one object or another but also all the ways and means by which such an object can be defined; not only certain problems, but also all possible methods of solving them.

To what has been said it is only necessary to add that also in the older branches of mathematics, the theory of numbers, Euclidean geometry, classical algebra and analysis, and the theory of probability, rapid development has continued throughout the whole period of modern mathematics so that these fields have been enriched by many new fundamental ideas and results; let us mention, for example, the results attained in the theory of numbers and in the geometry of everyday space by the Russian and Soviet mathematicians P. L. Čebyšev, E. S. Fedorov, I. M. Vinogradov, and others. The development on a wide front of the theory of probability has been connected with the extraordinarily important regularities observable in statistical physics and in contemporary problems of technology.

7. Characteristic features of modern mathematics. What are the most general characteristics of modern mathematics as a whole, distinguishing it from the earlier development of geometry, algebra, and analysis?

First of all is the immense extension of the subject matter of mathematics and of its applications. Such an extension of subject matter and range of application represents an enormous quantitative and qualitative growth, brought about by the appearance of powerful new theories and methods that allow us to solve problems completely inaccessible up to now. This extension of the subject matter of mathematics is characterized by the fact that contemporary mathematics conscientiously sets itself the task of studying all possible types of quantitative relationships and spatial forms.

A second characteristic feature of modern mathematics is the formation of general concepts on a new and higher level of abstraction. It is precisely this feature that guarantees preservation of the unity of mathematics, in spite of its immense growth in widely differing branches. Even in parts of mathematics that are extremely far from one another similarities of structure are brought to light by the general concepts and theories of the present day. They guarantee that contemporary mathematical methods will have great generality and breath of application; in particular, they produce a profound interpenetration of the fundamental branches of mathematics: geometry, algebra, and analysis.

As one of the characteristic features of modern mathematics, we must also mention the obvious dominance of the set-theoretical point of view. Of course, this point of view owes its significance to the fact that it summarizes in a certain sense the rich content of all the preceding developments of mathematics. Finally, one of the most characteristic features of modern mathematics is the profound analysis of its foundations, of the mutual influence of its concepts, of the structure of its separate theories, and of the methods of mathematical proof. Without such an analysis of foundations it would not be possible to improve or develop any further the principles and theories that have led to the present generalizations.

The characteristic feature of modern mathematics may be said to be that its subject matter consists not only of given quantitative relations and forms but of all possible ones. In geometry, we speak not only of spatial relations and forms but of all possible forms similar to spatial ones. In algebra, we speak of various abstract systems of objects with all possible laws of operation on them. In analysis, not only magnitudes are considered as variables but the very functions themselves. In a functional space all the functions of a given type (all the possible interdependences among the variables) are brought together. Summing up, it is possible to say that while elementary mathematics deals with constant magnitudes, and the next period with variable magnitudes, *contemporary mathematics is the mathematics of all possible (in general, variable) quantitative relations and interdependences among magnitudes.* This definition is, of course, incomplete, but it does emphasize the characteristic feature of modern mathematics which distinguishes it from the mathematics of preceding ages.*

Suggested Reading

Preliminary remark. The original Russian text of *Mathematics: its content, methods, and meaning* contains a list of recommended books at the end of each of its twenty chapters. In the present translation these books have been retained only if they have been translated into English. In compensation, the lists given here contain many other, readily available, works in the English language.

Books dealing with mathematics in general

E. T. Bell, *The development of mathematics*, 2d ed., McGraw-Hill, New York, 1945.

R. Courant and H. Robbins, *What is mathematics?* Oxford University Press, New York, 1941.

H. Eves and C. V. Newsom, *An introduction to the foundations and fundamental concepts of mathematics*, Rinehart, New York, 1958.

G. H. Hardy, *A mathematician's apology*, Macmillan, New York, 1940.

R. L. Wilder, *Introduction to the foundations of mathematics*, Wiley, New York, 1952.

Books of a historical character

R. C. Archibald, *Outline of the history of mathematics*, 5th ed., Mathematical Association of America, Oberlin, Ohio, 1941.

F. Cajori, *History of mathematics*, 2d ed., Macmillan, New York, 1919.

Euclid, The thirteen books of Euclid's *Elements* translated with an introduction and commentary by T. L. Heath, 2d ed., 3 vols., Dover, New York, 1956.

O. E. Neugebauer, *The exact sciences in antiquity*, Princeton University Press, Princeton, N. J., 1952.

D. E. Smith, *History of mathematics*, Vol. I. *General survey of the history of elementary mathematics*, Vol. II. *Special topics of elementary mathematics*, Dover, New York, 1958.

D. J. Struik, *A concise history of mathematics*, Dover, New York, 1948.

B. L. van der Waerden, *Science awakening*, P. Noordhoff, Groningen, 1954.

---

* This section is followed in the original Russian text by two sections entitled "The essential nature of mathematics" and "The laws of the development of mathematics." These sections are omitted in the present translation in view of the fact that they discuss in more detail, and in the more general philosophical setting of dialectical materialism, points of view already stated with great clarity in the preceding sections.

CHAPTER II

ANALYSIS

§1. Introduction

The rise at the end of the Middle Ages of new conditions of manufacture in Europe, namely the birth of capitalism, which at this time was replacing the feudal system, was accompanied by important geographical discoveries and explorations. In 1492, relying on the idea that the earth is spherical, Columbus discovered the New World. The discovery by Columbus greatly extended the boundaries of the known world and produced a revolution in the minds of men. The end of the 15th century and the beginning of the 16th saw the creative activity of the great artist-humanists Leonardo da Vinci, Raphael, and Michelangelo, which gave new meaning to art. In 1543 Copernicus published his work "On the revolution of the heavenly bodies," which completely changed the face of astronomy; in 1609 appeared the "New astronomy" of Kepler, containing his first and second laws for the motion of the planets around the sun, and in 1618 his book "Harmony of the world," containing the third law. Galileo, on the basis of his study of the works of Archimedes and his own bold experiments, laid the foundations for the new mechanics, an indispensable science for the newly arising technology. In 1609 Galileo directed his recently constructed telescope, though still small and imperfect, toward the night sky; the first glance in a telescope was enough to destroy the ideal celestial spheres of Aristotle and the dogma of the perfect form of celestial bodies. The surface of the moon was seen to be covered with mountains and pitted with craters. Venus displayed phases like the Moon, Jupiter was surrounded by four satellites and provided a miniature visual model of the solar system. The Milky Way fell apart into separate stars, and for the first time men felt the staggeringly immense distance

65

of the stars. No other scientific discovery has ever made such an impression on the civilized world.*

The further development of navigation, and consequently of astronomy, and also the new development of technology and mechanics necessitated the study of many new mathematical problems. The novelty of these problems consisted chiefly in the fact that they required mathematical study of the laws of motion in a broad sense of the word.

The state of rest and motionlessness is unknown in nature. The whole of nature, from the smallest particles up to the most massive bodies, is in a state of eternal creation and annihilation, in a perpetual flux, in unceasing motion and change. In the final analysis, every natural science studies some aspect of this motion. Mathematical analysis is that branch of mathematics that provides methods for the quantitative investigation of various processes of change, motion, and dependence of one magnitude on another. So it naturally arose in a period when the development of mechanics and astronomy, brought to life by questions of technology and navigation, had already produced a considerable accumulation of observations, measurements, and hypotheses and was leading science straight toward quantitative investigation of the simplest forms of motion.

The name "infinitesimal analysis" says nothing about the subject matter under discussion but emphasizes the method. We are dealing here with the special mathematical method of infinitesimals, or in its modern form, of limits. We now give some typical examples of arguments which make use of the method of limits and in one of the later sections we will define the necessary concepts.

Example 1. As was established experimentally by Galileo, the distance *s* covered in the time *t* by a body falling freely in a vacuum is expressed by the formula

$$s = \frac{gt^2}{2} \tag{1}$$

(*g* is a constant equal to 9.81 m/sec²).† What is the velocity of the falling body at each point in its path?

Let the body be passing through the point *A* at the time *t* and consider what happens in the short interval of time of length Δt; that is, in the time from *t* to $t + \Delta t$. The distance covered will be increased by a certain

---

* This section is based on the beautiful essay of Academician S. I. Vavilov "Galileo" (Great Soviet Encyclopedia, Volume 10, 1952).

† Nowadays formula (1) is deduced from the general laws of mechanics, but historically it was just this formula which, after being established experimentally by Galileo, served as a part of the accumulation of experience that was subsequently generalized by those laws.

increment Δs. The original distance is $s_1 = gt^2/2$; the increased distance is

$$s_2 = \frac{g(t + \Delta t)^2}{2} = \frac{gt^2}{2} + \frac{g}{2}(2t\Delta t + \Delta t^2).$$

From this we find the increment

$$\Delta s = s_2 - s_1 = \frac{g}{2}(2t\Delta t + \Delta t^2).$$

This represents the distance covered in the time from t to $t + \Delta t$. To find the average velocity over the section of the path Δs, we divide Δs by Δt:

$$v_{av} = \frac{\Delta s}{\Delta t} = gt + \frac{g}{2}\Delta t.$$

Letting Δt approach zero we obtain an average velocity which approaches as close as we like to the true velocity at the point A. On the other hand, we see that the second summand on the right-hand side of the equation becomes vanishingly small with decreasing Δt, so that the average v_{av} approaches the value gt, a fact which it is convenient to write as follows:

$$v = \lim_{\Delta t \to 0} v_{av} = \lim_{\Delta t \to 0} \frac{\Delta s}{\Delta t}$$

$$= \lim_{\Delta t \to 0} \left(gt + \frac{g}{2}\Delta t\right) = gt.$$

Consequently, gt is the true velocity at the time t.

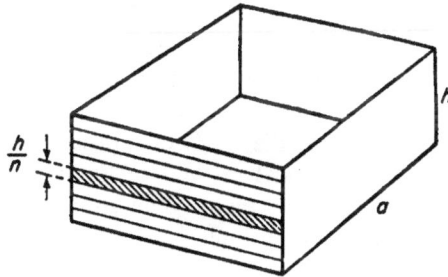

Example 2. A reservoir with a square base of side a and vertical walls of height h

FIG. 1.

is full to the top with water (figure 1). With what force is the water acting on one of the walls of the reservoir?

We divide the surface of the wall into n horizontal strips of height h/n. The pressure exerted at each point of the vessel is equal, by a well-known law, to the weight of the column of water lying above it. So at the lower edge of each of the strips the pressure, expressed in suitable units, will be equal respectively to

$$\frac{h}{n}, \frac{2h}{n}, \frac{3h}{n}, \ldots, \frac{(n-1)h}{n}, h.$$

We obtain an approximate expression for the desired force P, if we assume that the pressure is constant over each strip. Thus the approximate value of P is equal to

$$P \approx \frac{ah}{n} \cdot \frac{h}{n} + \frac{ah}{n} \cdot \frac{2h}{n} + \cdots + \frac{ah}{n} \frac{(n-1)\,h}{n} + \frac{ah}{n} h$$

$$= \frac{ah^2}{n^2} (1 + 2 + \cdots + n) = \frac{ah^2}{n^2} \cdot \frac{n(n+1)}{2} = \frac{ah^2}{2} \left(1 + \frac{1}{n}\right).$$

To find the true value of the force, we divide the side into narrower and narrower strips, increasing n without limit. With increasing n the magnitude $1/n$ in the above formula will become smaller and smaller and in the limit we obtain the exact formula

$$P = \frac{ah^2}{2}.$$

The idea of the method of limits is simple and amounts to the following. In order to determine the exact value of a certain magnitude, we first determine not the magnitude itself but some approximation to it. However, we make not one approximation but a whole series of them, each more accurate than the last. Then from examination of this chain of approximations, that is from examination of the process of approximation itself, we uniquely determine the exact value of the magnitude. By this method, which is in essence a profoundly dialectical one, we obtain a fixed constant as the result of a process or motion.

The mathematical method of limits was evolved as the result of the persistent labor of many generations on problems that could not be solved by the simple methods of arithmetic, algebra, and elementary geometry.

What were the problems whose solution led to the fundamental concepts of analysis, and what were the methods of solution that were set up for these problems? Let us examine some of them.

The mathematicians of the 17th century gradually discovered that a large number of problems arising from various kinds of motion with consequent dependence of certain variables on others, and also from geometric problems which had not yielded to former methods, could be reduced to two types. Simple examples of problems of the first type are: find the velocity at any time of a given nonuniform motion (or more generally, find the rate of change of a given magnitude), and draw a tangent to a given curve. These problems (our first example is one of them) led to a branch of analysis that received the name "differential calculus." The simplest examples of the second type of problem are:

find the area of a curvilinear figure (the problem of quadrature), or the distance traversed in a nonuniform motion, or more generally the total effect of the action of a continuously changing magnitude (compare the second of our two examples). This group of problems led to another branch of analysis, the "integral calculus." Thus two fundamental problems were singled out: the problem of tangents and the problem of quadratures.

In this chapter we will describe in detail the underlying ideas of the solution of these two problems. Particularly important here is the theorem of Newton and Leibnitz to the effect that the problem of quadratures is the inverse, in a well-known sense, of the problem of tangents. For solving the problem of tangents, and problems that can be reduced to it, there was worked out a suitable algorithm, a completely general method leading directly to the solution, namely the method of derivatives or of differentiation.

The history of the creation and development of analysis and of the role played in its growth by the analytic geometry of Descartes has already been described in Chapter I. We see that in the second half of the 17th century and the first half of the 18th a complete change took place in the whole of mathematics. To the divisions that already existed, arithmetic, elementary geometry, and the rudiments of algebra and trigonometry, were added such general methods as analytic geometry, differential and integral calculus, and the theory of the simplest differential equations. It was now possible to solve problems whose solutions up to now had been quite inaccessible.

It turned out that if the law for the formation of a given curve is not too complicated, then it is always possible to construct a tangent to it at an arbitrary point; it is only necessary to calculate, with the help of the rules of differential calculus, the so-called derivative, which in most cases requires a very short time. Up till then it had been possible to draw tangents only to the circle and to one or two other curves, and no one had suspected the existence of a general solution of the problem.

If we know the distance traversed by a moving point up to any desired instant of time, then by the same method we can at once find the velocity of the point at a given moment, and also its acceleration. Conversely, from the acceleration it is possible to find the velocity and the distance, by making use of the inverse of differentiation, namely integration. As a result, it was not very difficult, for example, to prove from the Newtonian laws of motion and the law of universal gravitation that the planets must move around the sun in ellipses according to the laws of Kepler.

Of the greatest importance in practical life is the problem of the greatest and least values of a magnitude, the so-called problem of maxima and minima. Let us take an example: From a log of wood with circular cross section of given radius we wish to cut a beam of rectangular cross section such that it will offer the greatest resistance to bending. What should be the ratio of the sides? A short argument on the stiffness of beams of rectangular cross section (applying simple concepts from the integral calculus), followed by the solving of a maximum problem (which involves calculating a derivative) provides the answer that the greatest stiffness is produced for a rectangular cross section whose height is in the ratio to its base of $\sqrt{2}:1$. The problems of maxima and minima are solved as simply as those of drawing tangents.

At various points of a curved line, if it is not a straight line or a circle, the curvature is in general different. How can we calculate the radius of a circle with the same curvature as the given line at the given point, the so-called radius of curvature of the curve at the point? It turns out that this is equally simple; it is only necessary to apply the operation of differentiation twice. The radius of curvature plays a great role in many questions of mechanics.

Before the invention of the new methods of calculation, it had been possible to find the area only of polygons, of the circle, of a sector or a segment of the circle, and of two or three other figures. In addition, Archimedes had already invented a way to calculate the area of a segment of a parabola. The extremely ingenious method which he used in this problem was based on special properties of the parabola and consequently gave rise to the idea that every new problem in the calculation of area would very likely require its own methods of investigation, even more ingenious and difficult than those of Archimedes. So mathematicians were greatly pleased when it turned out that the theorem of Newton and Leibnitz, to the effect that the inversion of the problem of tangents would solve the problem of quadrature, at once provided a method of calculating the areas bounded by curves of widely different kinds. It became clear that a general method exists, which is suitable for an infinite number of the most different figures. The same remark is true for the calculation of volumes, surfaces, the lengths of curves, the mass of inhomogeneous bodies, and so forth.

The new method accomplished even more in mechanics. It seemed that there was no problem in mechanics that the new calculations would not clarify and solve.

Not long before, Pascal had explained the increase in the size of the Torricelli vacuum with increasing altitude as a consequence of the decrease in atmospheric pressure. But exactly what is the law governing this decrease? The question is answered immediately by the investigation of a simple differential equation.

It is well known to sailors that they should take two or three turns of the mooring cable around the capstan if one man is to be able to keep a large vessel at its mooring. Why is this? It turned out that from a mathematical point of view the problem is almost completely identical with the preceding one and can be solved at once.

Thus, after the creation of analysis, there followed a period of tempestuous development of its applications to the most varied branches of technology and natural science. Since it is founded on abstraction from the special features of particular problems, mathematical analysis reflects the actual deep-lying properties of the material world; and this is the reason why it provides the means for investigation of such a wide range of practical questions. The mechanical motion of solid bodies, the motion of liquids and gases of their particular particles, their laws of flow in the mass, the conduction of heat and electricity, the course of chemical reactions, all these phenomena are studied in the corresponding sciences by means of mathematical analysis.

At the same time as its applications were being extended, the subject of analysis itself was being immeasurably enriched by the creation and development of various new branches, such as the theory of series, applications of geometry to analysis, and the theory of differential equations.

Among mathematicians of the 18th century, there was a widespread opinion that any problem of the natural sciences, provided only that one could find a correct mathematical description of it, could be solved by means of analytic geometry and the differential and integral calculus.

Mathematicians proceeded gradually to more complicated problems of natural science and technology, which demanded further development of their methods. For the solution of such problems it became necessary to create further branches of mathematics: the calculus of variations, the theory of functions of a complex variable, field theory, integral equations, and functional analysis. But all these new methods of calculation were essentially immediate extensions and generalizations of the remarkable methods discovered in the 17th century. The greatest mathematicians of the 18th century, David Bernoulli (1700–1782), Leonard Euler (1707–1783) and Lagrange (1736–1813), who blazed new paths in science, constantly took as their starting point the fundamental problems of the exact sciences. This energetic development of analysis was continued into the 19th century by such famous mathematicians as Gauss (1777–1855), Cauchy (1789–1857), M. V. Ostrogradskiĭ (1801–1861), P. L. Čebyšev (1821–1894), Riemann (1826–1866), Abel (1802–1829), Weierstrass (1815–1897), all of whom made truly remarkable contributions to the development of mathematical analysis.

The Russian mathematical genius, N. I. Lobačevskiĭ, had an influence on the development of certain questions of mathematical analysis, and we should also mention the leading mathematicians who were active at the turn of the 20th century: A. A. Markov (1856–1922), A. M. Lyapunov (1857–1918), H. Poincaré (1854–1912), F. Klein (1849–1925), D. Hilbert (1862–1943).

The second half of the 19th century witnessed a profound critical examination and clarification of the foundations of analysis. The various powerful methods that had accumulated were now put on a uniform systematic basis, corresponding to the advanced level of mathematical rigor. All these methods are the means by which, along with arithmetic, algebra, geometry and trigonometry, we give a mathematical interpretation to the world around us, describe the course of actual events, and solve the important practical problems connected with them.

Analytic geometry, differential and integral calculus, and the theory of differential equations are studied at all technical institutes, so that these branches of mathematics are known to millions of citizens; the elements of these sciences are also taught at many technical schools; there is also some question of their being introduced into the secondary schools.

In most recent times the general use of rapid calculating machines has introduced a new era in mathematics. These machines, in conjunction with the branches of mathematics just mentioned, open up strange new possibilities for mankind.

At the present time, analysis and the branches arising from it represent a widely diversified mathematical science, consisting of several broad independent disciplines closely connected with one another; each of these disciplines is being developed and perfected.

More than ever before, a significant role is being played in analysis by the requirements of daily life, by problems connected with the imposing development of technology. Of great importance are the aerodynamical problems of hypersonic velocities, which are being solved with constant success. The most difficult problems of mathematical physics have now reached the stage where they can be solved in practical numerical form. In contemporary physics such theories as quantum mechanics (which studies the problems peculiar to the microcosm of the atom) not only require the most advanced branches of contemporary mathematical analysis for solving their problems but could not even describe their fundamental concepts without the use of analysis.

The purpose of the present chapter is to give a popular presentation, suitable for a reader acquainted only with elementary mathematics, of the growth and the simplest applications of such basic concepts of analysis as function, limit, derivative, and integral. Since the various special branches of analysis will be dealt with in other chapters of the book, the present chapter has a more elementary character and a reader who has already studied a usual first course in analysis may omit it without harm to his understanding of the rest of the book.

§2. Function

The concept of a function. The various objects or phenomena that we observe in nature are organically connected with one another; they are interdependent. The simplest relations of this sort have long been known to mankind and information about them has been accumulated and formulated as physical laws. These laws indicate that the various magnitudes characterizing a given phenomenon are so closely related to one another that some of them are completely determined by the values of others. For example, the length of the sides of a rectangle completely determines its area, the volume of a given amount of gas at a given temperature is determined by the pressure, and the elongation of a given metallic rod is determined by its temperature. It was uniformities of this sort that served as the origin of the concept of *function*.

Consider an algebraic formula which, corresponding to each value of the literal magnitudes occurring in it, allows us to find the value of the magnitude expressed by the formula; the basic idea here is that of a function. Let us consider some examples of functions expressed by such formulas.

1. Let us suppose that at the beginning of a certain period of time a material point was at rest and that subsequently it began to fall as the result of gravity. Then the distance s traced out by the point up to time t is expressed by the formula

$$s = \frac{gt^2}{2}, \tag{1}$$

where g is the acceleration of gravity.

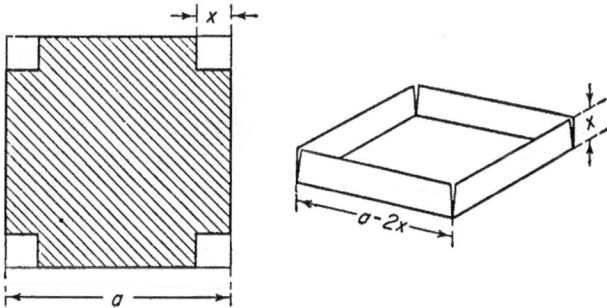

FIG. 2.

2. From a square of side a we construct an open rectangular box of height x (figure 2). The volume V of the box is calculated from the formula

$$V = x(a - 2x)^2. \tag{2}$$

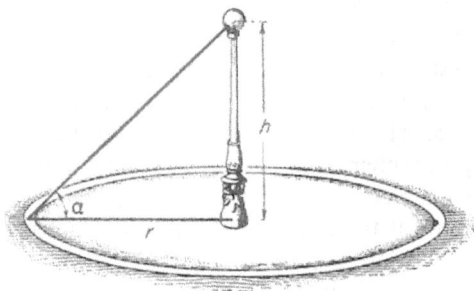

FIG. 3.

Formula (2) allows us, for every height x under the obvious restriction $0 \leqslant x \leqslant a/2$, to find the volume of the box.

3. Let a pillar (figure 3) be erected at the center of a circular skating rink with a light at height h. The illumination T at the edge of the circle may be expressed by the formula

$$T = \frac{A \sin \alpha}{h^2 + r^2}, \tag{3}$$

where r is the radius of the circle, $\tan \alpha = h/r$, and A is a certain magnitude characterizing the power of the light. If we know the height h we can calculate T from formula (3).

4. The roots of the quadratic equation

$$x^2 + px - 1 = 0 \tag{4}$$

are given by the formula

$$x = -\frac{p}{2} \pm \sqrt{1 + \frac{p^2}{4}}. \tag{5}$$

The characteristic feature of a formula in general, and of the examples just given in particular, is that the formula enables us, for any given value of one of the variables (the time t, the height x of the box, the height h of the pillar, the coefficient p of the quadratic equation), which is called the independent variable, to calculate the value of the other variable (the distance s, the volume V, the illumination T, the root x of the equation), which is called a dependent variable or a function of the first variable.

Each of the formulas introduced provides an example of a function: the distance s traced by the point is a function to the time t; the volume

V of the box is a function of height x; the illumination T of the edge of the rink is a function of the height h of the pillar; the two roots of the quadratic equation (4) are functions of the coefficient p.

It should be remarked that in some cases the independent variable may assume any desired numerical value, as in example 4 where the coefficient p of the quadratic equation (4) may be an arbitrary number. In other cases the independent variable may take an arbitrary value from some set (or collection) of numbers determined in advance; as in example 2, where the volume of the box is a function of its height x, which can take any value from the set of numbers x satisfying the inequality $0 < x < a/2$. Similarly, in example 3 the illumination T at the edge of the rink is a function of the height h of the pillar, which theoretically can take any value satisfying the inequality $h > 0$, but in practice h must satisfy the inequalities $0 < h \leqslant H$, where the magnitude H is determined by the technical facilities at the disposal of the administration of the rink.

Let us introduce other examples of this kind. The formula

$$y = \sqrt{1 - x^2}$$

determines a real function (expressing a relationship between the real numbers x and y) only for those values of x which satisfy the inequalities $-1 \leqslant x \leqslant +1$, and the formula $y = \log (1 - x^2)$ only for those x which satisfy the inequalities $-1 < x < 1$.

So it is necessary to take account of the fact that actual functions may not be defined for all numerical values of the independent variable but only for those values which belong to a certain set, which most often fills out an interval on the x-axis, with or without the end points.

We are now in the position to give the definition of a function accepted in present-day mathematics.

The (dependent) magnitude y is a function of the (independent) magnitude x if there exists a rule whereby to each value of x belonging to a certain set of numbers there corresponds a definite value of y.

The set of values x appearing in this definition is called the *domain of the function.*

Every new concept gives rise to a new symbolism. The transition from arithmetic to algebra was made possible by the construction of formulas which were valid for arbitrary numbers, and the search for general solutions gave rise to the literal symbolism of algebra.

The problem of analysis is the study of functions, that is of the dependence of one variable on another. Consequently, just as in algebra a transition took place from concrete numbers to arbitrary numbers, denoted by letters, so in analysis there was the corresponding transition from concrete formulas to arbitrary formulas. The phrase "*y* is a function of *x*" is conventionally written as

$$y = f(x).$$

Just as in algebra different letters are used for different numbers, so in analysis different notations are used for different types of dependence, that is for different functions: thus we write $y = F(x)$, $y = \phi(x)$, $\cdots$

Graphs of functions. One of the most fruitful and brilliant ideas of the second half of the 17th century was the idea of the connection between the concept of a function and the geometric representation of a line. This connection can be realized, for example, by means of a rectangular Cartesian system of coordinates, with which the reader is certainly familiar in a general way from his secondary school mathematics.

Let us set up on the plane a rectangular Cartesian system of coordinates. This means that on the plane we choose two mutually perpendicular lines (the axis of abscissas and the axis of ordinates), on each of which we fix a positive direction. Then to each point M of the plane we may assign two numbers (x, y), which are its coordinates, expressing in the given system of measurement the distance, taken with the proper sign,* of the point M from the axis of ordinates and the axis of abscissas respectively.

With such a system of coordinates we may represent functions graphically in the form of certain lines. Suppose we are given a function

$$y = f(x). \tag{6}$$

This means, as we know, that for every value of x belonging to the domain of definition of the given function, it is possible to determine by some means, for example by calculation, a corresponding value y. Let us give to x all possible numerical values, for each x determine y according to our rule (6), and construct on the plane the point with coordinates x and y. In this way, for every point M' on the x-axis (figure 4) there will correspond

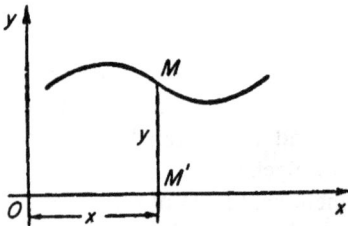

FIG. 4.

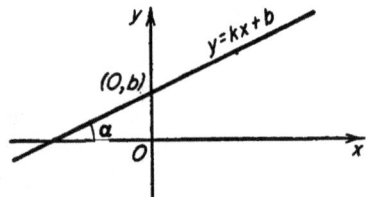

FIG. 5.

---

* The number x is the abscissa and y is the ordinate of the point M.

a point M with coordinates x and $y = f(x)$. The set of all points M forms a certain line, which we call the graph of the function $y = f(x)$.

Thus, *the graph of the function $f(x)$ is the geometric locus of the points whose coordinates satisfy equation* (6).

In school we became acquainted with the graphs of the simplest functions. Thus the reader probably knows that the function $y = kx + b$, where k and b are constants, is the graph (figure 5) of a straight line forming the angle α with the positive direction of the x-axis, where $\tan \alpha = k$, and intersecting the y-axis at the point $(0, b)$. This function is called a *linear function*.

Linear functions occur very frequently in the applications. Let us recall that many physical laws are represented, with considerable accuracy, by linear functions. For example, the length l of a body may be considered with good approximation as a linear function of its temperature

$$l = l_0 + \alpha l_0 t,$$

where α is the coefficient of linear expansion, and l_0 is the length of the body for $t = 0$. If x is the time and y is the distance covered by a moving point, then the linear function $y = kx + b$ obviously expresses the fact that the point is moving with uniform velocity k; and the number b denotes the distance, at time $x_0 = 0$, of the moving point from the fixed zero-point from which we measure our distances. Linear functions are extremely useful because of their simplicity and because it is possible to consider nonuniform changes as being approximately linear, even if only for small intervals.

But in many cases it is necessary to make use of nonlinear functional dependence. Let us recall for example the law of Boyle-Mariotte

$$v = \frac{c}{p},$$

where the magnitudes p and v are inversely proportional. The graph of such a relation represents a hyperbola (figure 6).

The physical law of Boyle-Mariotte corresponds actually to the case that p and v are positive; it represents a branch of the hyperbola lying in the first quadrant.

The general class of oscillatory processes includes periodic motions, which are usually described by the familiar trigonometric functions. For example, if we extend a hanging spring from its position of equilibrium, then, so long as we stay within the elastic limits of the spring, the point A will perform vertical oscillations which are quite accurately expressed by the law

$$x = a \cos(pt + \alpha),$$

where x is the displacement of the point A from its position of equilibrium, t is the time, and the numbers a, p and α are certain constants determined by the material, the dimensions, and the initial extension of the spring.

It should be kept in mind that a function may be defined in various domains by various formulas, determined by the circumstances of the

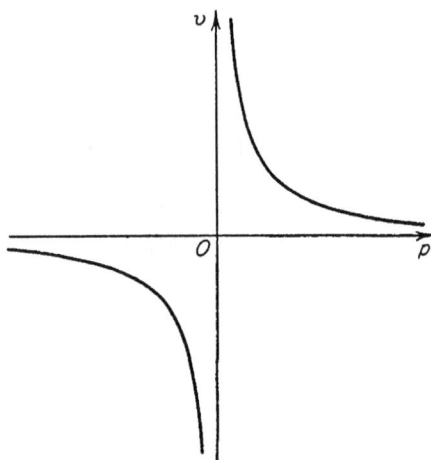

FIG. 6.

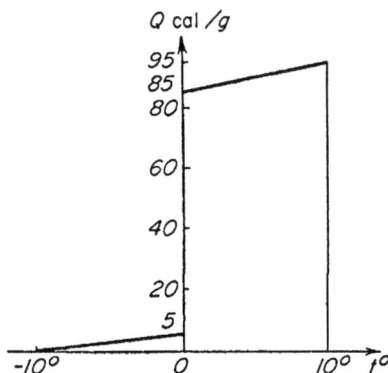

FIG. 7.

case. For example, the relation $Q = f(t)$ between the temperature t of a gram of water (or ice) and the quantity of heat Q in it, as t varies between $-10°$ and $+10°$, is a completely determined function which it is difficult to express in a single formula,* but it is easy to represent this function by two formulas. Since the specific heat of ice is equal to 0.5 and that of water is equal to 1, this function, if we agree that $Q = 0$ at $-10°$, is represented by the formula

$$Q = 0.5t + 5,$$

as t varies in the interval $-10° \leqslant t < 0°$ and by another formula

$$Q = t + 85,$$

as t varies in the interval $0° < t \leqslant 10°$. For $t = 0$ this function is indefinite or multiple-valued; for convenience, we may agree that at $t = 0$ it takes some well-defined value, for example $f(0) = 45$. The graph of the function $Q = f(t)$ is given in figure 7.

---

* This does not mean that such an expression is impossible. In Chapter XII we will show how to obtain a single formula.

We have introduced many examples of functions given by formulas. The possibility of representing a function by means of formulas is extremely important from the mathematical point of view, since such formulas provide very favorable conditions for investigating the properties of the functions by mathematical methods.

But one must not think that a formula is the only method of defining a function. There are many other methods; for example, the graph of the function, which gives a visual geometric picture of it. The following example gives a good illustration of another method.

To record variation of the temperature of the air during the course of 24 hours, meteorological stations make use of an instrument called the thermograph. A thermograph consists of a drum rotated about its axis by a clockwork mechanism, and of a curved brass framework that is extremely sensitive to changes of temperature. As a result, a pen fastened to the framework by a system of levers rises with rising temperature; and conversely, a fall in the temperature lowers the pen. On the drum is wound a ribbon of graph paper, on which the pen draws a continuous line, forming the graph of the function $T = f(t)$, which expresses the interdependence of the time and the temperature of the air. From this graph we may determine, without calculation, the value of the temperature at any moment of time t.

This example shows that a graph in itself determines a function independently of whether the function is given by a formula or not.

Incidentally, we shall return to this question (see Chapter XII) and shall prove the following important assertion: Every continuous graph can be represented by a formula, or, as it is still customary to say, by an analytic expression. This statement is also true for many discontinuous graphs.*

We remark that the truth of this statement, which is of great theoretical importance, was completely realized in mathematics only in the middle of the past century. Up to that time mathematicians understood by the term "function" only an analytic expression (formula). But they were under the mistaken impression that many discontinuous graphs did not correspond to any analytic expression, since they assumed that if a function was given by a formula, then its graph must possess certain particularly desirable properties in comparison with the other graphs.

But in the 19th century, it was discovered that every continuous graph may be represented by a more or less complicated formula. Thus the exceptional role of the analytic expression as a means of definition of

---

* Of course, the above statement will be completely clear to the reader only after we have given a precise definition of exactly what is meant in mathematics by the term "formula" and "analytic expression."

functions was weakened and there came into existence the new, more flexible definition given above for the concept of a function. By this definition a variable y is called a function of a variable x if there exists a rule whereby to every value of x in the domain of definition of the function there corresponds a completely determined value y, independent of the way in which this rule is given: by a formula, a graph, a table or in any other way.

We may remark here that in the mathematical literature the above definition of a function is often associated with the name of Dirichlet, but it is worth emphasizing that this definition was given simultaneously and independently by N. I. Lobačevskiĭ. Finally we suggest as an exercise that the reader sketch the graphs of the functions x^3, $\sqrt{x}$, $\sin x$, $\sin 2x$, $\sin (x + \pi/4)$, $\ln x$, $\ln(1 + x)$, $|x - 3|$, $(x + |x|)/2$.

We should also note that the graph of a function which for all values of x satisfies the relation

$$f(-x) = f(x)$$

is symmetric with respect to the y-axis and in the case

$$f(-x) = -f(x)$$

the graph is symmetric with respect to the origin of coordinates. Consider also how to obtain the graph of a function $f(a + x)$, when a is a constant, from the graph of $f(x)$. Finally, consider how, using the graphs of the functions $f(x)$ and $\phi(x)$, it is possible to find the values of the composite function $y = f[\phi(x)]$.

§3. Limits

In §1 it was stated that modern mathematical analysis uses a special method, which was worked out in the course of many centuries and serves now as its basic instrument. We are speaking here of the method of infinitesimals, or, as is essentially the same, of limits. We shall try to give some idea of these concepts. For this purpose we consider the following example.

We wish to calculate the area bounded by the parabola with equation $y = x^2$, by the x-axis and by the straight line $x = 1$ (figure 8). Elementary mathematics will not furnish us with a means for solving this problem. But here is how we may proceed.

We divide the interval $[0, 1]$ along the x-axis into n equal parts at the points

$$0, \frac{1}{n}, \frac{2}{n}, \cdots, \frac{n-1}{n}, 1$$

and on each of these parts construct the rectangle whose left side extends up to the parabola. As a result we obtain the system of rectangles shaded in figure 8, the sum S_n of whose areas is given by

$$S_n = 0 \cdot \frac{1}{n} + \left(\frac{1}{n}\right)^2 \frac{1}{n} + \left(\frac{2}{n}\right)^2 \frac{1}{n} + \cdots + \left(\frac{n-1}{n}\right)^2 \frac{1}{n}$$

$$= \frac{1^2 + 2^2 + \cdots + (n-1)^2}{n^3} = \frac{(n-1)\,n(2n-1)}{6n^3}. *$$

Let us express S_n in the following form:

$$S_n = \frac{1}{3} + \left(\frac{1}{6n^2} - \frac{1}{2n}\right) = \frac{1}{3} + \alpha_n. \qquad (7)$$

The quantity α_n, which depends on n, is admittedly rather unwieldy in appearance, but it possesses a certain remarkable property: If n is increased beyond all bounds, then α_n approaches 0. This property may also be expressed as follows: If we are given an arbitrary positive number ϵ, then it is possible to choose an integer N sufficiently large that for all n greater than N the number α_n will be less than the given ϵ in absolute value.†

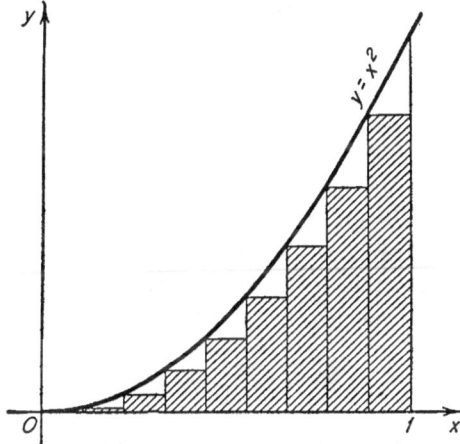

FIG. 8.

---

* If in the obvious equalities $(k+1)^3 - k^3 = 3k^2 + 3k + 1$, for the different values $k = 1, 2, \cdots, n-1$, we add the left and right sides separately, we obtain the equation

$$n^3 - 1 = 3\sigma_n + \frac{3(n-1)n}{2} + n - 1$$

where $\sigma_n = 1^2 + 2^2 + \cdots + (n-1)^2$. Solving this equation for σ_n, we get

$$\sigma_n = \frac{(n-1)n(2n-1)}{6}.$$

† For example, if $\epsilon = 0.001$, we may take $N = 500$. In fact, since

$$\frac{1}{6n^2} < \frac{1}{2n}$$

for positive integral n, therefore

$$|\alpha_n| = \left|\frac{1}{6n^2} - \frac{1}{2n}\right| = \frac{1}{2n} - \frac{1}{6n^2} < \frac{1}{2n} < 0.001$$

The magnitude α_n is an example of an infinitesimal in the sense in which that word is used in modern mathematics.

In figure 8 we see that if we increase the number n beyond all bounds, the sum S_n of the areas of the shaded rectangles will approach the desired area of the curvilinear figure. On the other hand, equation (7), in view of the fact that α_n approaches zero as n increases beyond all bounds, shows that the sum S_n at the same time approaches 1/3. From this it follows that the desired area S of the figure is equal to 1/3, and we have solved our problem.

So the method under discussion amounts to this, that in order to find a certain magnitude S we introduce another magnitude S_n, a *variable* magnitude which approaches S through particular values S_1, S_2, S_3, $\cdots$, which depend according to some law on the natural numbers $n = 1, 2, \cdots$. Then, from the fact that the variable S_n may be represented as the sum of a constant $\frac{1}{3}$ and an infinitesimal α_n, we conclude that S_n approaches $\frac{1}{3}$ and so $S = \frac{1}{3}$. In the language of the modern theory of limits we may say that for increasing n the variable magnitude S_n approaches a limit, which is equal to $\frac{1}{3}$.

Now let us give a precise definition of the concepts introduced here.

If a variable magnitude $\alpha_n (n = 1, 2, \cdots)$ has the property that for every arbitrarily small positive number ϵ it is possible to choose an integer N so large that for all $n > N$ we have $|\alpha_n| < \epsilon$, then we say that α_n is an *infinitesimal* and we write

$$\lim_{n \to \infty} \alpha_n = 0 \text{ or } \alpha_n \to 0.$$

On the other hand, if a variable x_n may be represented as a sum

$$x_n = a + \alpha_n,$$

where a is constant and α_n is an infinitesimal, then we say that the variable x_n, for n increasing beyond all bounds, approaches the number a and we write

$$\lim x_n = a \text{ or } x_n \to a.$$

The number a is called the *limit* of x_n. In particular the limit of an infinitesimal is obviously zero.

---

for arbitrary $n > 500$. In the same way it would be possible to assign arbitrarily small values ϵ, for example:

$$\epsilon_1 = 0.0001, \quad \epsilon_2 = 0.00001, \cdots,$$

and for each of them to choose, as above, appropriate values $N = N_1, N_2, \cdots$.

Let us consider the following examples of variable magnitudes

$$x_n = \frac{1}{n}, y_n = -\frac{1}{n^2}, z_n = \frac{(-1)^n}{n}, u_n = \frac{n-1}{n} = 1 - \frac{1}{n};$$

$$v_n = (-1)^n \ (n = 1, 2, \cdots).$$

It is clear that x_n, y_n, and z_n are infinitesimals, the first of them approaching zero through decreasing values, the second through increasing negative values, while the third takes on values which oscillate around zero. Further, $u_n \to 1$, while v_n does not have a limit at all, since with increasing n it does not approach any constant number but continually oscillates, taking on the values 1 and -1.

Another important concept in analysis is that of an *infinitely large* magnitude, which is defined as a variable x_n $(n = 1, 2, \cdots)$, with the property that after choice of an arbitrarily large positive number M it is possible to find a number N such that for all $n > N$

$$|x_n| > M.$$

The fact that the magnitude x_n is infinitely large is written thus

$$\lim x_n = \infty \text{ or } x_n \to \infty.$$

Such a magnitude x_n is said to approach infinity. If it is positive (negative) from some value on, this fact is expressed thus: $x_n \to +\infty (x_n \to -\infty)$. For example, for $n = 1, 2, \cdots$

$$\lim n^2 = +\infty, \lim (-n^3) = -\infty;$$

$$\lim \log \frac{1}{n} = -\infty, \lim \tan \left(\frac{\pi}{2} + \frac{1}{n}\right) = -\infty.$$

It is easy to see that if a magnitude α_n is infinitely large, then $\beta_n = 1/\alpha_n$ is infinitely small, and conversely.

Two variable magnitudes x_n and y_n may be added, subtracted, multiplied, and divided the one by the other so as to produce new magnitudes that are in general also variable: namely their sum $x_n + y_n$, their difference $x_n - y_n$, their product $x_n y_n$, and their quotient x_n/y_n. Correspondingly their particular values will be

$$x_1 \pm y_1, x_2 \pm y_2, x_3 \pm y_3, \cdots$$

$$x_1 y_1, x_2 y_2, x_3 y_3, \cdots$$

$$\frac{x_1}{y_1}, \frac{x_2}{y_2}, \frac{x_3}{y_3}, \cdots.$$

It is also possible to prove, as is fairly evident, that if the variables x_n and y_n approach finite limits, then their sum, difference, product, and quotient also approach limits which are correspondingly equal to the sum, difference, product, and quotient of these limits. This fact may be expressed thus:

$$\lim (x_n \pm y_n) = \lim x_n \pm \lim y_n; \lim (x_n y_n) = \lim x_n \lim y_n;$$

$$\lim \frac{x_n}{y_n} = \frac{\lim x_n}{\lim y_n}.$$

However, in the case of the quotient it is necessary to assume that the limit of the denominator ($\lim y_n$) is not equal to zero. If $\lim y_n = 0$ and $\lim x_n \neq 0$, then the ratio of x_n to y_n will not have a finite limit but will approach infinity.

Especially interesting, and at the same time important, is the case when the numerator and the denominator simultaneously approach zero. Here it is impossible to state in advance whether the ratio x_n/y_n will approach a limit, and if it does, what that limit will be, since the answer to this question depends entirely on the character of the approach of x_n and y_n to zero. For example, if

$$x_n = \frac{1}{n}, y_n = \frac{1}{n^2}, z_n = \frac{(-1)^n}{n} \ (n = 1, 2, \cdots),$$

then

$$\frac{y_n}{x_n} = \frac{1}{n} \to 0, \frac{x_n}{y_n} = n \to \infty.$$

On the other hand, the magnitude

$$\frac{x_n}{z_n} = (-1)^n$$

evidently does not approach any limit.

Thus the case when the numerator and the denominator of the fraction both approach zero cannot be dealt with in advance by general theorems, and for each particular fraction of this kind it is necessary to make a special investigation.

We shall see later that the fundamental problem of the differential calculus, which may be considered as the problem of determining the velocity of a nonuniform motion at a given moment, reduces to determining the limit of the ratio of two infinitesimal magnitudes, namely the increase of the distance covered and the increase in the time.

So far we have considered variables x_n which take on a sequence of numerical values $x_1, x_2, x_3, \cdots, x_n, \cdots$, while the index n runs through the sequence of natural numbers $n = 1, 2, 3, \cdots$. But it is also possible to consider the case that n varies continuously, like the time for example, and here also to determine the limit of the variable x_n. The properties of such limits are completely analogous to those formulated earlier for discrete (that is, discontinuous) variables. We also note that there is no special significance in the fact that n increases beyond all bounds. It is equally possible to consider the case that, while varying continuously, n approaches a given value n_0.

As an example let us investigate the variation in the magnitude of $(\sin x)/x$ as x approaches zero. Table 1 shows the values of this magnitude for certain values of x:

Table 1

x	$\dfrac{\sin x}{x}$
0.50	0.9589 ...
0.10	0.9983 ...
0.05	0.9996 ...

(it is assumed that the values of x are given in radian measure).

It is obvious that as x approaches zero the magnitude $(\sin x)/x$ approaches 1, but of course we must still give a rigorous proof of this fact. The proof may be obtained, for example, from the following inequality, which is valid for all nonzero angles in the first quadrant:

$$\sin x < x < \tan x.$$

If we divide both sides of this inequality by $\sin x$, we obtain

$$1 < \frac{x}{\sin x} < \frac{1}{\cos x},$$

from which follows

$$\cos x < \frac{\sin x}{x} < 1.$$

But as x decreases to zero $\cos x$ approaches 1, so that the magnitude $(\sin x)/x$, being contained in the interval between $\cos x$ and 1, also approaches 1, that is

$$\lim_{x \to 0} \frac{\sin x}{x} = 1.$$

We shall have occasion below to make use of this fact.

Our equation has been proved for the case that x approaches zero through positive values. But by changing the proof in an obvious way, it is possible to obtain the same result when x approaches zero through negative values.

Let us now discuss for a moment the following question. A variable magnitude may or may not have a limit and the question arises whether it is possible to give a criterion for determining the existence of a limit for a variable. We will confine ourselves to an important and sufficiently general case, for which such a criterion can be given. Let us suppose that the variable magnitude x_n increases or at least does not decrease; that is, it satisfies the inequalities

$$x_1 \leqslant x_2 \leqslant x_3 \leqslant \cdots,$$

and let us also suppose we have determined that none of its values exceeds a certain fixed number M; that is, $x_n \leqslant M$ $(n = 1, 2, \cdots)$. If we mark the values of x_n and the number M on the x-axis, we see that the variable point x_n moves along the axis to the right but constantly remains to the left of the point M. It is rather obvious that the variable point x_n must inevitably approach a certain limit point a, situated to the left of M or at most coinciding with M.

So, in the case under consideration, the limit

$$\lim x_n = a$$

of our variable exists.

The above argument has an intuitive character but we may consider it as a proof. In a course in modern analysis a complete proof of this fact is given on the basis of the theory of real numbers.

As an example let us consider the variable

$$u_n = \left(1 + \frac{1}{n}\right)^n \ (n = 1, 2, 3, \cdots).$$

The first few values are $u_1 = 2$, $u_2 = 2.25$, $u_3 \approx 2.37$, $u_4 \approx 2.44$, $\cdots$, which are seen to increase. From the binomial theorem of Newton it is possible to prove that this increase holds for arbitrary n. Moreover, it is also easy to prove that for all n the inequality $u_n < 3$ is valid. Consequently, our variable must have a limit which is not greater than 3. We shall see that this limit plays a very important role in mathematical physics and in a certain sense is the most natural base for logarithms of numbers.

It is customary to denote this limit by the letter e. It is equal to

$$e = \lim_{n \to \infty} \left(1 + \frac{1}{n}\right)^n = 2.718281828459045 \cdots$$

A more detailed analysis shows that the number e is not rational.*

It is also possible to show that the limit under consideration exists and is equal to e not only when $n \to +\infty$ but also when $n \to -\infty$. In both cases n may also take on noninteger values.

Let us mention an important application to physics of the concept of a limit. It consists of the remarkable fact that only by using the concept of a limit (passage to the limit) is it possible for us to give a complete definition of many of the concrete magnitudes encountered in physics.

Let us also consider for the moment the following geometric example. In elementary geometry the figures considered first are those bounded by straight line segments. But later there arises the more difficult task of finding the length of the circumference of a circle with given radius.

If we analyze the difficulties connected with the solution of this problem, we find that they reduce to the following.

We must give an answer to the question, what is meant by the length of the circumference; that is, we must give a precise definition of this length. It is essential that the definition should be expressible in terms of the lengths of straight-line segments and also that it should provide us with the possibility of effectively calculating the length of the circumference.

It is understood, of course, that the result of this calculation should be in agreement with practical experience. For example, if we consider a circumference consisting of an actual thread, then, if we cut the thread and stretch it out, we must obtain a segment whose length, within the limits of accuracy of measurement, coincides with our computed length.

As is known from elementary geometry, the solution of this problem reduces to the following definition. The length of a circumference is defined to be the limit approached by the perimeter of a regular[†] polygon inscribed in it as the number of sides of the polygon increases beyond all bounds. Thus the solution of the problem is based essentially on the concept of a limit.

The length of an arbitrary smooth curve is defined in the same way.

---

* In this connection we should remark that addition, subtraction, multiplication, and division (excluding division by zero) of rational numbers, that is numbers of the form p/q where p and q are integers, leads to rational numbers. But this is not necessarily the case for the operation of taking a limit. The limit of a sequence of rational numbers may be irrational number.

† It is not important that the polygon should be regular. The only essential feature is that the greatest side of the variable inscribed polygon should approach zero.

In the paragraphs just following, we will meet with a number of examples of geometric and physical magnitudes that can be defined only with the concept of a limit.

The concepts of limit and infinitesimal were given a definitive formulation at the beginning of the last century. The definitions introduced here are connected with the name of Cauchy, before whose time mathematicians operated with concepts that were less clear. The present-day concepts of a limit, of an infinitesimal as a variable magnitude, and of a real number, resulted from the development of mathematical analysis and were at the same time the means of stating and clarifying its many achievements.

§4. Continuous Functions

Continuous functions form the basic class of functions for the operations of mathematical analysis. The general idea of a continuous function may be obtained from the fact that its graph is continuous; that is, its curve may be drawn without lifting the pencil from the paper.

A continuous function gives the mathematical expression of a situation often encountered in practical life, namely that to a small increase in an independent variable there corresponds a small increase in the dependent variable, or function. Excellent examples of a continuous function are given by the various rules governing the motion of bodies $s = f(t)$, expressing the dependence of the distance s on the time t. Since the time and the distance are continuous, a law of motion of the body $s = f(t)$ sets up between them a definite continuous relation, characterized by the fact that to a small increase in the time corresponds a small increase in the distance.

Mankind arrived at the abstraction of continuity by observing the surrounding so-called dense media, namely solids, liquids, and gases; for example, metals, water, and air. In actual fact, as is well known now, every physical medium represents the accumulation of a large number of separate particles in motion. But these particles and the distances between them are so small in comparison with the dimensions of the media in which the phenomena of microscopic physics take place that many of these phenomena may be studied with sufficient accuracy if we consider the medium as being approximately without interstices, that is as continuously distributed over the occupied space. It is on such an assumption that many of the physical sciences are based, for example, hydrodynamics, aerodynamics, and the theory of elasticity. The mathematical concept of continuity naturally plays a large role in these sciences, and in many others as well.

Let us consider an arbitrary function $y = f(x)$ and some specific value of the independent variable x_0. If our function reflects a continuous process, then to values x which differ only slightly from x_0 will correspond values of the function $f(x)$ differing only slightly from the value $f(x_0)$ at the point x_0. Thus if the increment $x - x_0$ of the independent variable is small, then the corresponding increment $f(x) - f(x_0)$ of the function will also be small. In other words if the increment of the independent variable $x - x_0$ approaches zero, then the increment $f(x) - f(x_0)$ of the function must also approach zero, a fact which may be expressed in the following way:

$$\lim_{x - x_0 \to 0} [f(x) - f(x_0)] = 0. \tag{8}$$

This relation constitutes the mathematical definition of continuity of the function at the point x_0; namely, the function $f(x)$ is said to be *continuous at the point x_0*, if equality (8) holds.

Finally, we give the following definition. A function is said to be *continuous in a given interval*, if it is continuous at every point x_0 of this interval; that is, if at every such point equality (8) is fulfilled.

Thus, in order to introduce a mathematical definition of the property of a function reflected in the fact that its graph is continuous (in the everyday sense of this word), it was necessary first to define local continuity (continuity at the point x_0) and then on this basis to define continuity of the function in the whole interval.

This definition, first introduced at the beginning of the last century by Cauchy, is now generally adopted in contemporary mathematical analysis. The test of many concrete examples has shown that it corresponds very well to the practical notion we have of a continuous function, for instance, as represented by its continuous graph.

As examples of continuous functions, the reader may consider the elementary functions well known to him from school mathematics x^n, $\sin x$, $\cos x$, a^x, $\log x$, arc $\sin x$, arc $\cos x$. All these functions are continuous in the intervals for which they are defined.

If continuous functions are added, subtracted, multiplied, or divided (except for division by zero), the result is also a continuous function. But in the case of division the continuity is usually destroyed for those values x_0 for which the function in the denominator vanishes. The result of the division in that case is a function which is *discontinuous* at the point x_0.

The function $y = 1/x$ may serve as an example of a function which is discontinuous at the point $x = 0$. Other discontinuous functions are represented by the graphs in figure 9.

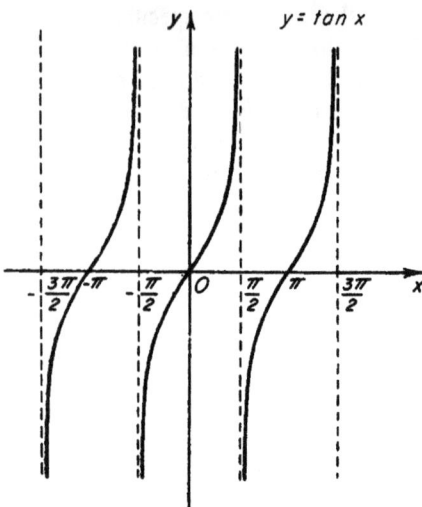

FIG. 9a.

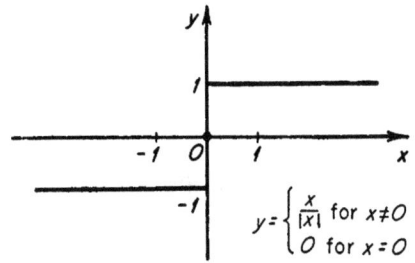

$$y = \begin{cases} \dfrac{x}{|x|} & \text{for } x \neq 0 \\ 0 & \text{for } x = 0 \end{cases}$$

FIG. 9b.

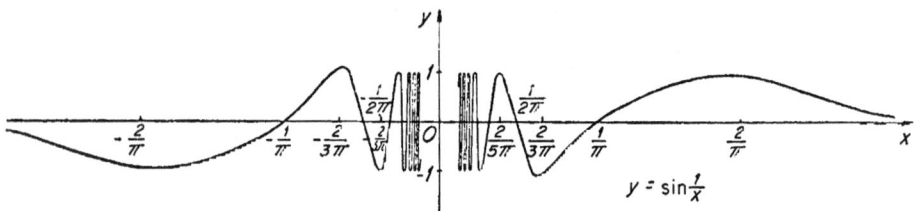

$y = \sin \dfrac{1}{x}$

FIG. 9c.

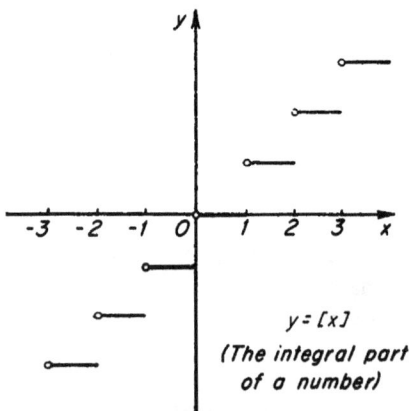

$y = [x]$
(The integral part
of a number)

FIG. 9d.

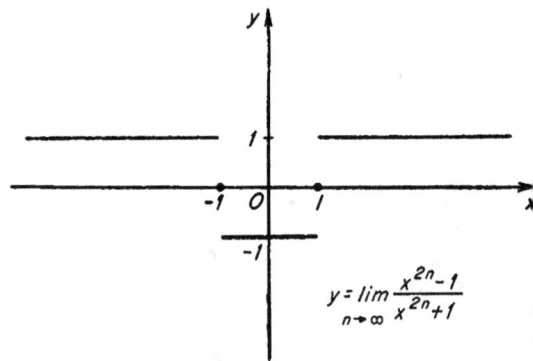

$y = \lim\limits_{n \to \infty} \dfrac{x^{2n} - 1}{x^{2n} + 1}$

FIG. 9e.

We recommend that the reader examine these graphs carefully. He will notice that the breaks in the functions are of different kinds: In some cases a limit $f(x)$ exists as x approaches the point x_0 where the function suffers a discontinuity, but this limit is different from $f(x_0)$. In other cases, as in figure 9c, the limit simply does not exist. It may also happen that as x approaches x_0 from one side $f(x) - f(x_0) \to 0$, but as $x \to x_0$ from the other side, $f(x) - f(x_0)$ does not approach zero. In this case, of course, the function has a discontinuity, but we may say that at such a point it is "continuous from one side." All these cases are represented in the graphs of figure 9.

As an exercise we recommend to the reader to consider the question, what value must be given to the functions

$$\frac{\sin x}{x}, \quad \frac{1 - \cos x}{x^2}, \quad \frac{x^3 - 1}{x - 1}, \quad \frac{\tan x}{x}$$

at those points where they are not defined (that is, at the points where the denominator is equal to zero), in order that they may be continuous at these points. Also, is it possible to find such numbers for the functions

$$\tan x, \quad \frac{1}{x - 1}, \quad \frac{x - 2}{(x^2 - 4)} \ ?$$

These discontinuous functions in mathematics represent the numerous jumplike processes to be met with in nature. In the case of a sudden blow, for example, the value of the velocity of a body changes in such a jumplike fashion. Many qualitative transitions take place with such jumps. In §2 we introduced the function $Q = f(t)$, expressing the way in which the quantity of heat in a given quantity of water (or ice) depends on the temperature. In the neighborhood of the melting point of ice the quantity of heat $Q = f(t)$ changes in a jumplike fashion with changing t.

Functions with isolated discontinuities are encountered quite often in analysis, along with the continuous functions. But as an example of a more complicated function, where the number of discontinuities is infinite, let us consider the so-called Riemann function, which is equal to zero at all irrational points and equal to $1/q$ at rational points of the form $x = p/q$ (where p/q is a fraction in its lowest terms). This function is discontinuous at all rational points and continuous at irrational points. By altering it slightly we may easily obtain an example of a function which is discontinuous at all points.* Let us remark by the way that even for such complicated functions modern analysis has discovered many in-

---

* It is sufficient to set the function equal to unity at the irrational points.

teresting laws, which are investigated in one of the independent branches
of analysis, the theory of functions of a real variable. This theory has
developed with extraordinary rapidity during the past 50 years.

§5. Derivative

The next fundamental concept of analysis is the concept of *derivative*.
Let us consider two problems from which it arose historically.

Velocity.   At the beginning of the present chapter we defined the velocity
of a freely falling body. To do so we made use of a passage to the limit
from the average velocity over short distances to the velocity at the given
point and the given time. The same procedure may be used to define the
instantaneous velocity for an arbitrary nonuniform motion. In fact, let
the function

$$s = f(t) \tag{9}$$

express the dependence of the distance s covered by the material point
in the time t. To find the velocity at the moment $t = t_0$, let us consider
the interval of time from t_0 to $t_0 + h$ ($h \neq 0$). During this time the point
will cover the distance

$$\Delta s = f(t_0 + h) - f(t_0) .$$

The average velocity v_{av} over this part of the path will depend on h

$$v_{av} = \frac{\Delta s}{h} = \frac{1}{h}[f(t_0 + h) - f(t_0)],$$

and will represent the actual velocity at the point t_0 with greater and
greater accuracy as h becomes smaller. It follows that the true velocity
at the time t_0 is equal to the limit

$$v = \lim_{h \to 0} \frac{f(t_0 + h) - f(t_0)}{h}$$

of the ratio of the increase in the distance to the increase in the time,
as the latter approaches zero without ever being actually equal to zero.
In order to calculate the velocity for different forms of motion, we must
discover how to find this limit for various functions $f(t)$.

Tangent.   We are led to investigate a precisely analogous limit by
another problem, this time a geometric one, namely the problem of
drawing a tangent to an arbitrary plane curve.

Let the curve C be the graph of a function $y = f(x)$, and let A be the point on the curve C with abscissa x_0 (figure 10). Which straight line shall we call the tangent to C at the point A? In elementary geometry this question does not arise. The only curve studied there, namely the circumference of a circle, allows us to define the tangent as a straight line which has only one point in common with the curve. But for other curves such a definition will clearly not correspond to our intuitive picture of "tangency." Thus, of the two straight lines L and M in figure 11, the first is obviously not tangent to the curve drawn there (a sinusoidal curve), although it has only one point in common with it; while the second straight line has many points in common with the curve, and yet it is tangent to the curve at each of these points.

To define the tangent, let us consider on the curve C (figure 10) another point A', distinct from A, with abscissa $x_0 + h$. Let us draw the secant AA' and denote the angle which it forms with the x-axis by β. We now allow the point A' to approach A along the curve C. If the secant AA' correspondingly approaches a limiting position, then the straight line T which has this limiting position is called the *tangent* at the point A. Evidently the angle α formed by the straight line T with the x-axis, must be equal to the limiting value of the variable angle β.

The value of $\tan \beta$ is easily determined from the triangle ABA' (figure 10):

$$\tan \beta = \frac{BA'}{AB} = \frac{f(x_0 + h) - f(x_0)}{h}.$$

For the limiting position we must have

$$\tan \alpha = \lim_{A' \to A} \tan \beta = \lim_{h \to 0} \frac{f(x_0 + h) - f(x_0)}{h},$$

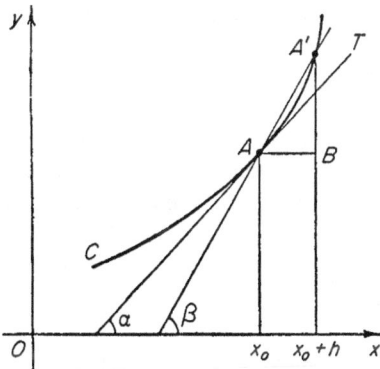

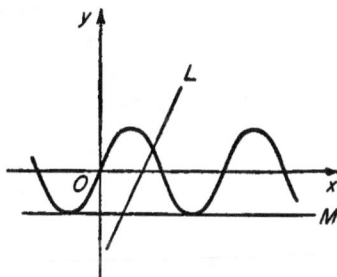

Fig. 10.   Fig. 11.

that is, the trigonometric tangent of the angle of inclination of the tangent line is equal to the limit of the ratio of the increase in the function $f(x)$ at the point x_0 to the corresponding increase in the independent variable, as the latter approaches zero without ever being actually equal to zero.

Let us give still another example leading to the calculation of an analogous limit. Let us suppose that a variable electric current is flowing through a conductor. Let us assume that we know the function $Q = f(t)$ expressing the quantity of electricity that has passed through a fixed cross section of the conductor up to time t. In the period from t_0 to $t_0 + h$, there will flow through this cross section a quantity of electricity ΔQ equal to $f(t_0 + h) - f(t_0)$. The average value of the current will therefore be equal to

$$I_{\mathrm{av}} = \frac{\Delta Q}{h} = \frac{f(t_0 + h) - f(t_0)}{h}.$$

The limit of this ratio as $h \to 0$ will give us the value of the current at the time t_0

$$I = \lim_{h \to 0} \frac{f(t_0 + h) - f(t_0)}{h}.$$

All the three problems discussed, in spite of the fact that they refer to different branches of science, namely mechanics, geometry, and the theory of electricity, have led to one and the same mathematical operation to be performed on a given function, namely to find the limit of the ratio of the increase of the function to the corresponding increase h of the independent variable as $h \to 0$. The number of such widely different problems could be increased at will, and their solution would lead to the same operation. To it we are led, for example, by the question of the rate of a chemical reaction, or of the density of a nonhomogeneous mass and so forth. In view of the exceptional role played by this operation on functions, it has received a special name, differentiation, and the result of the operation is called the *derivative* of the function.

Thus, the *derivative of the function* $y = f(x)$, or more precisely, the *value of the derivative at the given point* x is the limit* approached by the ratio of the increase $f(x + h) - f(x)$ of the function to the increase h of the independent variable, as the latter approaches zero. We often write $h = \Delta x$, and $f(x + \Delta x) - f(x) = \Delta y$, in which case the definition of the derivative is written in the concise form:

$$\lim_{\Delta x \to 0} \frac{\Delta y}{\Delta x}.$$

---

* It is understood that we are speaking here of the case where the limit in question actually exists. If this limit does not exist, then we say that at the point x the function does not have a derivative.

The value of the derivative obviously depends on the point x at which it is found. Thus the derivative of a function $y = f(x)$ is itself a function of x. It is customary to denote the derivative thus

$$f'(x) = \lim_{h \to 0} \frac{f(x+h) - f(x)}{h} = \lim_{\Delta x \to 0} \frac{\Delta y}{\Delta x}.$$

Certain other notations are also customary for the derivative:

$$\frac{df(x)}{dx}, \text{ or } \frac{dy}{dx}, \text{ or } y', \text{ or } y'_x.$$

We should also remark that the notation $\frac{dy}{dx}$ looks like a fraction, although it is read as a single symbol for the derivative. In the following sections the numerator and the denominator of this "fraction" will take on independent meaning, in such a way that their ratio will coincide with the derivative so that this manner of writing is completely justified.

The results of these examples may now be formulated as follows.

The velocity of a point for which the distance s is a given function of the time $s = f(t)$ is equal to the derivative of this function

$$v = s' = f'(t).$$

More concisely, the velocity is the derivative of the distance with respect to time.

The trigonometric tangent of the angle of inclination of the tangent line to the curve $y = f(x)$ at the point with abscissa x is equal to the derivative of the function $f(x)$ at this point:

$$\tan \alpha = y' = f'(x).$$

The strength of the current I at the time t, if $Q = f(t)$ is the quantity of electricity which up to time t has passed through a cross section of the conductor, is equal to the derivative

$$I = Q' = f'(t).$$

Let us make the following remark. The velocity of a nonuniform motion at a given time is a purely physical concept, arising from practical experience. Mankind arrived at it as the result of numerous observations on different concrete motions. The study of nonuniform motion of a body on different parts of its path, the comparison of different motions of this sort taking place simultaneously, and in particular the study of the phenomena of collisions of bodies, all represented an accumulation of practical experience that led to the setting up of the physical concept of the velocity of a nonuniform motion at a given time. But the exact definition of velocity necessarily depended upon the method of defining its numerical value, and to define this value was possible only with the concept of the derivative.

In mechanics the velocity of a body moving according to the rule $s = f(t)$ at the time t is defined as the derivative of the function $f(t)$ for this value of t.

The discussion at the beginning of the present section has shown, on the one hand, the advantages of introducing the operation of finding the derivative, and on the other has given a reasonable justification for the above formulated definition of the velocity at any given moment.

Thus, when we raised the question of finding the velocity of a point in nonuniform motion we had, properly speaking, only an empirical notion of its value but no exact definition. But now, as a result of our analysis, we have reached an exact definition of the value of the velocity at a given moment, namely the derivative of the distance with respect to the time. This result is extremely important from a practical point of view, since our empirical knowledge of the velocity has been greatly enriched by the fact that we can now make an exact numerical calculation.

What has just been said refers equally well, of course, to the strength of a current and to many other concepts expressing the rate of some process, physical, chemical, and so forth.

This situation may serve as an example for numerous others of a similar nature, where practical experience has led to the formation of a concept relating to the external world (velocity, work, density, area, and so forth) and then mathematics has enabled us to define this concept precisely, whereupon we can make use of the concept in practical calculations.

We have already noted at the beginning of the chapter that the concept of a derivative arose chiefly as the result of many centuries of effort directed towards the solving of two problems: drawing a tangent to a curve and finding the velocity of a nonuniform motion. These problems, and also the calculation of areas discussed later, interested mathematicians in ancient times. But until the 16th century the statement and the method of solution for each problem of this sort bore an extremely special character. The accumulation of all this extensive material was reduced to a theoretically complete system in the 17th century in the work of Newton and Leibnitz. An important contribution to the foundations of present-day analysis was also made by Euler.

But it must be said that Newton and Leibnitz and their contemporaries provided very little logical basis for their great mathematical discoveries;

in their methods of reasoning and in the concepts with which they operated there was much that is unclear from our point of view. Even at that time the mathematicians themselves were quite conscious of this, as is shown by the embittered discussions to be found in their correspondence with one another. However, these mathematicians of the 17th and 18th centuries carried on their purely mathematical activities in very close association with the research of other investigators, in the various branches of natural science (physics, mechanics, chemistry, technology). The statement of a mathematical problem usually arose from practical needs or from a wish to understand some phenomenon of nature, and as soon as the problem was solved, the solution was submitted in one way or another to a practical test. Consequently, in spite of a certain lack of logical basis, mathematics was able to advance in extremely useful directions.

Examples for the calculation of derivatives. The definition of the derivative as the limit

$$f'(x) = \lim_{h \to 0} \frac{f(x + h) - f(x)}{h}$$

allows us to calculate the derivative of any given concrete function.

Of course, it must be admitted that cases are possible where the function at one point or another or even at many points simply does not have a derivative; in other words, the ratio

$$\frac{f(x + h) - f(x)}{h}$$

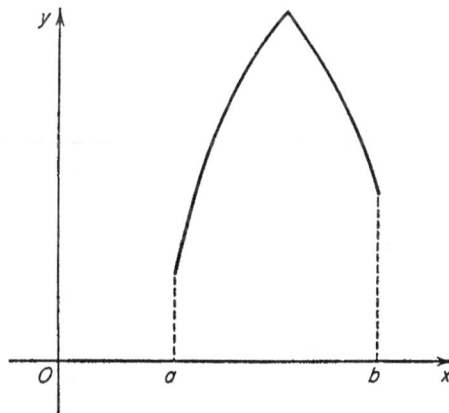

FIG. 12.

as $h \to 0$ does not approach a finite limit. This case obviously occurs at every point of discontinuity of the function $f(x)$, since here the ratio

$$\frac{f(x + h) - f(x)}{h} \tag{10}$$

has a numerator which does not approach zero while the denominator decreases without bound. The derivative may also fail to exist at a point where the function is continuous. A simple example is given by any

point where the graph of the function forms an angle (figure 12). At such
a point the curve of the graph has no definite tangent, and consequently
the function has no derivative. Often at such points the expression (10)
approaches different values, depending on whether h approaches zero
from the right or from the left, so that if h approaches zero in an arbitrary
manner, the ratio (10) simply has no limit. An example of a more complicated function without a derivative is given by

$$y = \begin{cases} x \sin \dfrac{1}{x} \text{ for } x \neq 0, \\ 0 \qquad \text{ for } x = 0. \end{cases}$$

The graph of this function is drawn in figure 13. At the point $x = 0$
it has no derivative because, as is evident from the graph, the secant OA
does not approach any definite position even when $A \to 0$ from one side.
In fact, the secant OA oscillates endlessly back and forth between the
straight line OM and the straight line OL. The corresponding ratio (10)
in this case has no limit, even if h preserves the same sign as it approaches
zero.

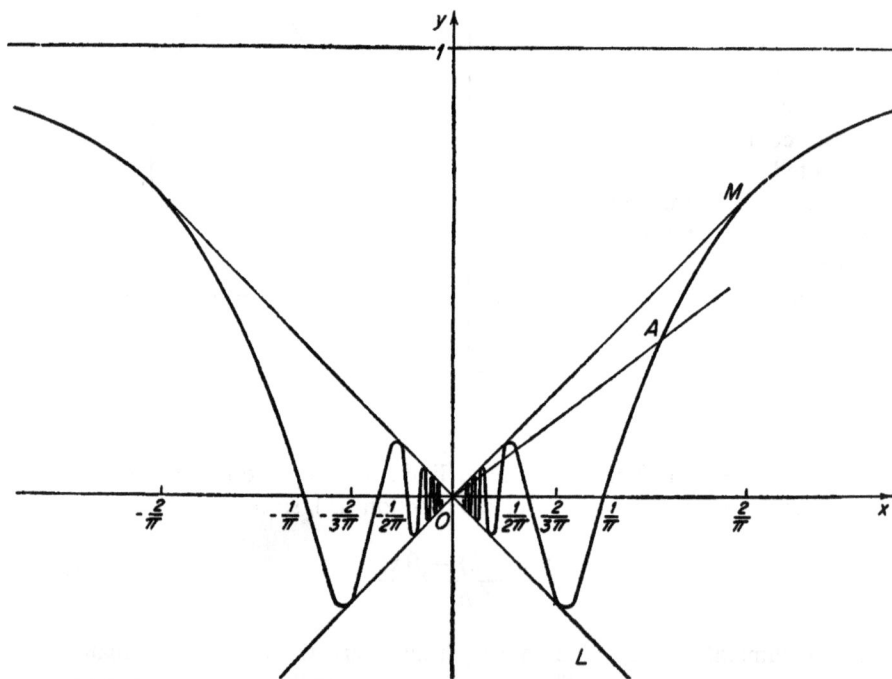

FIG. 13.

Let us remark finally that it is possible to define, in a purely analytic way by means of a formula, a continuous function which does not have a derivative at any point. An example of such a function was first given by the outstanding German mathematician of the last century, Weierstrass.

Consequently the class of differentiable functions is considerably narrower than that of continuous functions.

Let us pass now to the actual calculation of the derivatives of the simplest functions.

1. $y = c$, where c is a constant. A constant may be considered as a special case of a function that remains equal to the same number for arbitrary x. Its graph is a straight line parallel to the x-axis at a distance equal to c. This straight line forms with the x-axis an angle $\alpha = 0$, and obviously the derivative of a constant is identically equal to zero: $y' = (c)' = 0$. From the point of view of mechanics, this equation means that the velocity of a fixed point is equal to zero.

2. $y = x^2$

$$\frac{f(x + h) - f(x)}{h} = \frac{(x + h)^2 - x^2}{h} = 2x + h.$$

As $h \to 0$ we obtain* in the limit $2x$; consequently

$$y' = (x^2)' = 2x.$$

3. $y = x^n$ (n a positive integer).

$$\frac{f(x + h) - f(x)}{h} = \frac{(x + h)^n - x^n}{h}$$

$$= \frac{x^n + nx^{n-1}h + \dfrac{n(n - 1)}{2!} x^{n-2}h^2 + \cdots + h^n - x^n}{h}$$

$$= nx^{n-1} + \frac{n(n - 1)}{2!} x^{n-2}h + \cdots + h^{n-1}.$$

Every addend on the right side, beginning with the second, approaches zero as $h \to 0$; consequently

$$y' = (x^n)' = nx^{n-1}.$$

This formula remains true for arbitrary n positive or negative, fractional

---

* We always assume here that $h \neq 0$.

or even irrational, although the proof must then be different. We will make use of this fact without proving it. Thus for example

$$(\sqrt{x})' = (x^{\frac{1}{2}})' = \frac{1}{2}x^{-\frac{1}{2}} = \frac{1}{2\sqrt{x}}, (x > 0);$$

$$(\sqrt[3]{x})' = (x^{\frac{1}{3}})' = \frac{1}{3}x^{-\frac{2}{3}} = \frac{1}{3\sqrt[3]{x^2}}, (x \neq 0);$$

$$\left(\frac{1}{x}\right)' = (x^{-1})' = -1 \cdot x^{-2} = -\frac{1}{x^2}, (x \neq 0);$$

$$(x^{\pi})' = \pi x^{\pi-1}, \qquad\qquad (x > 0).$$

4. $y = \sin x$.

$$\frac{\sin(x+h) - \sin x}{h} = \frac{2\sin h/2 \cos(x+h/2)}{h} = \frac{\sin h/2}{h/2} \cdot \cos\left(x + \frac{h}{2}\right).$$

As explained earlier, the first fraction approaches unity as $h \to 0$, and $\cos(x + h/2)$ obviously approaches $\cos x$. Thus the derivative of the sine is equal to the cosine

$$y' = (\sin x)' = \cos x.$$

We suggest to the reader that by the same sort of argument he prove that

$$(\cos x)' = -\sin x.$$

5. Earlier (Chapter II, §3) we have already noted the existence of the limit

$$\lim_{n\to\infty} \left(1 + \frac{1}{n}\right)^n = e = 2.71828 \cdots .$$

We also remarked that for the calculation of this limit no essential role is played by the fact that n took on only positive integral values. It is important only that the infinitesimal $1/n$, which is being added to unity, and the exponent n, which is increasing beyond all bounds, should be reciprocal to each other.

Making use of this assertion, we can easily find the derivative of the logarithm $y = \log_a x$

$$\frac{\log_a(x+h) - \log_a x}{h} = \frac{1}{h}\log_a \frac{x+h}{x} = \frac{1}{x}\log_a \left(1 + \frac{h}{x}\right)^{x/h}.$$

The continuity of the logarithm allows us to replace the quantity under the log sign by its limit, which is equal to e; thus

$$\lim_{h \to 0} \left(1 + \frac{h}{x}\right)^{x/h} = e$$

(in this case the role of $n \to \infty$ is played by the increasing quantity x/h). As a result, we obtain the rule for differentiating a logarithm

$$(\log_a x)' = \frac{1}{x} \log_a e.$$

This rule becomes particularly simple if as the base of our logarithms we choose the number e. Logarithms taken to this base are called *natural logarithms* and are denoted by $\ln x$. We may write

$$(\log_e x)' = \frac{1}{x}$$

or again

$$(\ln x)' = \frac{1}{x}.$$

§6. Rules for Differentiation

From the examples given earlier it may appear that the calculation of the derivative of every new function demands the invention of new methods. This is not the case. The development of analysis was made possible to no small extent by the discovery of a simple unified method for finding the derivative of an arbitrary "elementary" function (that is, a function which may be expressed by a formula consisting of a finite combination of the fundamental algebraic operations, the trigonometric functions, the operation of raising to a power, and the taking of logarithms). At the basis of this method are the so-called *rules of differentiation*. They consist of a number of theorems that allow us to reduce more complicated problems to simpler ones.

We will explain here the rules of differentiation and will try to be very brief in deducing them. If the reader wishes to form merely a general idea of analysis, he may omit the present section, remembering only that there exists a means of actually finding the derivative of any elementary function. In this case it will be necessary, of course, for him to take on faith some of the calculations in our later examples.

Derivative of a sum. Assume that y is given as a function of x by the expression

$$y = \phi(x) + \psi(x),$$

where $u = \phi(x)$ and $v = \psi(x)$ are known functions of x. We assume moreover that we can find the derivatives of the functions u and v. How then are we to find the derivative of the function y? The answer is simple

$$y' = (u + v)' = u' + v'. \tag{11}$$

In fact, let us give x an increment Δx; then u, v, and y will each receive an increment Δu, Δv, and Δy, connected by the equation

$$\Delta y = \Delta u + \Delta v.$$

Thus*

$$\frac{\Delta y}{\Delta x} = \frac{\Delta u}{\Delta x} + \frac{\Delta v}{\Delta x},$$

and after the passage to the limit for $\Delta x \to 0$ we at once get formula (11), if, of course, the functions u and v have derivatives.

Analogously we may derive the formula for differentiating the difference of two functions

$$(u - v)' = u' - v'. \tag{12}$$

Derivative of a product. The rule for the differentiation of a product is somewhat more complicated. The derivative of the product of two functions, each of which has a derivative, exists, and is equal to the sum of the product of the first function by the derivative of the second and the product of the second by the derivative of the first; that is

$$(uv)' = uv' + vu'. \tag{13}$$

In fact, let us give x an increment Δx. Then the functions u, v and $y = uv$ will receive the increments Δu, Δv, Δy, satisfying the relation

$$\Delta y = (u + \Delta u)(v + \Delta v) - uv = u\,\Delta v + v\,\Delta u + \Delta u\,\Delta v,$$

from which

$$\frac{\Delta y}{\Delta x} = u\,\frac{\Delta v}{\Delta x} + v\,\frac{\Delta u}{\Delta x} + \Delta u\,\frac{\Delta v}{\Delta x}.$$

After passage to the limit for $\Delta x \to 0$ the first two summands on the right side produce the right side of formula (13) while the third summand vanishes.† Consequently, in the limit we obtain the rule (13).

---

* Here Δx is never equal to zero.

† The final summand here approaches zero for $\Delta x \to 0$, since $\Delta v/\Delta x$ approaches a finite number, equal to the derivative v', which was assumed from the beginning to exist, and $\Delta u \to 0$, since the function u, assumed to have a derivative, is continuous.

In the particular case $v = c =$ constant, we have

$$(cu)' = cu' + uc' = cu', \tag{14}$$

since the derivative of a constant is equal to zero.

Derivative of a quotient. Let $y = u/v$, where u and v have a derivative for a given x, with $v \neq 0$ for that value of x. Obviously

$$\Delta y = \frac{u + \Delta u}{v + \Delta v} - \frac{u}{v} = \frac{v \Delta u - u \Delta v}{(v + \Delta v) v},$$

from which

$$\frac{\Delta y}{\Delta x} = \frac{v \dfrac{\Delta y}{\Delta x} - u \dfrac{\Delta v}{\Delta x}}{(v + \Delta v) v} \to \frac{vu' - uv'}{v^2} \quad (\Delta x \to 0).$$

Here we have again made use of the fact that for a function v which has a derivative we necessarily have $\Delta v \to 0$, when $\Delta x \to 0$. Thus

$$\left(\frac{u}{v}\right)' = \frac{vu' - uv'}{v^2}. \tag{15}$$

Let us give some examples of the application of these rules

$$(2x^3 - 5)' = 2(x^3)' - (5)' = 2 \cdot 3x^2 - 0 = 6x^2;$$

$$(x^2 \sin x)' = x^2(\sin x)' + (x^2)' \sin x = x^2 \cos x + 2x \sin x;$$

$$(\tan x)' = \left(\frac{\sin x}{\cos x}\right)' = \frac{\cos x(\sin x)' - \sin x(\cos x)'}{\cos^2 x}$$

$$= \frac{\cos x \cdot \cos x - \sin x(-\sin x)}{\cos^2 x} = \frac{1}{\cos^2 x} = \sec^2 x.$$

We recommend to the reader to prove for himself the formula

$$(\cot x)' = -\csc^2 x.$$

Derivative of the inverse function. Let us consider a function $y = f(x)$, which is continuous and increasing (decreasing) on the interval $[a, b]$. By increasing (decreasing) we mean that to a greater value of x in the interval $[a, b]$ corresponds a greater (smaller) value of y (figure 14).

Let $c = f(a)$ and $d = f(b)$. In figure 14 it is evident that for each value of y from the interval $[c, d]$ (or $[d, c]$, respectively) there corresponds exactly one value of x from the interval $[a, b]$ such that $y = f(x)$. Thu on the interval $[c, d]$ (or $[d, c]$) we have a completely determined function $x = \phi(y)$, which is called the *inverse function* of $y = f(x)$. In figure 14 it is clear that the function $\phi(y)$ is continuous, a fact which is proved in modern analysis by strictly analytical methods. Now let Δx and Δy

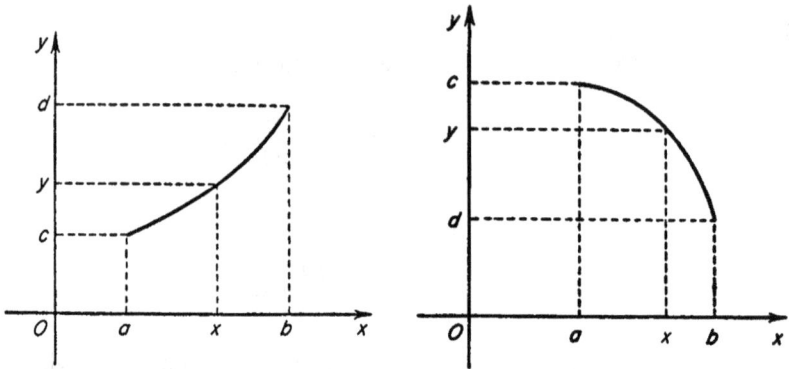

FIG. 14.

correspond respectively to the increments in x and y. It is evident that

$$\frac{\Delta y}{\Delta x} = \frac{1}{\Delta x / \Delta y}, \text{ if } \Delta y \neq 0.$$

In the limit this gives us a simple relation between derivatives of the direct and inverse functions

$$y'_x = \frac{1}{x'_y}. \tag{16}$$

Let us make use of this relation to find the derivative of the function $y = a^x$. The inverse function is $x = \log_a y$, which we are already able to differentiate, and so we may write

$$(a^x)'_x = \frac{1}{(\log_a y)'_y} = \frac{1}{1/y (\log_a e)} = y \log_e a = a^x \ln a. \tag{17}$$

In particular $(e^x)' = e^x$.

As another example let us take $y = $ arc sin x. The inverse function is $x = \sin y$. Thus

$$(\text{arc sin } x)'_x = \frac{1}{(\sin y)'_y} = \frac{1}{\cos y} = \frac{1}{\sqrt{1 - (\sin y)^2}} = \frac{1}{\sqrt{1 - x^2}}.$$

Table of derivatives. Let us tabulate the derivatives of the simplest elementary functions (Table 2).

Table 2

y	y'	y	y'	y	y'
c	0	$\ln x$	$\dfrac{1}{x}$	$\tan x$	$\sec^2 x$
x^a	ax^{a-1}	$\log_a x$	$\dfrac{1}{x}\log_a e$	$\arcsin x$	$\dfrac{1}{\sqrt{1-x^2}}$
e^x	e^x	$\sin x$	$\cos x$	$\arccos x$	$-\dfrac{1}{\sqrt{1-x^2}}$
a^x	$a^x \ln a$	$\cos x$	$-\sin x$	$\arctan x$	$\dfrac{1}{1+x^2}$

These formulas have been calculated and explained earlier, with the exception of the last two which the reader may, if he wishes, easily derive for himself by using the rule for differentiation of an inverse function.

Calculation of the derivative of a function of a function. It remains to consider the last and most difficult rule for differentiation. The reader in possession of this rule and of a set of tables may with perfect right consider that he is able to differentiate any elementary function.

In order to apply the rule we are about to give, it is necessary to be completely clear about how the function we wish to differentiate is constructed; that is, which operations must we perform on the independent variable x, and in which order, to produce the value of the dependent variable y.

For example, to calculate the function

$$y = \sin x^2,$$

it is necessary first of all to raise x to the second power and then to take the sine of the magnitude so obtained, a procedure which may be described in the following way: $y = \sin u$, where $u = x^2$.

On the other hand, in order to calculate the function

$$y = \sin^2 x,$$

it is necessary first of all to find the sine of x, and then to raise the value so found to the second power, a procedure which may be written thus: $y = u^2$, where $u = \sin x$.

Here are some examples:

1. $y = (3x + 4)^3, y = u^3, u = 3x + 4.$
2. $y = \sqrt{1 - x^2}, y = u^{\frac{1}{2}}, u = 1 - x^2.$
3. $y = e^{kx}; y = e^u, u = kx.$

In more complicated cases we have a chain of simple relations, which may have several links. For example,

4. $y = \cos^3 x^2; y = u^3; u = \cos v; v = x^2.$

If y is a function of the variable u

$$y = f(u), \qquad\qquad (18)$$

and u in its turn is a function of the variable x

$$u = \phi(x), \qquad\qquad (19)$$

then y, being a function of u, is also a certain function of x, which may be denoted as follows

$$y = F(x) = f[\phi(x)]. \qquad\qquad (20)$$

By considering more complicated cases we may form, for example, the function

$$y = \Phi(x) = f\{\phi[\psi(x)]\},$$

which is equivalent to the equations

$$y = f(u), u = \phi(v), v = \psi(x),$$

and we could form still longer chains.

We now show how to calculate the derivative of the function $F(x)$ defined by equation (20) if we know the derivative of $f(u)$ with respect to u and the derivative of $\phi(x)$ with respect to x.

Let us give to x the increment Δx; then by (19) u will receive a certain increment Δu and by (18) y will receive an increment Δy. Thus we may write

$$\frac{\Delta y}{\Delta x} = \frac{\Delta y}{\Delta u} \cdot \frac{\Delta u}{\Delta x}.$$

Now let Δx approach zero. Then $\Delta u/\Delta x \to u_x'$. Furthermore, from the continuity of u, the increase $\Delta u \to 0$, and therefore $\Delta y/\Delta u \to y_u'$ (the existence of the derivatives y_u' and u_x' was assumed).

Thus we have proved the important formula for the derivative of a function of a function*

$$y'_x = y'_u u'_x .$$ (21)

Let us calculate, from formula (21) and the fundamental table of derivatives given, the derivatives of the functions we have been considering:

1. $y = (3x + 4)^3 = u^3,\ y'_x = (u^3)'_u (3x + 4)'_x = 3u^2 \cdot 3 = 9(3x + 4)^2.$

2. $y = \sqrt{1 - x^2} = u^{\frac{1}{2}},\ y'_x = (u^{\frac{1}{2}})'_x (1 - x^2)'_x = \frac{1}{2} u^{-\frac{1}{2}} (-2x)$

$$= -\frac{x}{\sqrt{1 - x^2}}.$$

3. $y = e^{kx} = e^u,\ y'_x = (e^u)'_u \cdot u'_x = e^u \cdot k = ke^{kx} .$

If $y = f(u),\ u = \phi(v),\ v = \psi(x)$, then

$$y'_x = y'_u \cdot u_x = y'_u(u'_v \cdot v'_x) = y'_u \cdot u'_v \cdot v'_x .$$

It is clear how to generalize this formula for the case of an arbitrary (finite) number of functions in the chain. For example,

4. $y = \cos^3 x^2;\ y'_x = (u^3)'_u (\cos v)'_v \cdot (x^2)'_x = 3u^2(-\sin v) \cdot 2x$

$$= -6x \cos^2 x^2 \sin x^2.$$

In our explanation of how to calculate the derivative of a function of a function, we have introduced intermediate variables $u, v, \cdots$. But in fact, after a little practice one may dispense with them, simply keeping in mind the functions they denote.

The elementary functions. To close the present section let us remark that the functions whose derivatives were listed in tabular form (Table 2) may be used to define the so-called elementary functions. These *elementary functions* are defined as those functions that may be obtained from the preceding simple functions by the four arithmetical operations and the operation of taking a function of a function, each of these operations being performed a finite number of times.

For example, the polynomial $x^2 - 2x^2 + 3x - 5$ is an elementary function since it is obtained by arithmetic operations from a number of functions of the form x^k. The function $\ln \sqrt{1 - x^2}$ is also elementary,

---

* In deducing this formula we have tacitly assumed that, as Δx approaches zero, Δu is never equal to zero. But the formula remains true even when this assumption does not hold.

since it is obtained from the polynomial $u = 1 - x^2$ by the operation $v = u^{1/2}$, and subsequently the operation $\ln v$.

The rules for differentiation discussed earlier are sufficient to obtain the derivative of any elementary function, as soon as we know the derivatives of the simplest elementary functions.

§7. Maximum and Minimum; Investigation of the Graphs of Functions

One of the simplest and most important applications of the derivative is in the theory of maxima and minima. Let us suppose that on a certain interval $a \leqslant x \leqslant b$ we are given a function $y = f(x)$ which is not only continuous but also has a derivative at every point. Our ability to calculate the derivative enables us to form a clear picture of the graph of the function. On an interval on which the derivative is always positive the tangent to the graph will be directed upward. On such an interval the

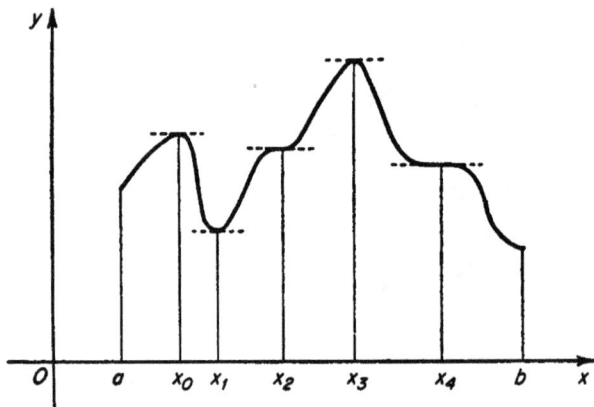

FIG. 15.

function will increase; that is, to a greater value of x will correspond a greater value of $f(x)$. On the other hand, on an interval where the derivative is always negative, the function will decrease; the graph will run downward.

Maximum and minimum. In figure 15 we have drawn the graph of a function $y = f(x)$ defined on the interval (a, b). Of a special interest are the points of this graph whose abcissas are x_0, x_1, x_3.

At the point x_0 the function $f(x)$ is said to have a local maximum; by this we mean that at this point $f(x)$ is greater than at neighboring points; more precisely $f(x_0) \geqslant f(x)$ for every x in a certain interval around the point x_0.

A local minimum is defined analogously.

For our function a local maximum occurs at the points x_0 and x_3, and a local minimum at the point x_1.

At every maximum or minimum point, if it is inside the interval $[a, b]$, i.e., if it does not coincide with one of the end points a or b, the derivative must be equal to zero.

This last statement, a very important one, follows immediately from the definition of the derivative as the limit of the ratio $\Delta y/\Delta x$. In fact, if we move a short distance from the maximum point, then $\Delta y \leqslant 0$. Thus for positive Δx the ratio $\Delta y/\Delta x$ is nonpositive, and for negative Δx the ratio $\Delta y/\Delta x$ is nonnegative. The limit of this ratio, which exists by hypothesis, can therefore be neither positive nor negative and there remains only the possibility that it is zero. By inspection of the diagram it is seen that this means that at maximum or minimum points (it is customary to leave out the word "local," although it is understood) the tangent to the graph is horizontal. In figure 15 we should remark that at the points x_2 and x_4 also the tangent is horizontal, just as it is at the points x_0, x_1, x_3, although at these points the function has neither maximum nor minimum. In general, there may be more points at which the derivative of the function is equal to zero (stationary points) than there are maximum or minimum points.

Determination of the greatest and least values of a function. In numerous technical questions it is necessary to find the point x at which a given function $f(x)$ attains its greatest or its least value on a given interval.

In case we are interested in the greatest value, we must find x_0 on the interval $[a, b]$ for which among all x on $[a, b]$ the inequality $f(x_0) \geqslant f(x)$ is fulfilled.

But now the fundamental question arises, whether in general there exists such a point. By the methods of modern analysis it is possible to prove the following existence theorem: If the function $f(x)$ is continuous on a finite interval, then there exists at least one point on the interval for which the function attains its maximum (minimum) value on the interval $[a, b]$.

From what has been said already, it follows that these maximum or minimum points must be sought among the "stationary" points. This fact is the basis for the following well-known method for finding maxima and minima.

First we find the derivative of $f(x)$ and then solve the equation obtained by setting it equal to zero

$$f'(x) = 0.$$

If x_1, x_2, $\cdots$, x_n are the roots of this equation, we then compare the numbers $f(x_1)$, $f(x_2)$, $\cdots$, $f(x_n)$ with one another. Of course, it is necessary to take into account that the maximum or minimum of the function may be found not within the interval but at the end (as is the case with the minimum in figure 15) or at a point where the function has no derivative (as in figure 12). Thus to the points x_1, x_2, $\cdots$, x_n we must add the ends a and b of the interval and also those points, if they exist, at which there is no derivative. It only remains to compare the values of the function at all these points and to choose among them the greatest or the least.

With respect to the stated existence theorem, it is important to add that this theorem ceases, in general, to hold in the case that the function $f(x)$ is continuous only on the interval (a, b); that is, on the set of points x satisfying the inequalities $a < x < b$. We leave it to the reader to consider the fact that the function $1/x$ has neither a maximum nor a minimum on the interval $(0, 1)$.

Let us look at some examples.

From a square piece of tin of side a it is required to make a rectangular open box of maximum volume. If from the corners of the original square we take away squares of side x (see §2, example 2) we get a box with the volume

$$V = x(a - 2x)^2.$$

Our problem then becomes to find the value of x for which the function $V(x)$ attains its greatest value on the interval $0 \leqslant x \leqslant a/2$. In accordance with the rule, we find the derivative and set it equal to zero

$$V'(x) = (a - 2x)^2 - 4x(a - 2x) = 0.$$

Solving this equation, we find the two roots

$$x_1 = \frac{a}{2}, x_2 = \frac{a}{6}.$$

To these we adjoin the left end of the interval (the right end is identical with x_1) and compare the values of the function at these points

$$V(0) = 0; V\left(\frac{a}{6}\right) = \frac{2}{27} a^3, V\left(\frac{a}{2}\right) = 0.$$

Thus the box will have the greatest volume, equal to $2/27\, a^3$, for the height $x = a/6$.

As a second example, let us examine the problem of the lamp at the skating rink (see §2, example 3). At whàt height h should we place the lamp in order that the edge of the rink may receive the greatest illumination?

For formula (3) §2, our problem reduces to determining the value of h for $T = A \sin \alpha / h^2 + r^2$ takes on its greatest value. Instead of h it is more convenient here to find the angle α (figure 3, Chapter I). We have

$$h = r \tan \alpha,$$

so that

$$T = \frac{A}{r^2} \frac{\sin \alpha}{1 + \tan^2 \alpha} = \frac{A}{r^2} \sin \alpha \cos^2 \alpha.$$

Then it is required to find the maximum of the function $T(\alpha)$ among those values of α which satisfy the inequality $0 < \alpha < \pi/2$. To do this, we find the derivative and set it equal to zero

$$T'(\alpha) = \frac{A}{r^2} (\cos^3 \alpha - 2 \sin^2 \alpha \cos \alpha) = 0.$$

This equation splits into two

$$\cos \alpha = 0, \cos^2 \alpha - 2 \sin^2 \alpha = 0.$$

The first equation has the root $\alpha = \pi/2$, which coincides with the end of the interval $(0, \pi/2)$. The second equation may be put in the form

$$\tan^2 \alpha = \tfrac{1}{2}.$$

But since $0 < \alpha < \pi/2$, we have the result $\alpha \approx 35°15'$. So this is the value for which the function $T(\alpha)$ attains its maximum (at the ends of the interval, $T = 0$). The desired height h is thus equal to

$$h = r \tan \alpha = \frac{r}{\sqrt{2}} \approx 0.7r.$$

For best illumination of the edge of the rink the lamp should be placed at a height equal to about 0.7 times the radius.

But now let us suppose that the facilities at our disposal do not allow us to raise the lamp to a height greater than a certain H. Then the angle α may vary not from 0 to $\pi/2$ but only within the narrower limits $0 < \alpha \leqslant \text{arc tan } (H/r)$. For example, let $r = 12$ meters and $H = 9$ meters. In this case, it is in fact possible to raise the lamp to the height $h = r/\sqrt{2}$, which amounts to somewhat more than 8 meters, so that this is what we ought to do. But if H is less than 8 meters (for example, if we have at our disposal only a pole of length 6 meters), then it turns out that the derivative of the function $T(\alpha)$ in the interval $[0, \text{arc tan } (H/r)]$ is nowhere equal to zero. In this case the maximum is attained at the end of the interval, and the lamp should be raised to the greatest possible height $H = 6$ meters.

Up to now we have considered a function on a finite interval. If the interval is infinite in length, then even a continuous function may fail to attain its greatest or least value but may, for example, continue to grow or to decrease as x approaches infinity.

Thus the functions $y = kx + b$ (see figure 5, Chapter I), $y = $ arc tan x (figure 16a), $y = \ln x$ (figure 16b) nowhere attain either a

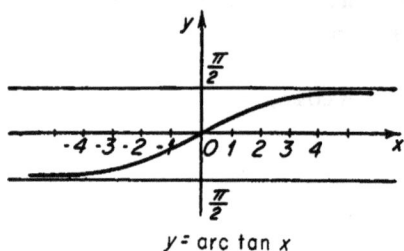

$y = $ arc tan x

FIG. 16a.

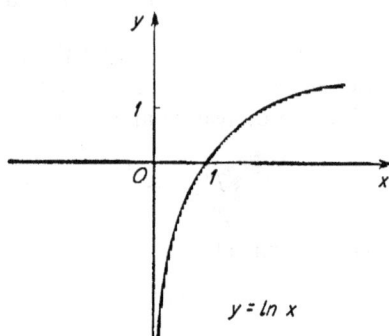

$y = \ln x$

FIG. 16b.

maximum or a minimum. The function $y = e^{-x^2}$ (figure 16c) attains its maximum at the point $x = 1$, but nowhere attains a minimum. As for the function $y = x/(1 + x^2)$ (figure 16d), it reaches its minimum at the point $x = -1$ and its maximum at the point $x = 1$.

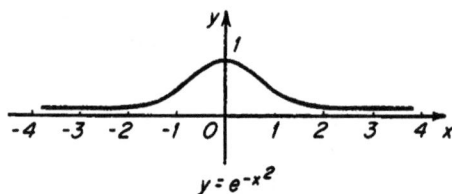

$y = e^{-x^2}$

FIG. 16c.

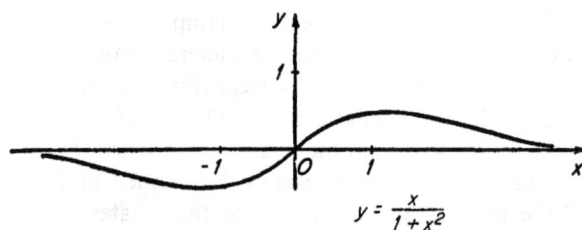

$y = \dfrac{x}{1 + x^2}$

FIG. 16d.

In the case of an interval of infinite length the investigation may be reduced to the ordinary rules. It is only necessary to consider in place of $f(a)$ and $f(b)$ the limits

$$A = \lim_{x \to -\infty} f(x), \quad B = \lim_{x \to +\infty} f(x).$$

Derivatives of higher orders. We have just seen how, for closer study of the graph of a function, we must examine the changes in its derivative $f'(x)$. This derivative is a function of x, so that we may in turn find its derivative.

The derivative of the derivative is called the second derivative and is denoted by

$$[y']' = y'' \quad \text{or} \quad [f'(x)]' = f''(x).$$

Analogously, we may calculate the third derivative

$$[y'']' = y''' \quad \text{or} \quad [f''(x)]' = f'''(x)$$

and more generally the nth derivative or, as it is also called, the derivative of nth order

$$y^{(n)} = f^{(n)}(x).$$

Of course, it must be kept in mind that, for a certain value of x (or even for all values of x) this sequence may break off at the derivative of some order, say the kth; it may happen that $f^{(k)}(x)$ exists but not $f^{(k+1)}(x)$. Derivatives of arbitrary order will appear later in §9 in connection with the Taylor formula. For the moment we confine ourselves to the second derivative.

Significance of the second derivative; convexity and concavity. The second derivative has a simple significance in mechanics. Let $s = f(t)$ be a law of motion along a straight line; then s' is the velocity and s'' is the "velocity of the change in the velocity" or more simply the "acceleration" of the point at time t. For example, for a falling body under the force of gravity

$$s = \frac{gt^2}{2} + v_0 t + s_0 \, ,$$

$$s' = gt + v_0 \, ,$$

$$s'' = g,$$

that is, the acceleration of falling bodies is constant.

The second derivative also has a simple geometric meaning. Just as the sign of the first derivative determines whether the function is increasing or decreasing, so the sign of the second derivative determines the side toward which the graph of the function will be curved.

Suppose, for example, that on a given interval the second derivative is everywhere positive. Then the first derivative increases and therefore

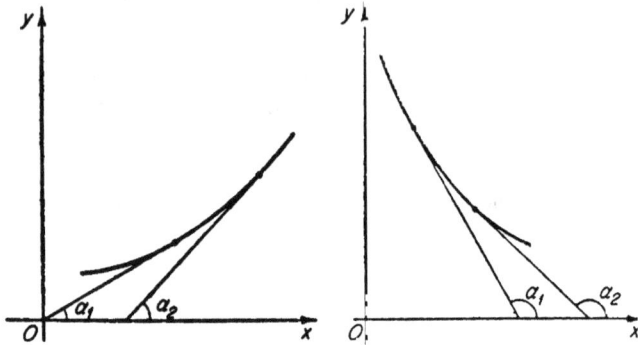

FIG. 17.

$f'(x) = \tan \alpha$ increases and the angle α of inclination of the tangent line itself increases (figure 17). Thus as we move along the curve it keeps turning constantly to the same side, namely upward, and is thus, as they say, "convex downward."

On the other hand, in a part of a curve where the second derivative is constantly negative (figure 18) the graph of the function is "convex upward."*

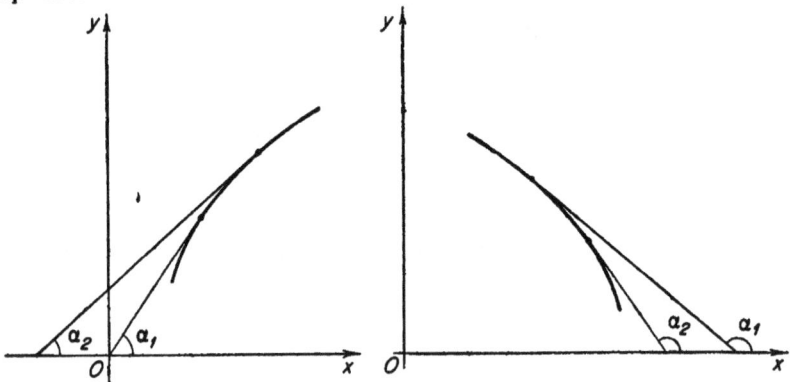

FIG. 18.

* Strictly defined, the "convexity upward" is that property of the curve that consists of its lying above (more precisely "not below") the chord joining any two of its points; analogously, for "convexity downward" (which is also simply called "concavity"), the curve does not lie above its chords.

Criteria for maxima and minima; study of the graphs of curves. If throughout the whole interval over which x varies the curve is convex upward and if at a certain point x_0 of this interval the derivative is equal to zero, then at this point the function necessarily attains its maximum; and its minimum in the case of convexity downward. This simple consideration often allows us, after finding a point at which the derivative is equal to zero, to decide thereupon whether at this point the function has a local maximum or minimum.*

Example 1. Let us study the appearance of the graph of the function

$$f(x) = \frac{x^3}{3} - \frac{5x^2}{2} + 6x - 2.$$

We take its first derivative and set it equal to zero,

$$f'(x) = x^2 - 5x + 6 = 0.$$

The roots of the equation obtained in this way are $x_1 = 2$, $x_2 = 3$. The corresponding values of the function are

$$f(2) = 2\tfrac{2}{3}, \; f(3) = 2\tfrac{1}{2}.$$

We then mark these two points on the diagram. Along with these we may also mark the point with coordinates $x = 0$ and $y = f(0) = -2$ where the graph intersects the y-axis. The second derivative is $f''(x) = 2x - 5$. This reduces to zero for $x = \tfrac{5}{2}$, so that

$$f''(x) > 0 \text{ for } x > \tfrac{5}{2},$$

$$f''(x) < 0 \text{ for } x < \tfrac{5}{2}.$$

The point

$$x = \frac{5}{2}, y = f\left(\frac{5}{2}\right) = 2\tfrac{7}{12}$$

is a *point of inflection* of the graph. To the left of this point the curve is convex upward, and to the right it is convex downward.

It is now evident that the point $x = 2$ is a maximum point and the point $x = 3$ is a minimum point for the function.

---

* In more complicated cases, where the second derivative itself changes sign, the problem of determining the character of the stationary point is solved by means of the Taylor formula (§9).

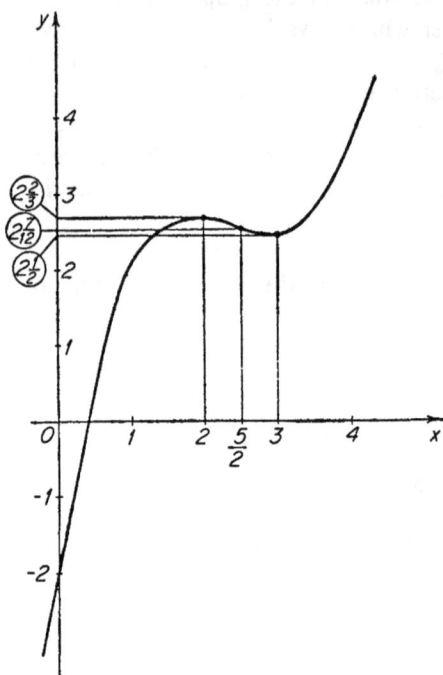

On the basis of these results we conclude that the graph of the function $y = f(x)$ has the appearance sketched in figure 19. To the right of the point $(0, -2)$ the curve rises with increasing x, is convex upwards, and attains its maximum at the point $(2, 2\frac{2}{3})$, after which it begins to fall. At the point $(2\frac{1}{2}, 2\frac{7}{12})$, where $f''(x) = 0$, the convexity changes to concavity. Then at the point $(3, 2\frac{1}{2})$ the function attains its minimum and from there on rises to infinity. The final statement comes from the fact that the first term of the function, the one containing the highest (third) power of x, approaches infinity faster than the second and third terms. For the same reason the graph of the function approaches $-\infty$ as x assumes numerically larger negative values.

Example 2. We shall prove the inequality $e^x \geqslant 1 + x$ for arbitrary x. For this purpose we consider the function $f(x) = e^x - x - 1$. Its first derivative is $f'(x) = e^x - 1$, which reduces to zero only for $x = 0$. The second derivative $f''(x) = e^x > 0$ for all x. Consequently the graph of the function $f(x)$ is convex downward. The number $f(0) = 0$ is a minimum for the function and $e^x - x - 1 \geqslant 0$ for all x.

The study of graphs has many different purposes. They often show very clearly, for example, the number of real roots of a given equation. Thus, in order to demonstrate that the equation

$$xe^x = 2$$

has a single real root, we may study the graphs of the functions $y = e^x$ and $y = 2/x$ (as sketched in figure 20). It is easy to see that these graphs intersect at only one point, so that the equation $e^x = 2/x$ has exactly one root.

The methods of analysis are extensively applied to questions of approximate calculation of the roots of an equation. On this subject, see Chapter IV, §5.

§8. Increment and Differential of a Function

The differential of a function. Let us consider a function $y = f(x)$ that has a derivative. The increment of this function

$$\Delta y = f(x + \Delta x) - f(x),$$

corresponding to the increment Δx, has the property that the ratio $\Delta y/\Delta x$, as $\Delta x \rightarrow 0$, approaches a finite limit, equal to the derivative

$$\frac{\Delta y}{\Delta x} \rightarrow f'(x).$$

This fact may be written as an equality

$$\frac{\Delta y}{\Delta x} = f'(x) + \alpha,$$

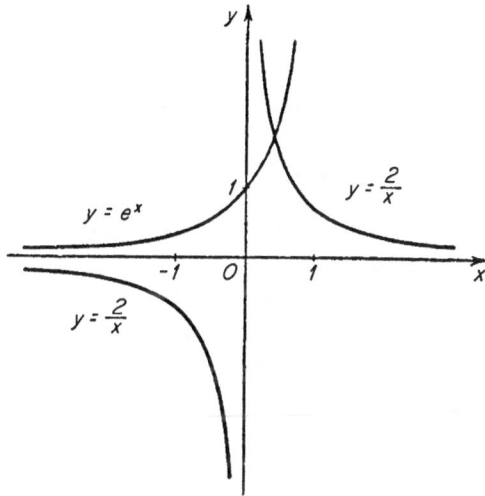

FIG. 20.

where the value of α depends on Δx in such a way that as $\Delta x \rightarrow 0$, α also approaches zero. Thus the increment of a function may be represented in the form

$$\Delta y = f'(x)\Delta x + \alpha \Delta x,$$

where $\alpha \rightarrow 0$, if $\Delta x \rightarrow 0$.

The first summand on the right side of this equality depends on Δx in a very simple way, namely it is proportional to Δx. It is called the *differential* of the function, at the point x, corresponding to the given increment Δx, and is denoted by

$$dy = f'(x)\Delta x.$$

The second summand has the characteristic property that, as $\Delta x \rightarrow 0$, it approaches zero more rapidly than Δx, as a result of the presence of the factor α. It is therefore said to be an infinitesimal of higher order than Δx and, in case $f'(x) \neq 0$, it is also of higher order than the first summand.

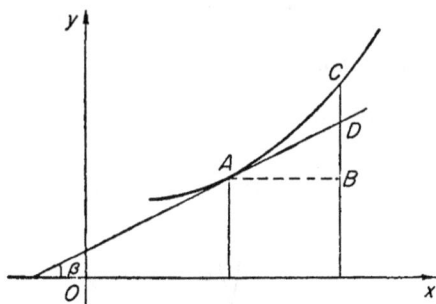

FIG. 21.

By this we mean that for sufficiently small Δx the second summand is small in itself and its ratio to Δx is also arbitrarily small.

The decomposition of Δy into two summands, of which the first (the principal part) depends linearly on Δx and the second is negligible for small Δx, may be illustrated by figure 21. The segment $BC = \Delta y$, where $BC = BD + DC$, $BD = \tan \beta \cdot \Delta x = f'(x) \Delta x = dy$, and DC is an infinitesimal of higher order than Δx.

In practical problems the differential is often used as an approximate value for the increment in the function. For example, suppose we have the problem of determining the volume of the walls of a closed cubical box whose interior dimensions are $10 \times 10 \times 10$ cm and the thickness of whose walls is 0.05 cm. If great accuracy is not required, we may argue as follows. The volume of all the walls of the box represents the increment Δy of the function $y = x^3$ for $x = 10$ and $\Delta x = 0.1$. So we find approximately

$$\Delta y \approx dy = (x^3)' \Delta x = 3x^2 \Delta x = 3 \cdot 10^2 \cdot 0.1 = 30 \text{ cm}^3.$$

For symmetry in the notation it is customary to denote the increment of the independent variable by dx and to call it also a differential. With this notation the differential of the function may be written thus:

$$dy = f'(x) \, dx.$$

Then the derivative is the ratio $f'(x) = dy/dx$ of the differential of the function to the differential of the independent variable.

The differential of a function originated historically in the concept of an "indivisible." This concept, which from a modern point of view was never very clearly defined, was in its time, in the 18th century, a fundamental one in mathematical analysis. The ideas concerning it have undergone essential changes in the course of several centuries. The indivisible, and later the differential of a function, were represented as actual infinitesimals, as something in the nature of an extremely small constant magnitude, which however was not zero. The definition given in this section is the one accepted in present-day analysis. According to this definition the differential is a finite magnitude for each increment Δx and is at the same time proportional to Δx. The other fundamental property of the differential, the character of its difference from Δy, may be recognized only in motion, so to speak: if we consider an increment Δx which is approaching zero (which is infinitesimal), then the difference between dy and Δy will be arbitrarily small even in comparison with Δx.

This substitution of the differential in place of small increments of the function forms the basis of most of the applications of infinitesimal analysis to the study of nature. The reader will see this in a particularly clear way in the case of differential equations, dealt with in this book in Chapters V and VI.

Thus, in order to determine the function that represents a given physical process, we try first of all to set up an equation that connects this function in some definite way with its derivatives of various orders. The method of obtaining such an equation, which is called a differential equation, often amounts to replacing increments of the desired functions by their corresponding differentials.

As an example let us solve the following problem. In a rectangular system of coordinates $Oxyz$, we consider the surface obtained by rotation of the parabola whose equation (in the Oyz plane) is $z = y^2$. This surface is called a paraboloid of revolution (figure 22). Let v denote the volume of the body bounded by the paraboloid and the plane parallel to the Oxy plane at a distance z from it. It is evident that v is a function of z ($z > 0$).

To determine the function v, we attempt to find its differential dv.

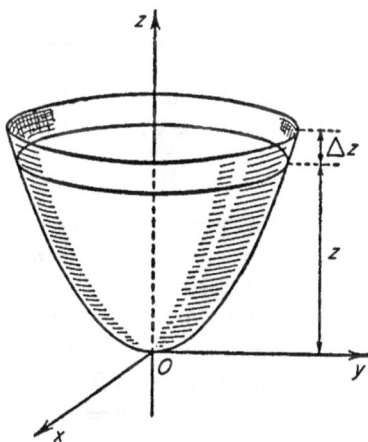

FIG. 22.

The increment Δv of the function v at the point z is equal to the volume bounded by the paraboloid and by two planes parallel to the Oxy plane at distances z and $z + \Delta z$ from it.

It is easy to see that the magnitude of Δv is greater than the volume of the circular cylinder of radius $\sqrt{z}$ and height Δz but less than that of the circular cylinder with radius $\sqrt{z + \Delta z}$ and height Δz.

Thus

$$\pi z\, \Delta z < \Delta v < \pi(z + \Delta z)\, \Delta z$$

and so

$$\Delta v = \pi(z + \theta\, \Delta z)\, \Delta z = \pi z\, \Delta z + \pi\theta\, \Delta z^2\,,$$

where θ is some number depending on Δz and satisfying the inequality $0 < \theta < 1$.

So we have succeeded in representing the increment Δv in the form of a sum, the first summand of which is proportional to Δz, while the second is an infinitesimal of higher order than Δz (as $\Delta z \to 0$). It follows that the first summand is the differential of the function v

$$dv = \pi z\, \Delta z,$$

or

$$dv = \pi z\, dz,$$

since $\Delta z = dz$ for the independent variable z.

The equation so obtained relates the differentials dv and dz (of the variables v and z) to each other and thus is called a differential equation.

If we take into account that

$$\frac{dv}{dz} = v',$$

where v' is the derivative of v with respect to the variable z, our differential equation may also be written in the form

$$v' = \pi z.$$

To solve this very simple differential equation we must find a function of z whose derivative is equal to πz. Problems of this sort are treated in a general way in §§10 and 11, but for the moment we urge the reader to verify that a solution of our equation is given by $v = \pi z^2/2 + C$, where for C we may choose an arbitrary number.* In our case the volume of the body is obviously zero for $z = 0$ (see figure 22), so that $C = 0$. Thus our function is given by $v = \pi z^2/2$.

The mean value theorem and examples of its application. The differential expresses the approximate value of the increment of the function in terms of the increment of the independent variable and of the derivative at the initial point. So for the increment from $x = a$ to $x = b$, we have

$$f(b) - f(a) \approx f'(a)(b - a).$$

---

* This formula gives all the solutions.

It is possible to obtain an exact equation of this sort if we replace the derivative $f'(a)$ at the initial point by the derivative at some intermediate point, suitably chosen in the interval (a, b). More precisely: *If $y = f(x)$ is a function which is differentiable on the interval $a \leqslant x \leqslant b$, then there exists a point ξ, strictly within this interval, such that the following exact equality holds*

$$f(b) - f(a) = f'(\xi)(b - a). \tag{22}$$

The geometric interpretation of this "mean-value theorem" (also called Lagrange's formula or the finite-difference formula) is extraordinarily simple. Let A, B be the points on the graph of the function $f(x)$ which correspond to $x = a$ and $x = b$, and let us join A and B by the chord AB (figure 23). Now let us move the straight line AB, keeping it constantly

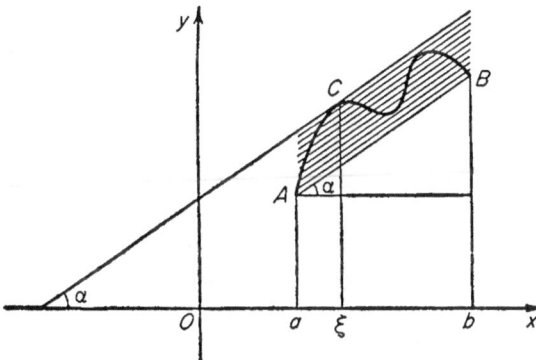

FIG. 23.

parallel to itself, up or down. At the moment when this straight line cuts the graph for the last time, it will be tangent to the graph at a certain point C. At this point (let the corresponding abscissa be $x = \xi$), the tangent line will form the same angle of inclination α as the chord AB. But for the chord we have

$$\tan \alpha = \frac{f(b) - f(a)}{b - a}.$$

On the other hand at the point C

$$\tan \alpha = f'(\xi).$$

This equation

$$\frac{f(b) - f(a)}{b - a} = f'(\xi)$$

is exactly the mean-value theorem.*

Formula (22) has the peculiar feature that the point ξ appearing in it is unknown to us; we know only that it lies "somewhere in the interval (a, b)." But in spite of this indeterminacy, the formula has great theoretical significance and is part of the proof of many theorems in analysis. The immediate practical importance of this formula is also very great, since it enables us to estimate the increase in a function when we know the limits between which its derivative can vary. For example,

$$|\sin b - \sin a| = |\cos \xi| (b - a) \leqslant b - a.$$

Here a, b and ξ are angles, expressed in radian measure; ξ is some value between a and b; ξ itself is unknown, but we know that $|\cos \xi| \leqslant 1$.

From formula (22) it is clear that a function whose derivative is everywhere equal to zero must be a constant; at no part of the interval can it receive an increment different from zero. Analogously, the reader will easily prove that a function whose derivative is everywhere positive must everywhere increase, and if its derivative is negative, the function must decrease. We give here without proof one of the many generalizations of the mean-value theorem.

For arbitrary functions $\phi(x)$ and $\psi(x)$ differentiable in the interval $[a, b]$, provided only that $\psi'(x) \neq 0$ in (a, b), the following equation[†] holds

$$\frac{\phi(b) - \phi(a)}{\psi(b) - \psi(a)} = \frac{\phi'(\xi)}{\psi'(\xi)} , \tag{23}$$

where ξ is some point in the interval (a, b).[‡]

From this theorem we can derive a general method for calculating the limits of an expression like

$$\lim_{x \to 0} \frac{\phi(x)}{\psi(x)} , \tag{24}$$

---

* Of course these arguments only give a geometric interpretation of the theorem and by no means form a rigorous proof.

† Formula (23) can be derived by a simple application of the mean-value theorem to the function

$$f(x) = \phi(x) - \frac{\phi(b) - \phi(a)}{\psi(b) - \psi(a)} \psi(x).$$

‡ By the symbols $[a, b]$ and (a, b) we denote the sets of values of x satisfying the inequalities $a \leqslant x \leqslant b$ and $a < x < b$ respectively.

if $\phi(0) = \psi(0) = 0$. From formula (23) we have

$$\frac{\phi(x)}{\psi(x)} = \frac{\phi(x) - \phi(0)}{\psi(x) - \psi(0)} = \frac{\phi'(\xi)}{\psi'(\xi)},$$

where ξ is between 0 and x, and therefore $\xi \to 0$ together with x. This allows us to calculate the limit

$$\lim_{x \to 0} \frac{\phi'(x)}{\psi'(x)},$$

instead of the limit (24), which is in many cases very much easier.*

Example. Let us find the $\lim\limits_{x \to 0} \dfrac{x - \sin x}{x^3}$. By making use of the rule three times, we have successively

$$\lim_{x \to 0} \frac{x - \sin x}{x^3} = \lim_{x \to 0} \frac{1 - \cos x}{3x^2} = \lim_{x \to 0} \frac{\sin x}{6x} = \lim_{x \to 0} \frac{\cos x}{6} = \frac{1}{6}.$$

§9. Taylor's Formula

The function

$$p(x) = a_0 + a_1 x + a_2 x^2 + \cdots + a_n x^n,$$

where the coefficients a_k are constants, is called a polynomial of degree n. In particular, $y = ax + b$ is a polynomial of the first degree and $y = ax^2 + bx + c$ is a polynomial of the second degree. Polynomials may be considered as the simplest of all functions. In order to calculate their value for a given x, we require only the operations of addition, subtraction, and multiplication; not even division is needed. Polynomials are continuous for all x and have derivatives of arbitrary order. Also, the derivative of a polynomial is again a polynomial, of degree lower by one, and the derivatives of order $n + 1$ and higher of a polynomial of degree n are equal to zero.

If to the polynomials we adjoin functions of the form

$$y = \frac{a_0 + a_1 x + \cdots + a_n x^n}{b_0 + b_1 x + \cdots + b_m x^m},$$

---

* The same rule is valid for finding the limit of a fractional expression in which the numerator and the denominator both approach infinity. This method, which is very convenient for finding such limits (or, as we say, for the removal of indeterminacies), will be used, for example, in §3 of Chapter XII.

for the calculation of which we also need division, and also the functions $\sqrt{x}$ and $\sqrt[3]{x}$ and, finally, arithmetical combinations of these functions, we obtain essentially all the functions whose values can be calculated by methods learned in the secondary school.

While we were still in school, we formed some notion of a number of other functions, like

$$\sqrt[5]{x}, \log x, \sin x, \text{arc} \tan x,$$

But though we became acquainted with the most important properties of these functions, we found no answer in elementary mathematics to the question: How can we calculate them? What sort of operations, for example, is it necessary to perform on x in order to obtain $\log x$ or $\sin x$? The answer to this question is given by methods that have been worked out in analysis. Let us examine one of these methods.

Taylor's formula.   On an interval containing the point a, let there be given a function $f(x)$ with derivatives of every order. The polynomial of first degree

$$p_1(x) = f(a) + f'(a)(x - a)$$

has the same value as $f(x)$ at the point $x = a$ and also, as is easily verified, has the same derivative as $f(x)$ as this point. Its graph is a straight line, which is tangent to the graph of $f(x)$ to the point a. It is possible to choose a polynomial of the second degree, namely

$$p_2(x) = f(a) + f'(a)(x - a) + \frac{f''(a)}{2}(x - a)^2,$$

which at the point of $x = a$ has with $f(x)$ a common value and a common first and second derivative. Its graph at the point a will follow that of $f(x)$ even more closely. It is natural to expect that if we construct a polynomial which at $x = a$ has the same first n derivatives as $f(x)$ at the same point, then this polynomial will be a still better approximation to $f(x)$ at points x near a. Thus we obtain the following approximate equality, which is Taylor's formula

$$f(x) \approx f(a) + f'(a)(x - a) + \frac{f''(a)}{2!}(x - a)^2 + \cdots + \frac{f^{(n)}(a)}{n!}(x - a)^n \quad (25)$$

The right side of this formula is a polynomial of degree n in $(x - a)$. For each x the value of this polynomial can be calculated if we know the values of $f(a), f'(a), \cdots, f^{(n)}(a)$.

For functions which have an $(n + 1)$th derivative, the right side of this formula, as is easy to show, differs from the left side by a small quantity which approaches zero more rapidly than $(x - a)^n$. Moreover, it is the only possible polynomial of degree n that differs from $f(x)$, for x close to a, by a quantity that approaches zero, as $x \to a$, more rapidly than $(x - a)^n$. If $f(x)$ itself is an algebraic polynomial of degree n, then the approximate equality (25) becomes an exact one.

Finally, and this is particularly important, we can give a simple expression for the difference between the right side of formula (25) and the actual value of $f(x)$. To make the approximate equality (25) exact, we must add to the right side a further term, called the "remainder term"

$$f(x) = f(a) + f'(a)(x - a) + \cdots + \frac{f^{(n)}(a)}{n!}(x - a)^n + \frac{f^{(n+1)}(\xi)}{(n + 1)!}(x - a)^{n+1} \tag{26}$$

This final supplementary term*

$$R_{n+1}(x) = \frac{f^{(n+1)}(\xi)}{(n + 1)!}(x - a)^{n+1}$$

has the peculiarity that the derivative appearing in it is to be calculated in each case not at the point a but at a suitably chosen point ξ, which is unknown but lies somewhere in the interval between a and x.

The proof of equality (26) is rather cumbersome but quite simple in essence. We shall give here a somewhat artificial version of the proof, which has the merit of being concise.

In order to find out by how much the left side in the approximate formula (25) differs from the right, let us consider the ratio of the difference between the two sides in equality (25) to the quantity $-(x - a)^{n+1}$

$$\frac{f(x) - \left[f(a) + f'(a)(x - a) + \cdots + \frac{f^{(n)}(a)}{n!}(x - a)^n \right]}{-(x - a)^{n+1}}. \tag{27}$$

We also introduce the function

$$\phi(u) = f(u) + f'(u)(x - u) + \cdots + \frac{f^{(n)}(u)}{n!}(x - u)^n$$

of a variable u, taking x to be fixed (constant). Then the numerator in (27) will represent the increase of this function as we pass from $u = a$ to $u = x$, and the denominator will be the increase over the same interval of the function

$$\psi(u) = (x - u)^{n+1}.$$

---

* This is only one of the possible forms for the remainder term $R_{n+1}(x)$.

We now make use of the generalized mean-value theorem quoted earlier

$$\frac{\phi(x) - \phi(a)}{\psi(x) - \psi(a)} = \frac{\phi'(\xi)}{\psi'(\xi)}.$$

Differentiating the functions $\phi(u)$ and $\psi(u)$ with respect to u (it must be recalled that the value of x has been fixed) we find that

$$\frac{\phi'(\xi)}{\psi'(\xi)} = -\frac{f^{(n+1)}(\xi)}{(n+1)!}.$$

The equality of this last expression with the original quantity (27) gives Taylor's formula in the form (26).

In the form (26) Taylor's formula not only provides a means of approximate calculation of $f(x)$ but also allows us to estimate the error. Let us consider the simple example

$$y = \sin x.$$

The values of the function $\sin x$ and of its derivatives of arbitrary order are known for $x = 0$. Let us make use of these values to write Taylor's formula for $\sin x$, choosing $a = 0$ and limiting ourselves to the case $n = 4$. We find successively

$$f(x) = \sin x, \qquad f'(x) = \cos x, \qquad f''(x) = -\sin x,$$
$$f'''(x) = -\cos x, \qquad f^{\mathrm{IV}}(x) = \sin x, \qquad f^{\mathrm{V}}(x) = \cos x;$$
$$f(0) = 0 \qquad f'(0) = 1, \qquad f''(0) = 0,$$
$$f'''(0) = -1. \qquad f^{\mathrm{IV}}(0) = 0, \qquad f^{\mathrm{V}}(\xi) = \cos \xi.$$

Therefore

$$\sin x = x - \frac{x^3}{6} + R_5, \quad \text{where} \quad R_5 = \frac{x^5}{120} \cos \xi.$$

Although the exact value R_5 is unknown, still we can easily estimate it from the fact that $|\cos \xi| \leqslant 1$. For all values of x between 0 and $\pi/4$ we have

$$|R_5| = \left| \frac{x^5}{120} \cos \xi \right| < \frac{1}{120} \left(\frac{\pi}{4}\right)^5 < \frac{1}{400}.$$

Consequently, on the interval $[0, \pi/4]$ the function $\sin x$ may be considered, with accuracy up to $\frac{1}{400}$, as equal to the polynomial of third degree

$$\sin x = x - \frac{1}{6} x^3.$$

If we were to take more terms in Taylor's expansion for sin x, we would obtain a polynomial of higher degree which would approximate sin x still more closely.

The tables for trigonometric and other functions are calculated by similar methods.

The laws of nature, as a rule, can be expressed with good approximation by functions that may be differentiated as often as we like and that in their turn may be approximated by polynomials, the degree of the polynomial being determined by the accuracy desired.

Taylor's series.   If in formula (25) we take a larger and larger number of terms, then the difference between the right side and $f(x)$, expressed by the remainder term $R_{n+1}(x)$, may tend to zero. Of course this will not always occur: neither for all functions nor for all values of x. But there exists a broad class of functions (the so-called *analytic* functions) for which the remainder term $R_{n+1}(x)$ does in fact approach zero as $n \to \infty$, at least for all values of x within a certain interval around the point a. For these functions the Taylor formula allows us to calculate $f(x)$ with any desired degree of accuracy. Let us examine such functions more closely.

If $R_{n+1}(x) \to 0$ as $n \to \infty$, then from (26) it follows that

$$f(x) = \lim_{n \to \infty} \left[f(a) + f'(a)(x - a) + \cdots + \frac{f^{(n)}(a)}{n!} (x - a)^n \right]$$

In this case we say that $f(x)$ has been expanded in a convergent infinite series

$$f(x) = f(a) + f'(a)(x - a) + \frac{f''(a)}{2!} (x - a)^2 + \cdots,$$

in increasing powers of $(x - a)$. This series is called a *Taylor series*, and $f(x)$ is said to be the sum of the series. Let us consider some examples (with $a = 0$):

1. $(1 + x)^n = 1 + nx + \dfrac{n(n - 1)}{2!} x^2 + \dfrac{n(n - 1)(n - 2)}{3!} x^3 + \cdots$

   (valid for $|x| < 1$ and for arbitrary real n).

2. $\quad \sin x = x - \dfrac{x^3}{3!} + \dfrac{x^5}{5!} - \dfrac{x^7}{7!} + \cdots$ $\qquad$ (valid for all x).

3. $\cos x = 1 - \dfrac{x^2}{2!} + \dfrac{x^4}{4!} - \dfrac{x^6}{6!} + \cdots$ (valid for all x).

4. $e^x = 1 + x + \dfrac{x^2}{2!} + \dfrac{x^3}{3!} +$ (valid for all x).

5. $\arctan x = x - \dfrac{x^3}{3} + \dfrac{x^5}{5} - \cdots$ (valid for $|x| \leqslant 1$).

The first of these examples is the famous binomial theorem of Newton, which was obtained by Newton for all n but completely proved in his time only for integral n. This example served as a model for the establishment of the general Taylor formula. The last two formulas allow us, for $x = 1$, to calculate with arbitrarily good approximation the numbers e and π.

The Taylor formula, which opens up the way for most of the calculations in applied analysis, is extremely important from the practical point of view.

Many of the laws of nature, physical and chemical processes, the motion of bodies, and the like, are expressed with great accuracy by functions which may be expanded in a Taylor series. The theory of such functions can be formulated in a clearer and more complete way if we consider them as functions of a complex variable (see Chapter IX).

The idea of approximating a function by polynomials or of representing it as the sum of an infinite number of simpler functions underwent far-reaching developments in analysis, where it now forms an independent branch, the theory of approximation of functions (see Chapter XII).

§10. Integral

From Chapter I and from §1 of the present chapter the reader already knows that the concept of the integral, and more generally of the integral calculus, had its historical origin in the need for solving concrete problems, a characteristic example of which is the calculation of the area of a curvilinear figure. The present section is devoted to these questions. In it we will also discuss the aforementioned connection between the problems of the differential and the integral calculus, which was not fully cleared up until the 18th century.

Area. Let us suppose that a curve above the x-axis forms the graph of the function $y = f(x)$. We attempt to find the area S of the segment bounded by the line $y = f(x)$, by the x-axis and by the straight lines drawn through the points $x = a$ and $x = b$ parallel to the y-axis.

To solve this problem we proceed as follows. We divide the interval $[a, b]$ into n parts, not necessarily equal. We denote the length of the first part by Δx_1, of the second by Δx_2, and so forth up to the final part Δx_n. In each segment we choose points $\xi_1, \xi_2, \cdots, \xi_n$ and set up the sum

$$S_n = f(\xi_1)\,\Delta x_1 + f(\xi_1)\,\Delta x_2$$
$$\dotplus \cdots \dotplus f(\xi_n)\,\Delta x_n . \quad (28)$$

The magnitude S_n is obviously equal to the sum of the areas of the rectangles shaded in figure 24.

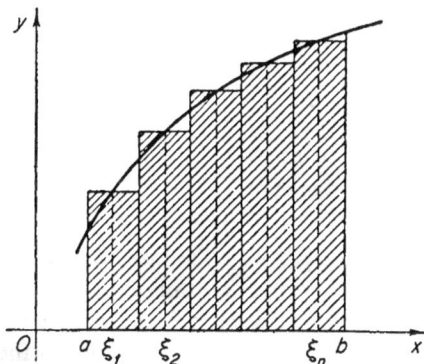

FIG. 24.

The finer we make the subdivision of the segment $[a, b]$, the closer S_n will be to the area S. If we carry out a sequence of such constructions, dividing the interval $[a, b]$ into successively smaller and smaller parts, then the sums S_n will approach S.

The possibility of dividing $[a, b]$ into unequal parts makes it necessary for us to define what we mean by "successively smaller" subdivisions. We assume not only that n increases beyond all bounds but also that the length of the greatest Δx_i in the nth subdivision approaches zero. Thus

$$S = \lim_{\max \Delta x_i \to 0} [f(\xi_1)\,\Delta x_1 + f(\xi_2)\,\Delta x_2 + \cdots + f(\xi_n)\,\Delta x_n]$$

$$= \lim_{\max \Delta x_i \to 0} \sum_{i=1}^{n} f(\xi_i)\,\Delta x_i . \quad (29)$$

The calculation of the desired area has in this way been reduced to finding the limit (29).

We note that when we first set up the problem, we had only an empirical idea of what we mean by the area of our curvilinear figure, but we had no precise definition. But now we have obtained an exact definition of the concept of area: It is the limit (29). We now have not only an intuitive notion of area but also a mathematical definition, on the

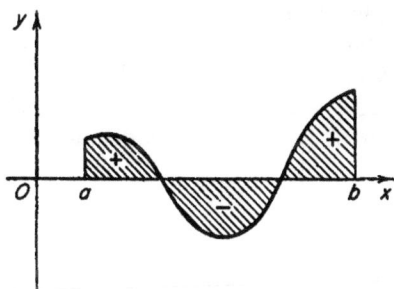

FIG. 25.

basis of which we can calculate the area numerically (compare the remarks at the end of §3, concerning velocity and the length of a circumference).

We have assumed that $f(x) \geqslant 0$. If $f(x)$ changes sign, then in figure 25, the limit (29) will give us the algebraic sum of the areas of the segments lying between the curve $y = f(x)$ and the x-axis, where the segments above the x-axis are taken with a plus sign and those below with a minus sign.

Definite integral. The need to calculate the limit (29) arises in many other problems. For example, suppose that a point is moving along a straight line with variable velocity $v = f(t)$. How are we to determine the distance s covered by the point in the time from $t = a$ to $t = b$?

Let us assume that the function $f(t)$ is continuous; that is, in small intervals of time the velocity changes only slightly. We divide the interval $[a, b]$ into n parts, of length $\Delta t_1, \Delta t_2, \cdots, \Delta t_n$. To calculate an approximate value for the distance covered in each interval Δt_i, we will suppose that the velocity in this period of time is constant, equal throughout to its actual value at some intermediate point ξ_1. The whole distance covered will then be expressed approximately by the sum

$$s_n = \sum_{i=1}^{n} f(\xi_i)\, \Delta t_i ,$$

and the exact value of the distance s covered in the time from a to b, will be the limit of such sums for finer and finer subdivisions; that is, it will be the limit (29)

$$s = \lim_{\max \Delta t_i \to 0} \sum_{i=1}^{n} f(\xi_i)\, \Delta t_i .$$

It would be easy to give many examples of practical problems leading to the calculation of such a limit. We will discuss some of them later, but for the moment the examples already given will sufficiently indicate the importance of this idea. The limit (29) is called the *definite integral* of the function $f(x)$ taken over the interval $[a, b]$, and it is denoted by

$$\int_a^b f(x)\, dx.$$

The expression $f(x)\, dx$ is called the integrand, a and b are the limits of integration; a is the lower limit, b is the upper limit.

The connection between differential and integral calculus. As an example of the direct calculation of a definite integral, we may take example 2, §1. We may now say that the problem considered there reduces to calculation of the definite integral

$$\int_0^h ax\,dx.$$

Another example was considered in §3, where we solved the problem of finding the area bounded by the parabola $y = x^2$. Here the problem reduces to calculation of the integral

$$\int_0^1 x^2\,dx.$$

We were able to calculate both these integrals directly, because we have simple formulas for the sum of the first n natural numbers and for the sum of their squares. But for an arbitrary function $f(x)$, we are far from being able to add up the sum (28) (that is, to express the result in a simple formula) if the points ξ_i and the increments Δx_i are given to suit some particular problem. Moreover, even when such a summation is possible, there is no general method for carrying it out; various methods, each of a quite special character, must be used in the various cases.

So we are confronted by the problem of finding a general method for the calculation of definite integrals. Historically this question interested mathematicians for a long period of time, since there were many practical aspects involved in a general method for finding the area of curvilinear figures, the volume of bodies bounded by a curved surface, and so forth.

We have already noted that Archimedes was able to calculate the area of a segment and of certain other figures. The number of special problems that could be solved, involving areas, volumes, centers of gravity of solids, and so forth, gradually increased, but progress in finding a general method was at first extremely slow. The general method could not be discovered until sufficient theoretical and computational material had been accumulated through the demands of practical life. The work of gathering and generalizing this material proceeded very gradually until the end of the Middle Ages; and its subsequent energetic development was a direct consequence of the rapid growth in the productive powers of Europe resulting from the breakup of the former (feudal) methods of manufacturing and the creation of new ones (capitalistic).

The accumulation of facts connected with definite integrals proceeded alongside of the corresponding investigations of problems related to the derivative of a function. The reader already knows from §1 that this immense preparatory labor was crowned with success in the 17th century by the work of Newton and Leibnitz. It is in this sense that Newton and Leibnitz are the creators of the differential and integral calculus.

One of the fundamental contributions of Newton and Leibnitz consists of the fact that they finally cleared up the profound connection between differential and integral calculus, which provides us, in particular, with a general method of calculating definite integrals for an extremely wide class of functions.

To explain this connection, we turn to an example from mechanics.

We suppose that a material point is moving along a straight line with velocity $v = f(t)$, where t is the time. We already know that the distance σ covered by our point in the time between $t = t_1$ and $t = t_2$ is given by the definite integral

$$\sigma = \int_{t_1}^{t_2} f(t)\, dt.$$

Now let us assume that the law of motion of the point is known to us; that is, we know the function $s = F(t)$ expressing the dependence on the time t of the distance s calculated from some initial point A on the straight line. The distance σ covered in the interval of time $[t_1, t_2]$ is obviously equal to the difference

$$\sigma = F(t_2) - F(t_1).$$

In this way we are led by physical considerations to the equality

$$\int_{t_1}^{t_2} f(t)\, dt = F(t_2) - F(t_1),$$

which expresses the connection between the law of motion of our point and its velocity.

From a mathematical point of view the function $F(t)$, as we already know from §5, may be defined as a function whose derivative for all values of t in the given interval is equal to $f(t)$, that is

$$F'(t) = f(t).$$

Such a function is called a *primitive* for $f(t)$.

We must keep in mind that if the function $f(t)$ has at least one primitive, then along with this one it will have an infinite number of others; for if $F(t)$ is a primitive for $f(t)$, then $F(t) + C$, where C is an arbitrary constant, is also a primitive. Moreover, in this way we exhaust the whole set of primitives for $f(t)$, since if $F_1(t)$ and $F_2(t)$ are primitives for the same function $f(t)$, then their difference $\phi(t) = F_1(t) - F_2(t)$ has a derivative

$\phi'(t)$ that is equal to zero at every point in a given interval so that $\phi(t)$ is a constant.*

From a physical point of view the various values of the constant C determine laws of motion which differ from one another only in the fact that they correspond to all possible choices for the initial point of the motion.

We are thus led to the result that for an extremely wide class of functions $f(x)$, including all cases where the function $f(x)$ may be considered as the velocity of a point at the time x, we have the following equality†

$$\int_a^b f(x)\, dx = F(b) - F(a), \tag{30}$$

where $F(x)$ is an arbitrary primitive for $f(x)$.

This equality is the famous *formula of Newton and Leibnitz*, which reduces the problem of calculating the definite integral of a function to finding a primitive for the function and in this way forms a link between the differential and the integral calculus.

Many particular problems that were studied by the greatest mathematicians are automatically solved by this formula, stating that the definite integral of the function $f(x)$ on the interval $[a, b]$ is equal to the difference between the values of any primitive at the left and right ends of the interval.‡ It is customary to write the difference (30) thus:

$$F(x)\Big|_a^b = F(b) - F(a).$$

Example 1. The equality

$$\left(\frac{x^3}{3}\right)' = x^2$$

shows that the function $x^3/3$ is a primitive for the function x^2. Thus, by the formula of Newton and Leibnitz,

$$\int_0^a x^2\, dx = \frac{x^3}{3}\Big|_0^a = \frac{a^3}{3} - \frac{0}{3} = \frac{a^3}{3}.$$

---

* By the mean value theorem
$$\phi(t) - \phi(t_0) = \phi'(v)(t - t_0) = 0,$$
when v lies between t and t_0. Thus $\phi(t) = \phi(t_0) = \text{const}$ for all t.

† It is possible to prove mathematically, without recourse to examples from mechanics, that if the function $f(x)$ is continuous (and even if it is discontinuous but Lebesgue-summable; see Chapter XV) on the interval $[a, b]$, then there exists a primitive $F(x)$ satisfying equality (30).

‡ This formula has been generalized in various ways (see for example §13, the formula of Ostrogradskiĭ).

Example 2. Let c and c' be two electric charges, on a straight line at distance r from each other. The attraction F between them is directed along this straight line and is equal to

$$F = \frac{a}{r^2}$$

($a = kcc'$, where k is a constant). The work W done by this force, when the charge c remains fixed but c' moves along the interval $[R_1, R_2]$, may be calculated by dividing the interval $[R_1, R_2]$ into parts Δr_i. On each of these parts we may consider the force to be approximately constant, so that the work done on each part is equal to $a/r_i^2\, \Delta r_i$. Making the parts smaller and smaller, we see that the work W is equal to the integral

$$W = \lim_{n\to\infty} \sum_{i=1}^{n} \frac{a}{r_i^2}\, \Delta r_i = \int_{R_1}^{R_2} \frac{a}{r^2}\, dr.$$

The value of this integral can be calculated at once, if we recall that

$$\frac{a}{r^2} = \left(-\frac{a}{r}\right)',$$

so that

$$W = -\frac{a}{r}\Big|_{R_1}^{R_2} = a\left(\frac{1}{R_1} - \frac{1}{R_2}\right).$$

In particular, the work done by a force F as the charge c', initially at a distance R_1 from c, moves out to infinity, is equal to

$$W = \lim_{R_2\to\infty} a\left(\frac{1}{R_1} - \frac{1}{R_2}\right) = \frac{a}{R_1}.$$

From the arguments given above for the formula of Newton and Leibnitz, it is clear that this formula gives mathematical expression to an actual tie existing in the objective world. It is a beautiful and important example of how mathematics gives expression to objective laws. We should remark that in his mathematical investigations, Newton always took a physical point of view. His work on the foundations of differential and integral calculus cannot be separated from his work on the foundations of mechanics.

The concepts of mathematical analysis, such as the derivative or the integral, as they presented themselves to Newton and his contemporaries, had not yet completely "broken away" from their physical and geometric origins, such as velocity and area. In fact, they were half mathematical in character and half physical. The conditions existing at that time were not yet suitable for producing a purely mathematical definition of these concepts. Consequently, the investigator could handle them correctly in complicated situations only if he remained in close contact with the practical aspects of his problem even during the intermediate (mathematical) stages of his argument.

From this point of view the creative work of Newton was different in character from that of Leibnitz.* Newton was guided at all stages by a physical way of looking at the problem. But the investigations of Leibnitz do not have such an immediate connection with physics, a fact that in the absence of clear-cut mathematical definitions sometimes led him to mistaken conclusions. On the other hand, the most characteristic feature of the creative activity of Leibnitz was his striving for generality, his efforts to find the most general methods for the problems of mathematical analysis.

The greatest merit of Leibnitz was his creation of a mathematical symbolism expressing the essence of the matter. The notations for such fundamental concepts of mathematical analysis as the differential dx, the second differential d^2x, the integral $\int y\,dx$, and the derivative d/dx were proposed by Leibnitz. The fact that these notations are still used shows how well they were chosen.

One advantage of a well-chosen symbolism is that it makes our proofs and calculations shorter and easier; also, it sometimes protects us against mistaken conclusions. Leibnitz, who was well aware of this, paid especial attention in all his work to the choice of notation.

The evolution of the concepts of mathematical analysis (derivative, integral, and so forth) continued, of course, after Newton and Leibnitz and is still continuing in our day; but there is one stage in this evolution that should be mentioned especially. It took place at the beginning of the last century and is related particularly to the work of Cauchy.

Cauchy gave a clear-cut formal definition of the concept of a limit and used it as the basis for his definitions of continuity, derivative, differential, and integral. These definitions have been introduced at the corresponding places in the present chapter. They are widely used in present-day analysis.

The great importance of these achievements lies in the fact that it is now possible to operate in a purely formal way not only in arithmetic,

---

* The discoveries of Newton and Leibnitz were made independently.

algebra, and elementary geometry, but also in this new and very extensive branch of mathematics, in mathematical analysis, and to obtain correct results in so doing.

Regarding practical application of the results of mathematical analysis, it is now possible to say: If the original data are verified in the actual world, then the results of our mathematical arguments will also be verified there. If we are properly assured of the accuracy of the original data, then there is no need to make a practical check of the correctness of the mathematical results; it is sufficient to check only the correctness of the formal arguments.

This statement naturally requires the following limitation. In mathematical arguments the original data, which we take from the actual world, are true only up to a certain accuracy. This means that at every step of our mathematical argument the results obtained will contain certain errors, which may accumulate as the number of steps in the argument increases.*

Returning now to the definite integral, let us consider a question of fundamental importance. For what functions $f(x)$, defined on the interval $[a, b]$, is it possible to guarantee the existence of the definite integral $\int_a^b f(x)\, dx$, namely a number to which the sum $\sum_1^n f(\xi_i)\, \Delta x_i$ tends as limit as max $\Delta x_i \to 0$? It must be kept in view that this number is to be the same for all subdivisions of the interval $[a, b]$ and all choices of the points ξ_i.

Functions for which the definite integral, namely the limit (29), exists are said to be *integrable* on the interval $[a, b]$. Investigations carried out in the last century show that all continuous functions are integrable.

But there are also discontinuous functions which are integrable. Among them, for example, are those functions which are bounded and either increasing or decreasing on the interval $[a, b]$.

The function that is equal to zero at the rational points in $[a, b]$ and equal to unity at the irrational points, may serve as an example of a non-integrable function, since for an arbitrary subdivision the integral sum s_n will be equal to zero or unity, depending on whether we choose the points ξ_i as rational numbers or irrational.

Let us note that in many cases the formula of Newton and Leibnitz provides an answer to the practical question of calculating a definite integral. But here arises the problem of finding a primitive for a given

---

* For example, it follows formally from $a = b$ and $b = c$ that $a = c$. But in practice this relation appears as follows: From the facts that $a = b$ is known with accuracy up to ϵ and $b = c$ is known with the same accuracy, it follows that $a = c$ is known with accuracy up to 2ϵ.

function; that is, of finding a function that has the given function for its derivative. We now proceed to discuss this problem. Let us note by the way that the problem of finding a primitive has great importance in other branches of mathematics also, particularly in the solution of differential equations.

§11. Indefinite Integrals; the Technique of Integration

An arbitrary primitive of a given function $f(x)$ is usually called an *indefinite integral* of $f(x)$ and is written in the form

$$\int f(x)\, dx.$$

In this way, if $F(x)$ is a completely determined primitive of $f(x)$, then the indefinite integral of $f(x)$ is given by

$$\int f(x)\, dx = F(x) + C, \tag{31}$$

where C is an arbitrary constant.

Let us also note that if the function $f(x)$ is given on the interval $[a, b]$ and, if $F(x)$ is a primitive for $f(x)$ and x is a point in the interval $[a, b]$, then by the formula of Newton and Leibnitz we may write

$$F(x) = F(a) + \int_a^x f(t)\, dt.$$

Here the integral on the right side differs from the primitive $F(x)$ only by the constant $F(a)$. In such a case this integral, if we consider it as a function of its upper limit x (for variable x), is a completely determined primitive of $f(x)$. Consequently, an indefinite integral of $f(x)$ may also be written as follows:

$$\int f(x)\, dx = \int_a^x f(t)\, dt + C,$$

where C is an arbitrary constant.

Let us set up a fundamental table of indefinite integrals, which can be obtained directly from the corresponding table of derivatives (see §6):

$$\int x^a \, dx = \frac{x^{a+1}}{a+1} + C(a \neq -1),$$

$$\int \frac{dx}{x} = \ln |x| + C,^*$$

$$\int a^x \, dx = \frac{a^x}{\ln a} + C,$$

$$\int e^x \, dx = e^x + C,$$

$$\int \sin x \, dx = -\cos x + C,$$

$$\int \cos x \, dx = \sin x + C,$$ (32)

$$\int \sec^2 x \, dx = \tan x + C,$$

$$\int \frac{dx}{\sqrt{1 - x^2}} = \arcsin x + C$$
$$= -\arccos x + C_1 \left(C_1 - C = \frac{\pi}{2} \right),$$

$$\int \frac{dx}{1 + x^2} = \arctan x + C.$$

The general properties of indefinite integrals may also be deduced from the corresponding properties of derivatives. For example, from the rule for the differentiation of a sum we obtain the formula

$$\int [f(x) \pm \phi(x)] \, dx = \int f(x) \, dx \pm \int \phi(x) \, dx + C,$$

and from the corresponding rule expressing the fact that a constant factor k may be taken outside the sign of differentiation we get

$$\int kf(x) \, dx = k \int f(x) \, dx + C.$$

For example,

$$\int \left(3x^2 + 2x - \frac{3}{\sqrt{x}} + \frac{4}{x} - 1 \right) dx$$

$$= 3\frac{x^3}{3} + \frac{2x^2}{2} - 3\frac{x^{-1/2+1}}{-\frac{1}{2}+1} + 4 \ln |x| - x + C.$$

---
* For $x > 0$, $(\ln |x|)' = (\ln x)' = 1/x$; for $x < 0$, $(\ln |x|)' = [\ln(-x)]' = 1/-x(-1) = 1/x$.

There are a number of methods for calculating indefinite integrals. Let us consider one of them, namely the *method of substitution* or change of variable, which is based on the following equality

$$\int f(x)\, dx = \int f[\phi(t)]\, \phi'(t)\, dt + C, \tag{33}$$

where $x = \phi(t)$ is a differentiable function. The relation (33) is to be understood in the sense that if in the function

$$F(x) = \int f(x)\, dx,$$

on the left side of equality (33), we set $x = \phi(t)$, we thereby obtain a function $F[\phi(t)]$ whose derivative with respect to t is equal to the expression under the sign of integration on the right side of equality (33). This fact follows immediately from the theorem on the derivative of a function of a function.

Let us give some examples of this method of substitution

$$\int e^{kx}\, dx = \int e^{t} \frac{1}{k}\, dt = \frac{1}{k} \int e^{t}\, dt = \frac{1}{k} e^{t} + C = \frac{e^{kx}}{k} + C$$

(substitution of $kx = t$, from which $k\, dx = dt$).

$$\int \frac{x\, dx}{\sqrt{a^2 - x^2}} = -\int dt = -t + C = -\sqrt{a^2 - x^2} + C$$

$$\left(\text{substitution of } t = \sqrt{a^2 - x^2}, \text{ from which } dt = -\frac{x\, dx}{\sqrt{a^2 - x^2}} \right).$$

$$\int \sqrt{a^2 - x^2}\, dx = \int \sqrt{a^2 - a^2 \sin^2 u}\, a \cos u\, du = a^2 \int \cos^2 u\, du$$

$$= a^2 \int \frac{1 + \cos 2u}{2}\, du = \frac{a^2}{2} \left(u + \frac{\sin 2u}{2} \right) + C$$

$$= \frac{a^2}{2} (u + \sin u \cos u) + C$$

$$= \frac{a^2}{2} \left(\arcsin \frac{x}{a} + \frac{x}{a^2} \sqrt{a^2 - x^2} \right) + C$$

(substitution of $x = a \sin u$).

As can be seen from these examples, the method of substitution or change of variables greatly extends the class of elementary functions that we are able to integrate; that is, for which we can find primitives that are themselves elementary functions. But it must be noted that from the point of view of actually calculating the result, we are in a much worse position, generally speaking, with respect to integration than for differentiation.

From §6 we know that the derivative of an arbitrary elementary function is itself an elementary function, which we may effectively calculate by making use of the rules of differentiation. But the converse statement is in general untrue, since there exist elementary functions whose indefinite integrals are not elementary functions. Examples are e^{-x^2}, $1/(\ln x)$, $(\sin x)/x$ and so forth. To obtain integrals of these functions we must make use of approximative methods and also introduce new functions which can not be reduced to elementary ones. We can not spend more time here on this question but must simply note that even in elementary mathematics it is possible to find many examples in which a direct operation can be carried out on a certain class of numbers, while the inverse operation can not be carried out on the same class; thus, a square of an arbitrary rational number is again a rational number, but the square root of a rational number is by no means always rational. Analogously, differentiation of elementary functions produces a function that is again elementary, but integration may lead us outside the class of elementary functions.

Some of the integrals that cannot be expressed in terms of elementary functions have great importance in mathematics and its applications. An example is

$$\int_0^x e^{-t^2}\, dt,$$

which plays a very important role in the theory of probability (see Chapter XI). Other examples are the integrals

$$\int_0^\phi \frac{d\theta}{\sqrt{1 - k^2 \sin^2 \theta}} \quad \text{and} \quad \int_0^\phi \sqrt{1 - k^2 \sin^2 \theta}\, d\theta\ (k^2 < 1),$$

which are called *elliptic integrals* of the first and second kind respectively. We are led to the calculation of these integrals by a large number of problems in physics (see Chapter V, §1, example 3). Detailed tables of these integrals for various values of the arguments x and ϕ have been calculated by approximate methods but with great accuracy.

It must be emphasized that the proof of the very fact that a given elementary function cannot be integrated in terms of elementary functions is in each case quite difficult. Such questions occupied the attention of outstanding mathematicians in the last century and have played an important role in the development of analysis. Fundamental results were obtained here by Čebyšev, who gave a complete answer to the question of expressing in terms of elementary functions the integrals of the form

$$\int x^m(a + bx^s)^p \, dx,$$

where m, s, and p are rational numbers. Up to his time three relations, obtained by Newton, were known for the exponents m, s, and p, which implied the integrability of this integral in terms of elementary functions. Čebyšev proved that in all other cases the integral cannot be expressed in terms of elementary functions.

We introduce here another method of integration, namely integration by parts. It is based on the formula we already know

$$(uv)' = uv' + u'v,$$

for the derivative of the product of the functions u and v. This formula may also be written

$$uv' = (uv)' - u'v.$$

Let us now integrate the left and right sides, keeping in mind that

$$\int (uv)' \, dx = uv + C.$$

We now finally obtain the equality

$$\int uv' \, dx = uv - \int u'v \, dx,$$

which is also called the *formula of integration by parts*. We have not written the constant C since we may consider that it is included in one of the indefinite integrals occurring in this equation.

Let us introduce some applications of this formula. Suppose we have to calculate $\int xe^x \, dx$. Here we will take $u = x$ and $v' = e^x$, and thus $u' = 1$, $v = e^x$, and consequently

$$\int xe^x \, dx = xe^x - \int 1 \cdot e^x \, dx = xe^x - e^x + C.$$

In the integral $\int \ln x \, dx$ it is convenient to take $u = \ln x$, $v' = 1$, so that $u' = 1/x$, $v = x$ and

$$\int \ln x \, dx = x \ln x - \int dx = x \ln x - x + C.$$

In the following characteristic example it is necessary to integrate twice by parts and then to find the desired integral from the equations so obtained:

$$\int e^x \sin x \, dx = e^x \sin x - \int e^x \cos x \, dx$$

$$= e^x \sin x - e^x \cos x - \int e^x \sin x \, dx,$$

from which

$$\int e^x \sin x \, dx = \frac{e^x}{2} (\sin x - \cos x) + C.$$

We end this section here; from it the reader will have obtained only a superficial idea of the theory of integration. We have not given any attention to many different methods in this theory. In particular we have not touched here on the very interesting question of the integration of rational fractions, a theory in which an important contribution was made by the well-known mathematician and mechanician, Ostrogradskiĭ.

§12. Functions of Several Variables

Up to now we have spoken only of functions of one variable, but in practice it is often necessary to deal also with functions depending on two, three, or in general many variables. For example, the area of a rectangle is a function

$$S = xy$$

of its base x and its height y. The volume of a rectangular parallelepiped is a function

$$v = xyz$$

of its three dimensions. The distance between two points A and B is a function

$$r = \sqrt{(x_1 - x_2)^2 + (y_1 - y_2)^2 + (z_1 - z_2)^2}$$

of the six coordinates of these points. The well-known formula

$$pv = RT$$

expresses the dependence of the volume v of a definite amount of gas on the pressure p and absolute temperature T.

Functions of several variables, like functions of one variable, are in many cases defined only on a certain region of values of the variables themselves. For example, the function

$$u = \ln(1 - x^2 - y^2 - z^2) \qquad (34)$$

is defined only for values of x, y and z that satisfy the condition

$$x^2 + y^2 + z^2 < 1. \qquad (35)$$

(For other x, y, z its values are not real numbers.) The set of points of space whose coordinates satisfy the inequality (35) obviously fills up a sphere of unit radius with its center at the origin of coordinates. The points on the boundary are not included in this sphere; the surface of the sphere has been so to speak "peeled off." Such a sphere is said to be open. The function (34) is defined only for such sets of three numbers (x, y, z) as are coordinates of points in the open sphere G. It is customary to state this fact concisely by saying that the function (34) is defined on the sphere G.

Let us give another example. The temperature of a nonuniformly heated body V is a function of the coordinates x, y, z of the points of the body. This function is not defined for all sets of three numbers x, y, z but only for such sets as are coordinates of points of the body V.

Finally, as a third example, let us consider the function

$$u = \phi(x) + \phi(y) + \phi(z),$$

where ϕ is a function of one variable defined on the interval $[0, 1]$. Obviously the function u is defined only for sets of three numbers (x, y, z) which are coordinates of points in the cube:

$$0 \leqslant x \leqslant 1, \, 0 \leqslant y \leqslant 1, \, 0 \leqslant z \leqslant 1.$$

We now give a formal definition of a function of three variables. Suppose that we are given a set E of triples of numbers (x, y, z) (points of space). If to each of these triples of numbers (points) of E there corresponds a definite number u in accordance with some law, then u is said to be a function of x, y, z (of the point), defined on the set of triples of numbers (on the points) E, a fact which is written thus:

$$u = F(x, y, z).$$

In place of F we may also write other letters: f, ϕ, ψ.

In practice the set E will usually be a set of points, filling out some geometrical body or surface: sphere, cube, annulus, and so forth, and then we simply say that the function is defined on this body or surface. Functions of two, four, and so forth, variables are defined analogously.

Implicit definition of a function. Let us note that functions of two variables may serve, under certain circumstances, as a useful means for the definition of functions of one variable. Given a function $F(x, y)$ of two variables let us set up the equation

$$F(x, y) = 0. \tag{36}$$

In general, this equation will define a certain set of points (x, y) of the surface on which our function is equal to zero. Such sets of points usually represent curves that may be considered as the graphs of one or several one-valued functions $y = \phi(x)$ or $x = \psi(y)$ of one variable. In such a case these one-valued functions are said to be defined implicitly by the equation (36). For example, the equation

$$x^2 + y^2 - r^2 = 0$$

gives an implicit definition of two functions of one variable

$$y = +\sqrt{r^2 - x^2} \quad \text{and} \quad y = -\sqrt{r^2 - x^2}.$$

But it is necessary to keep in mind that an equation of the form (36) may fail to define any function at all. For example, the equation

$$x^2 + y^2 + 1 = 0$$

obviously does not define any real function, since no pair of real numbers satisfies it.

Geometric representation. Functions of two variables may always be visualized as surfaces by means of a system of space coordinates. Thus the function

$$z = f(x, y) \tag{37}$$

is represented in a three-dimensional rectangular coordinate system by a surface, which is the geometric locus of points M whose coordinates x, y, z satisfy equation (37) (figure 26).

There is another, extremely useful method, of representing the function (37), which has found wide application in practice. Let us choose a sequence of numbers z_1, z_2, $\cdots$, and then draw on one and the same plane Oxy the curves

$$z_1 = f(x, y), \quad z_2 = f(x, y),$$

which are the so-called level lines of the function $f(x, y)$. From a set of level lines, if they correspond to values of z that are sufficiently close to one another, it is possible to form a very good image of the variation of the function $f(x, y)$, just as from the level lines of a topographical map one may judge the variation in altitude of the locality.

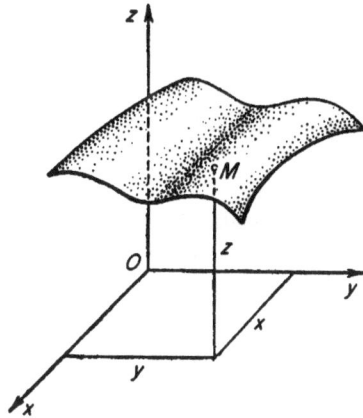

FIG. 26.

Figure 27 shows a map of the level lines of the function $z = x^2 + y^2$, the diagram at the right indicating how the function is built up from its level lines. In Chapter III, figure 50, a similar map is drawn for the level lines of the function $z = xy$.

Partial derivatives and differential. Let us make some remarks about the differentiation of the functions of several variables. As an example we take the arbitrary function

$$z = f(x, y)$$

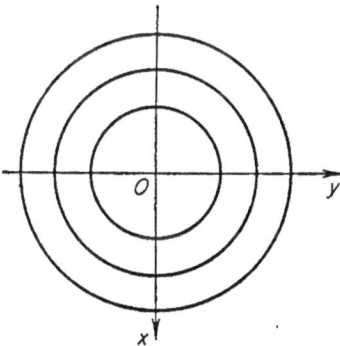
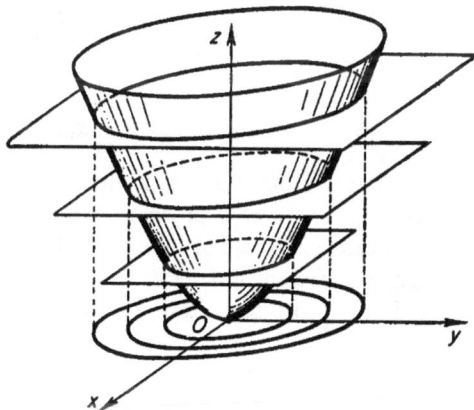

FIG. 27.

of two variables. If we fix the value of y, that is if we consider it as not varying, then our function of two variables becomes a function of the one variable x. The derivative of this function with respect to x, if it exists, is called the *partial derivative* with respect to x and is denoted thus:

$$\frac{\partial z}{\partial x}, \quad \text{or} \quad \frac{\partial f}{\partial x}, \quad \text{or} \quad f_x'(x, y).$$

The last of these three notations indicates clearly that the partial derivative with respect to x is in general a function of x and y. The partial derivative with respect to y is defined similarly.

Geometrically the function $f(x, y)$ represents a surface in a rectangular three-dimensional system of coordinates. The corresponding function of x for fixed y represents a plane curve (figure 28) obtained from the intersection of the surface with a plane parallel to the plane Oxz and at a distance y from it. The partial derivative $\partial z/\partial x$ is obviously equal to the trigonometric tangent of the angle between the tangent to the curve at the point (x, y) and the positive direction of the x-axis.

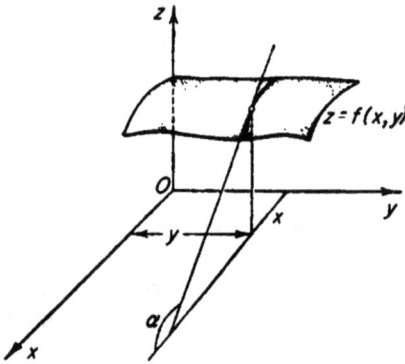

FIG. 28.

More generally, if we consider a function $z = f(x_1, x_2, \cdots, x_n)$ of the n variables $x_1, x_2, \cdots, x_n$, the partial derivative $\partial z/\partial x_i$ is defined as the derivative of this function with respect to x_i, calculated for fixed values of the other variables:

$$x_1, x_2, \ldots, x_{i-1}, x_{i+1}, \ldots, x_n.$$

We may say that the partial derivative of a function with respect to the variable x_i is the rate of change of this function in the direction of the change in x_i. It would also be possible to define a derivative in an arbitrary assigned direction, not necessarily coinciding with any of the coordinate axis, but we will not take the time to do this.

Examples.

1. $z = \dfrac{x}{y}, \dfrac{\partial z}{\partial x} = \dfrac{1}{y}, \dfrac{\partial z}{\partial y} = -\dfrac{x}{y^2}.$

2. $u = \dfrac{1}{\sqrt{x^2 + y^2 + z^2}}$,

$$\frac{\partial u}{\partial x} = -\frac{1}{x^2 + y^2 + z^2} \cdot \frac{2x}{2\sqrt{x^2 + y^2 + z^2}} = -\frac{x}{(x^2 + y^2 + z^2)^{3/2}} \cdot$$

It is sometimes necessary to form the partial derivatives of these partial derivatives; that is; the so-called partial derivatives of second order. For functions of two variables there are four of them

$$\frac{\partial^2 u}{\partial x^2}, \ \frac{\partial^2 u}{\partial x\, \partial y}, \ \frac{\partial^2 u}{\partial y\, \partial x}, \ \frac{\partial^2 u}{\partial y^2} \cdot$$

However, if these derivatives are continuous, then it is not hard to prove that the second and third of these four (the so-called mixed derivatives) coincide:

$$\frac{\partial^2 u}{\partial x\, \partial y} = \frac{\partial^2 u}{\partial y\, \partial x} \cdot$$

For example, in the case of first function considered,

$$\frac{\partial^2 z}{\partial x^2} = 0, \ \frac{\partial^2 z}{\partial x\, \partial y} = -\frac{1}{y^2}, \ \frac{\partial^2 z}{\partial y\, \partial x} = -\frac{1}{y^2}, \ \frac{\partial^2 z}{\partial y^2} = \frac{2x}{y^3},$$

the two mixed derivatives are seen to coincide.

For functions of several variables, just as was done for functions of one variable, we may introduce the concept of a differential.

For definiteness let us consider a function

$$z = f(x, y)$$

of two variables. If it has continuous partial derivatives, we can prove that its increment

$$\Delta z = f(x + \Delta x, y + \Delta y) - f(x, y),$$

corresponding to the increments Δx and Δy of its arguments, may be put in the form

$$\Delta z = \frac{\partial f}{\partial x} \Delta x + \frac{\partial f}{\partial y} \Delta y + \alpha \sqrt{\Delta x^2 + \Delta y^2},$$

where $\partial f / \partial x$ and $\partial f / \partial y$ are the partial derivatives of the function at the point (x, y) and the magnitude α depends on Δx and Δy in such a way that $\alpha \to 0$ as $\Delta x \to 0$ and $\Delta y \to 0$.

The sum of the first two components

$$dz = \frac{\partial f}{\partial x} \Delta x + \frac{\partial f}{\partial y} \Delta y$$

is linearly dependent* on Δx and Δy and is called the *differential of the function*. The third summand, because of the presence of the factor α, tending to zero with Δx and Δy, is an infinitesimal of higher order than the magnitude

$$\rho = \sqrt{\Delta x^2 + \Delta y^2},$$

describing the change in x and y.

Let us give an application of the concept of differential. The period of oscillation of a pendulum is calculated from the formula

$$T = 2\pi \sqrt{\frac{l}{g}} ,$$

where l is its length and g is the acceleration of gravity. Let us suppose that l and g are known with errors respectively equal to Δl and Δg. Then the error in the calculation of T will be equal to the increment ΔT corresponding to the increments of the arguments Δl and Δg. Replacing ΔT approximately by dT, we will have

$$\Delta T \approx dT = \pi \left(\frac{\Delta l}{\sqrt{lg}} - \frac{\sqrt{l}\, \Delta g}{\sqrt{g^3}} \right).$$

The signs of Δl and Δg are unknown, but we may obviously estimate ΔT by the inequality

$$|\Delta T| \leqslant \pi \left(\frac{|\Delta l|}{\sqrt{lg}} + \sqrt{\frac{l}{g^3}} |\Delta g| \right),$$

from which after division by T we get

$$\frac{|\Delta T|}{T} < \left(\frac{|\Delta l|}{l} + \frac{|\Delta g|}{g} \right).$$

Thus we may consider in practice that the relative error for T is equal to the sum of the relative errors for l and g.

---

* In general a function $Ax + By + C$, where A, B, C are constants, is called a linear function of x and y. If $C = 0$, it is called a homogeneous linear function. Here we omit the word "homogeneous."

For symmetry of notation, the increments of the independent variables Δx and Δy are usually denoted by the symbols dx and dy and are also called differentials. With this notation the differential of the function $u = f(x, y, z)$ may be written thus:

$$du = \frac{\partial f}{\partial x}\,dx + \frac{\partial f}{\partial y}\,dy + \frac{\partial f}{\partial z}\,dz.$$

Partial derivatives play a large role whenever we have to do with functions of several variables, as happens in many of the applications of analysis to technology and physics. We shall be dealing in Chapter VI with the problem of reconstructing a function from the properties of its partial derivatives.

In the following paragraphs, we give some simple examples of applications of partial derivatives in analysis.

Differentiation of implicit functions. Suppose we wish to find the derivative of y, where y is a function of x defined implicitly by the relation

$$F(x, y) = 0 \tag{38}$$

between these variables. If x and y satisfy the relation (38) and we give x the increment Δx, then y will receive an increment Δy such that $x + \Delta x$ and $y + \Delta y$ again satisfy (38). Consequently*

$$F(x + \Delta x, y + \Delta y) - F(x, y) = \frac{\partial F}{\partial x}\Delta x + \frac{\partial F}{\partial y}\Delta y + \alpha\sqrt{\Delta x^2 + \Delta y^2} = 0.$$

Thus, provided $\partial F/\partial y \neq 0$, it follows that

$$\lim_{\Delta x \to 0}\frac{\Delta y}{\Delta x} = y'_x = -\frac{\dfrac{\partial F}{\partial x}}{\dfrac{\partial F}{\partial y}}.$$

In this way we have obtained a method for finding the derivative of an implicit function y without first solving the equation (38) for y.

Maximum and minimum problems. If a function, let us say of two variables $z = f(x, y)$, attains its maximum at the point (x_0, y_0), that is if $f(x_0, y_0) \geqslant f(x, y)$ for all points (x, y) close to (x_0, y_0), then this point must also be the point of maximum altitude for any line formed by the

---

* We assume that $F(x, y)$ has continuous derivatives with respect to x and y.

intersection of the surface $z = f(x, y)$ with a plane parallel to Oxz or Oyz. So at such a point we must have

$$f'_x(x, y) = 0, f'_y(x, y) = 0. \tag{39}$$

The same equations must also hold for a point of local minimum. Consequently, the greatest or least values of the function are to be sought first of all at points where the conditions (39) are satisfied, but we must also not forget about points on the boundary of the domain of definition of the function and points where the function fails to have a derivative, if such points exist.

To establish whether a point (x, y) satisfying (39) is actually a maximum or minimum point, use is frequently made of various indirect arguments. For example, if for any reason it is clear that the function is differentiable and attains its minimum inside the region and that there is only one point where the conditions (39) are fulfilled, then obviously the minimum must be attained at this point.

For example, let it be required to make a rectangular tin box (without lid) with assigned volume V, using the smallest possible amount of material. If the sides of the base of this box are denoted by x and y, then its height h will be equal to V/xy, and consequently the surface S will be given by the function

$$S = xy + \frac{V}{xy}(2x + 2y) = xy + 2V\left(\frac{1}{x} + \frac{1}{y}\right) \tag{40}$$

of x and y. Since x and y by the terms of the problem must be positive, the question has been reduced to finding the minimum of the function $S(x, y)$ for all possible points (x, y) in the first quadrant of the plane (x, y), which we will denote by the letter G.

If the minimum is attained at some point of the region G, then the partial derivatives must be equal to zero

$$\frac{\partial S}{\partial x} = y - \frac{2V}{x^2} = 0,$$

$$\frac{\partial S}{\partial y} = x - \frac{2V}{y^2} = 0,$$

that is $yx^2 = 2V, xy^2 = 2V$, from which we find as the dimensions of the box:

$$x = y = \sqrt[3]{2V} \quad \text{and} \quad h = \sqrt[3]{\frac{V}{4}}. \tag{41}$$

We have solved the problem but have not altogether proved that our

solution is correct. A rigorous mathematician will say to us: "You have supposed from the very beginning that under the given conditions the box with minimum surface actually exists and, proceeding from this assumption, you have found its dimensions. So you have really obtained only the following result: If there exists a point (x, y) in G for which the function S attains its minimum, then the coordinates of this point must necessarily be determined by the equation (41). But now you must show that the minimum of S does exist for some point in G and then I will admit the correctness of your result." This remark is a very reasonable one, since, for example, our function S, as we shall soon see, does not possess any maximum in the region G. But let us show how it is possible to convince ourselves that in the given case the function actually does attain its minimum at a certain point (x, y) of the region G.

The fundamental theorem on which we shall base our argument is one that is proved in analysis with complete rigor; it amounts to the following. If the function f of one or several variables is everywhere continuous in a certain finite region H which is bounded and includes its boundary, then there always exists in H at least one point at which the function attains its minimum (maximum). With this theorem we can easily complete our analysis of the problem.

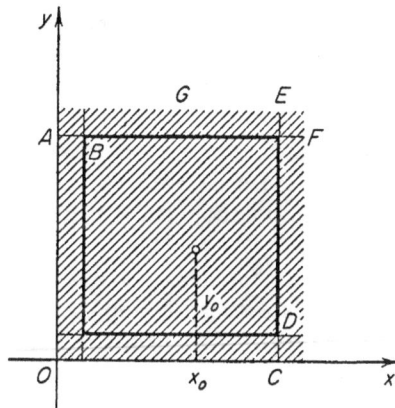

FIG. 29.

Let us consider an arbitrary point (x_0, y_0) of the region G; at this point let $S(x_0, y_0) = N$. Let us also choose a number R satisfying the two inequalities $R > N$, $2VR > N$ and construct a square Ω_R with side R^2, as in figure 29, where $AB = CD = 1/R$.

We now give a lower bound for the values of our function $S(x, y)$ at points of the region G lying outside the square Ω_R. If the point of the region G has abscissa $x < 1/R$, then

$$S(x, y) = xy + 2V\left(\frac{1}{x} + \frac{1}{y}\right) > 2V\frac{1}{x} > 2VR > N.$$

Analogously, if the point of the region G has its ordinate $y < 1/R$, then also $S > N$. Also, if the point of the region G has its abscissa $x > 1/R$

and if it lies above the straight line AF or has its ordinate $y > 1/R$ and
lies to the right of the straight line CE, then

$$S(x, y) > xy > \frac{1}{R} R^2 = R > N.$$

Thus, for all points (x, y) of the region G lying outside the square Ω_k ,
the inequality $S(x, y) > N$ holds, and since $S(x_0, y_0) = N$, the point
(x_0, y_0) must belong to the square and consequently the minimum of our
function on G is equal to its minimum on the square.

But the function $S(x, y)$ is continuous in this square and on its boundary,
so that by the theorem stated earlier there exists in the square a point
(x, y) where our function assumes its minimum for points in the square
and consequently for the entire region G. Thus the existence of a minimum
has been proved.

This argument may serve as an example of the way that it is possible
to discuss the existence of a maximum or a minimum for a function
defined on an unbounded domain.

The Taylor formula. Like functions of one variable, functions of
several variables may be represented by a Taylor formula. For example,
an expansion of the function

$$u = f(x, y)$$

in the neighborhood of the point (x_0, y_0) has the following form, if we
confine ourselves to the first and second powers of $x - x_0$ and $y - y_0$:

$$f(x, y) = f(x_0, y_0) + [f_x'(x_0, y_0)(x - x_0) + f_y'(x_0, y_0)(y - y_0)]$$

$$+ \frac{1}{2!} [f_{xx}''(x_0, y_0)(x - x_0)^2 + 2f_{xy}''(x_0, y_0)(x - x_0)(y - y_0)$$

$$+ f_{yy}''(x_0, y_0)(y - y_0)^2] + R_3 .$$

If the function $f(x, y)$ has continuous partial derivatives of the second
order, the remainder term here will approach zero faster than

$$r^2 = (x - x_0)^2 + (y - y_0)^2,$$

that is, faster than the square of the distance between the points (x, y)
and (x_0, y_0), as $r \to 0$. The Taylor formula provides a widely used method
of defining and approximately calculating the values of various functions.

Let us note that with the help of this formula we can also answer the
question asked earlier, whether a given function actually has a maximum

or minimum at a point where $\partial f/\partial x = \partial f/\partial y = 0$. In fact, if these conditions are satisfied at the point (x_0, y_0), then for points (x, y) close to (x_0, y_0), the value of the function will, by the Taylor formula, differ from $f(x_0, y_0)$ by the amount

$$f(x, y) - f(x_0, y_0)$$

$$= \frac{1}{2!} [A(x - x_0)^2 + 2B(x - x_0)(y - y_0) + C(y - y_0)^2] + R_3, \qquad (42)$$

where A, B, and C denote respectively the second partial derivatives $f''_{xx}, f''_{xy}, f''_{yy}$ at the point (x_0, y_0).

If it turns out that the function

$$\Phi(x, y) = A(x - x_0)^2 + 2B(x - x_0)(y - y_0) + C(y - y_0)^2$$

is positive for arbitrary values of $(x - x_0)$ and $(y - y_0)$ not both equal to zero, then the right side of equation (42) will also be positive for small values of $(x - x_0)$ and $(y - y_0)$, since for sufficiently small $(x - x_0)$ and $(y - y_0)$ the quantity R_3 is known to be less in absolute value than $\frac{1}{2} \Phi(x, y)$. Thus it will follow that at the point (x_0, y_0) the function f attains its minimum. On the other hand, if the function $\Phi(x, y)$ is negative for arbitrary $(x - x_0)$ and $(y - y_0)$ the right side of (42) will be negative for $(x - x_0)$ and $(y - y_0)$, so that at the point (x_0, y_0) the function will have a maximum. In more complicated cases it is necessary to consider the succeeding terms in the Taylor formula.

Problems concerning the maximum or the minimum of functions of three or more variables may be treated in a completely analogous fashion. As an exercise the reader may prove that if given masses

$$m_1, m_2, \cdots, m_n$$

are arranged in space at given points

$$P_1(x_1, y_1, z_1), P_2(x_2, y_2, z_2), \cdots, P_n(x_n, y_n, z_n),$$

the moment (of inertia) M of this system of masses about the point $P(x, y, z)$, defined as the sum of the products of the masses and the squares of their distances from the point P,

$$M(x, y, z) = \sum_{i=1}^{n} m_i[(x - x_i)^2 + (y - y_i)^2 + (z - z_i)^2],$$

will be a minimum if the point P is at the so-called center of gravity of the system, with the coordinates

$$x = \frac{\sum_{i=1}^{n} m_i x_i}{\sum_{i=1}^{n} m_i}, \, y = \frac{\sum_{i=1}^{n} m_i y_i}{\sum_{i=1}^{n} m_i}, \, z = \frac{\sum_{i=1}^{n} m_i z_i}{\sum_{i=1}^{n} m_i}$$

Maxima and minima with subsidiary conditions. For functions of several variables we may set up various problems concerning maximum and minimum. Let us illustrate with a simple example. Suppose that among all rectangles inscribed in a circle of radius R, we wish to find the one with greatest area. The area of a rectangle is equal to the product xy of its sides, where x and y are positive numbers connected in this case by the relation $x^2 + y^2 = (2R)^2$, as is clear from figure 30. Thus we are required to find the maximum of the function $f(x, y) = xy$ for all x and y satisfying the relation $x^2 + y^2 = 4R^2$.

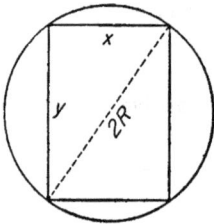

Problems of this sort, where it is necessary to find the maximum (or minimum) of a function $f(x, y)$ for those values only of x and y that satisfy a certain relation that $\phi(x, y) = 0$ are very common in practice.

Of course, it would be possible to solve the equation $\phi(x, y) = 0$ for y, to substitute the solution into the function $f(x, y)$ and in this way to seek the ordinary maximum for a function of one variable x. But this method is usually complicated and sometimes impossible.

Fig. 30.

For the solution of such problems in analysis, a much more convenient procedure called the method of Lagrange multipliers, has been worked out. The idea behind it is extremely simple. Let us consider the function

$$F(x, y) = f(x, y) + \lambda\phi(x, y),$$

where λ is an arbitrary positive number. Obviously, for x, y satisfying the condition $\phi(x, y) = 0$, the values of $F(x, y)$ coincide with those of $f(x, y)$.

For function $F(x, y)$ let us seek a maximum without conditions of any kind on x and y. At the maximum point the conditions $\partial F/\partial x = \partial F/\partial y = 0*$ must hold; in other words

$$\frac{\partial f}{\partial x} + \lambda \frac{\partial \phi}{\partial x} = 0; \tag{43}$$

$$\frac{\partial f}{\partial y} + \lambda \frac{\partial \phi}{\partial y} = 0. \tag{44}$$

---

* We are speaking here, of course, of a maximum attained in the domain of definition of the function $F(x, y)$. The functions $f(x, y)$ and $\phi(x, y)$ are assumed to be differentiable.

The values of x and y at the maximum point for $F(x, y)$, being a solution of the system (43) and (44), depend on the coefficient λ in these equations. Let us now suppose that we have succeeded in choosing the number λ in such a way that the coordinates of the maximum point satisfy the condition

$$\phi(x, y) = 0. \qquad (45)$$

Then this point will be an exact local maximum for the original problem.

In fact, we may consider the problem geometrically as follows. The function $f(x, y)$ is defined on a certain region G (figure 31). The condition $\phi(x, y) = 0$ will ordinarily be satisfied by the points of some curve Γ. We are required to find the greatest value of x and y on points of the line Γ. If $F(x, y)$ attains its maximum on the curve Γ, then $F(x, y)$ does not increase for small shifts in an arbitrary direction from this point, and in particular for shifts along the curve Γ. But for shifts along Γ, the values of $F(x, y)$, coincide

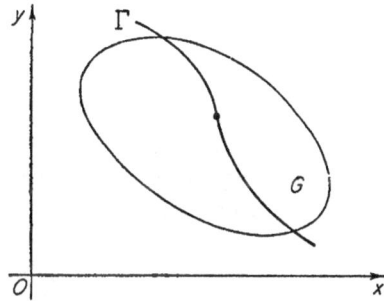

FIG. 31.

with those of $f(x, y)$ which means that for small shifts along the curve the function $f(x, y)$ does not increase, or in other words it has the local maximum at the point.

These arguments indicate a simple method of solving the problem. We solve equations (43), (44), (45) for the unknowns x, y, and λ, obtaining one or more solutions

$$(x_1, y_1, \lambda_1), \ (x_2, y_2, \lambda_2), \ \cdots. \qquad (46)$$

To the points $(x_1, y_1), (x_2, y_2), \cdots$ so determined we adjoin those points of the boundary of G where the curve Γ leaves the region G. Then from all these points we choose that one at which $f(x, y)$ takes on its greatest (or smallest) value.

Of course, the arguments here are far from proving the correctness of the method. In fact, we have not yet even proved that the points of local maximum for $f(x, y)$ on the curve Γ can be obtained as maximum points for the function $F(x, y)$ for some value of λ. However, it is possible to prove, as is done in the textbooks in analysis, that every point (x_0, y_0) where $f(x, y)$ has a local maximum on the curve will be obtained by the method indicated, provided only that at this point the partial derivatives $\phi'_x(x_0, y_0)$ and $\phi'_y(x_0, y_0)$ are not both equal to zero.*

Let us use the method of Lagrange to solve the problem at the beginning of the present section. In this case $f(x, y) = xy$; $\phi(x, y) = x^2 + y^2 - 4R^2$. We set up the equations (43), (44), (45)

$$y + 2\lambda x = 0,$$

$$x + 2\lambda y = 0,$$

$$x^2 + y^2 = 4R^2,$$

for which, taking into account that x and y are positive, we find the unique solution

$$x = y = R\sqrt{2}\left(\lambda = -\frac{1}{2}\right).$$

For these values of x and y, which are equal to one another so that the inscribed rectangle is a square, the area is in fact a maximum.

The method of Lagrange may be extended to deal with functions of three or more variables. There may be any number of subsidiary conditions (smaller than the number of variables) of the type of condition (45), and we will introduce the corresponding number of auxiliary multipliers.

Let us give some examples of problems involving maxima or minima with subsidiary conditions.

Example 1. For what height h and radius r will an open cylindrical tank of given volume V require the least amount of sheet metal for its manufacture; that is, the area of its sides and circular base will be a minimum?

The problem obviously reduces to finding the minimum of the function of the variables r and h

$$f(r, h) = 2\pi rh + \pi r^2$$

under the condition $\pi r^2 h = V$, which may be written in the form

$$\phi(r, h) = \pi r^2 h - V = 0.$$

---

* In the course in higher mathematics of V. I. Smirnov, the reader will find a simple example where this particular feature of the situation would lead to the loss of a solution if we apply the method of Lagrange mechanically and do not consider, in addition to the points mentioned above, a point where not only (45) holds but also:

$$\phi'_x(x_0, y_0) = 0, \qquad \phi'_y(x_0, y_0) = 0$$

Example 2. A moving point is required to pass from A to B (figure 32). On the path AM it moves with the velocity of v_1, and on MB with the velocity v_2. Where should the point M be placed on the line DD' so that the entire path from A to B may be covered as quickly as possible?

Let us take as unknowns the angles α and β marked in figure 32. The lengths a and b of the perpendiculars from the points A and B to the straight line DD' and the distance c between them are known. The time

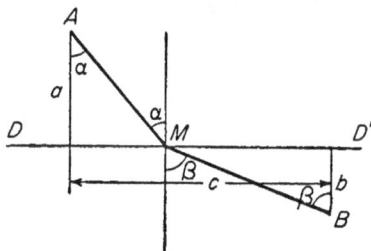

FIG. 32.

required for covering the entire path is represented, as can easily be seen, by the formula

$$f(\alpha, \beta) = \frac{a}{v_1 \cos \alpha} + \frac{b}{v_2 \cos \beta}.$$

It is required to find the minimum of this expression, taking into account the fact that α and β are connected by the relation

$$a \tan \alpha + b \tan \beta = c.$$

The reader may solve these examples by the Lagrange method. In the second example he will find that the best position for M is given by the condition

$$\frac{\sin \alpha}{\sin \beta} = \frac{v_1}{v_2}.$$

This is the well-known law for the refraction of light. Consequently, a ray of light will be refracted in its passage from one medium to another in such a way that the time for its passage from a point in one medium to a point in the other is a minimum. Conclusions of this sort are interesting not only for computational purposes but also from a general philosophical point of view; they have inspired researchers in the exact sciences to penetrate further and further into the profound and general laws of nature.

Finally let us note that the multipliers λ, introduced in the solution of problems by the method of Lagrange, are not merely auxiliary numbers. In each case they are closely connected with the essential nature of the particular problem and have a concrete interpretation.

§13. Generalizations of the Concept of Integral

In §10 we defined the definite integral of the function $f(x)$ on the interval $[a, b]$ as the limit of the sum

$$\sum_{i=1}^{n} f(\xi_i)\, \Delta x_i$$

when the length of the greatest segment Δx_i in the subdivision of $[a, b]$ approaches zero. In spite of the fact that the class of functions $f(x)$ for which this limit actually exists (the class of integrable functions) is a very wide one, and in particular includes all continuous and even many discontinuous functions, this class of functions has a serious shortcoming. If we add, subtract, or multiply, or under certain conditions divide the values of two integrable functions $f(x)$ and $\phi(x)$, we obtain functions which, as may easily be proved, are again integrable. For $f(x)/\phi(x)$ this will be true in all cases in which $1/\phi(x)$ remains bounded on $[a, b]$. But if a function is obtained as a result of a limiting process from a sequence of approximating integrable functions $f_1(x), f_2(x), f_3(x), \cdots$ such that for all values of x in the interval $[a, b]$

$$f(x) = \lim_{n \to \infty} f_n(x),$$

then the limit function $f(x)$ is not necessarily integrable.

In many cases this and other circumstances give rise to considerable complication, since the process of passing to a limit is widely used.

A way out of the difficulty was discovered by making further generalizations of the concept of an integral. The most important of these is the integral of Lebesgue, with which the reader will become acquainted in Chapter XV on the theory of functions of a real variable. But here we will confine ourselves to generalizations of the integral in other directions, which are also of the greatest importance in practice.

Multiple integrals. We have already studied the process of integration for functions of one variable defined on a one-dimensional region, namely an interval. But the analogous process may be extended to functions of two, three, or more variables, defined on corresponding regions.

For example, let us consider a surface

$$z = f(x, y)$$

defined in a rectangular system of coordinates, and on the plane Oxy let
there be given a region G bounded
by a closed curve Γ. It is required to
find the volume bounded by the
surface, by the plane Oxy and by the
cylindrical surface passing through
the curve Γ with generators parallel
to the Oz axis (figure 33). To solve
this problem we divide the plane
region G into subregions by a network of straight lines parallel to the
axes Ox and Oy and denote by

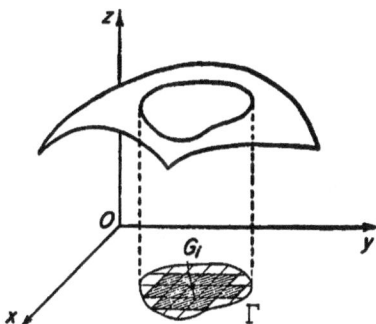

FIG. 33.

$$G_1, G_2, \cdots, G_n$$

those subregions which consist of
complete rectangles. If the net is sufficiently fine, then practically the
whole of the region G will be covered by the enumerated rectangles. In
each of them we choose at will a point

$$(\xi_1, \eta_1), (\xi_2, \eta_2), \cdots, (\xi_n, \eta_n)$$

and, assuming for simplicity that G_i denotes not only the rectangle but
also its area, we set up the sum

$$S_n = f(\xi_1, \eta_1) G_1 + f(\xi_2, \eta_2) G_2 + \cdots + f(\xi_n, \eta_n) G_n = \sum_{i=1}^{n} f(\xi_i, \eta_i) G_i. \tag{47}$$

It is clear that, if the surface is continuous and the net is sufficiently
fine, this sum may be brought as near as we like to the desired volume V.
We will obtain the desired volume exactly if we take the limit of the sum
(47) for finer and finer subdivisions (that is, for subdivisions such that
the greatest of the diagonals of our rectangles approaches zero)

$$\lim_{\max d(G_i) \to 0} \sum_{i=1}^{n} f(\xi_i, \eta_i) G_i = V. \tag{48}$$

From the point of view of analysis it is therefore necessary, in order to
determine the volume V, to carry out a certain mathematical operation
on the function $f(x, y)$ and its domain of definition G, an operation
indicated by the left side of equality (48). This operation is called the
integration of the function f over the region G, and its result is the integral
of f over G. It is customary to denote this result in the following way:

$$\iint\limits_{G} f(x, y) \, dx \, dy = \lim_{\max d(G_i) \to 0} \sum_{i=1}^{n} f(\xi_i, \eta_i) G_i. \tag{49}$$

Similarly, we may define the integral of a function of three variables over a three-dimensional region G, representing a certain body in space. Again we divide the region G into parts, this time by planes parallel to the coordinate planes. Among these parts we choose the ones which represent complete parallelepipeds and enumerate them

$$G_1, G_2, \cdots, G_n.$$

In each of these we choose an arbitrary point

$$(\xi_1, \eta_1, \zeta_1), (\xi_2, \eta_2, \zeta_2), \cdots, (\xi_n, \eta_n, \zeta_n)$$

and set up the sum

$$S = \sum_{i=1}^{n} f(\xi_i, \eta_i, \zeta_i) G_i, \tag{50}$$

where G_i denotes the volume of the parallelepiped G_i. Finally we define the integral of $f(x, y, z)$ over the region G as the limit

$$\lim_{\max d(G_i) \to 0} \sum_{i=1}^{n} f(\xi_i, \eta_i, \zeta_i) G_i = \iiint\limits_{G} f(x, y, z) \, dx \, dy \, dz, \tag{51}$$

to which the sum (50) tends when the greatest diagonal $d(G_i)$ approaches zero.

Let us consider an example. We imagine the region G is filled with a nonhomogeneous mass whose density at each point in G is given by a known function $\rho(x, y, z)$. The density $\rho(x, y, z)$ of the mass at the point (x, y, z) is defined as the limit approached by the ratio of the mass of an arbitrary small region containing the point (x, y, z) to the volume of this region as its diameter approaches zero.* To determine the mass of the body G it is natural to proceed as follows. We divide the region G into parts by planes parallel to the coordinate planes and enumerate the complete parallelepipeds formed in this way

$$G_1, G_2, \cdots, G_n.$$

Assuming that the dividing planes are sufficiently close to one another, we will make only a small error if we neglect the irregular regions of the body and define the mass of each of the regular regions G_i (the complete parallelepipeds) as the product

$$\rho(\xi_i, \eta_i, \zeta_i)G_i,$$

------

* The diameter of a region is defined as the least upper bound of the distance between two points of the region.

where (ξ_i, η_i, ζ_i) is an arbitrary point G_i. As a result the approximate value of the mass M will be expressed by the sum

$$S_n = \sum_{i=1}^{n} \rho(\xi_i, \eta_i, \zeta_i)\, G_i,$$

and its exact value will clearly be the limit of this sum as the greatest diagonal G_i approaches zero; that is

$$M = \iiint_G \rho(x, y, z)\, dx\, dy\, dz = \lim_{\max d(G_i) \to 0} \sum_{i=1}^{n} \rho(\xi_i, \eta_i, \zeta_i)\, G_i.$$

The integrals (49) and (51) are called double and triple integrals respectively.

Let us examine a problem which leads to a double integral. We imagine that water is flowing over a plane surface. Also, on this surface the underground water is seeping through (or soaking back into the ground) with an intensity $f(x, y)$ which is different at different points. We consider a region G bounded by a closed contour (figure 34) and assume that at every point of G we know the intensity $f(x, y)$, namely the amount of underground water seeping through per minute per cm^2 of surface; we will have $f(x, y) > 0$ where

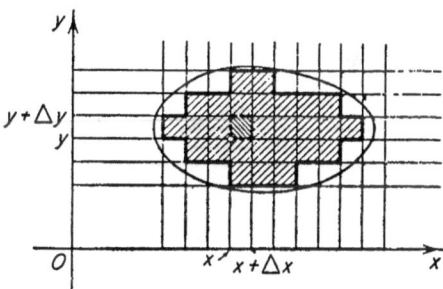

FIG. 34.

the water is seeping through and $f(x, y) < 0$ where it is soaking into the ground. How much water will accumulate on the surface G per minute?

If we divide G into small parts, consider the rate of seepage as approximately constant in each part and then pass to the limit for finer and finer subdivisions, we will obtain an expression for the whole amount of accumulated water in the form of an integral

$$\iint_G f(x, y)\, dx\, dy.$$

Double (two-fold) integrals were first introduced by Euler. Multiple integrals form an instrument which is used everyday in calculations and investigations of the most varied kind.

It would also be possible to show, though we will not do it here, that calculation of multiple integrals may be reduced, as a rule, to iterated calculation of ordinary one-dimensional integrals.

Contour and surface integrals.   Finally, we must mention that still other generalizations of the integral are possible. For example, the problem of defining the work done by a variable force applied to a material point, as the latter moves along a given curve, naturally leads to a so-called curvilinear integral, and the problem of finding the general charge on a surface on which electricity is continuously distributed with a given surface density leads to another new concept, an integral over a curved surface.

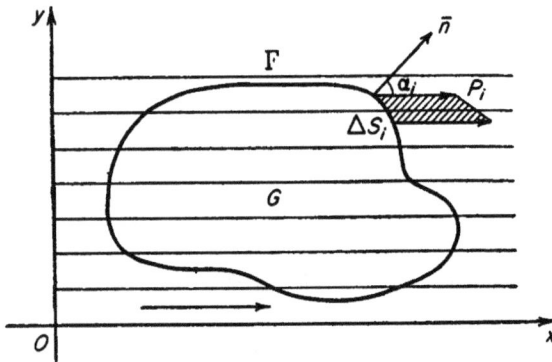

FIG. 35.

For example, suppose that a liquid is flowing through space (figure 35) and that the velocity of a particle of the liquid at the point (x, y) is given by a function $P(x, y)$, not depending on z. If we wish to determine the amount of liquid flowing per minute through the contour Γ,* we may reason in the following way. Let us divide Γ up into segments $\varDelta s_i$. The amount of water flowing through one segment $\varDelta s_i$ is approximately equal to the column of liquid shaded in figure 35; this column may be considered as the amount of liquid forcing its way per minute through that segment of the contour. But the area of the shaded parallelogram is equal to

$$P_i(x, y) \cdot \varDelta s_i \cdot \cos \alpha_i ,$$

where α_i is the angle between the direction $\bar{x}$ of the x-axis and the outward normal of the surface bounded by the contour Γ; this normal is the perpendicular $\bar{n}$ to the tangent, which we may consider as defining the direction of the segment $\varDelta s_i$. By summing up the areas of such parallelograms and passing to the limit for finer and finer subdivisions

---

   * More precisely, through a cylindrical surface with the contour for its base and with height equal to unity.

of the contour Γ, we determine the amount of water flowing per minute through the contour Γ; it is denoted thus:

$$\int_{\Gamma} P(x, y) \cos (\bar{n}, \bar{x}) \, ds$$

and is called a curvilinear integral. If the flow is not everywhere parallel, then its velocity at each point (x, y) will have a component $P(x, y)$ along the x-axis and a component $Q(x, y)$ along the y-axis. In this case we can show by an analogous argument that the quantity of water flowing through the contour will be equal to

$$\int_{\Gamma} [P(x, y) \cos (\bar{n}, \bar{x}) + Q(x, y) \cos (\bar{n}, \bar{y})] \, ds. *$$

When we speak of an integral over a curved surface G for a function $f(M)$ of its points $M(x, y, z)$, we mean the limit of sums of the form

$$\lim \sum_{i=1}^{n} f(M_i) \, \Delta\sigma_i = \iint_{G} f(x, y, z) \, d\sigma$$

for finer and finer subdivisions of the region G into segments whose areas are equal to $\Delta\sigma_i$.

General methods exist for transforming multiple, curvilinear, and surface integrals into other forms and for calculating their values, either exactly or approximately.

Formula of Ostrogradskiĭ. Several important and very general formulas relating an integral over a volume to an integral over its surface (and also an integral over a surface, curved or plane, to an integral around its boundary) were discovered in the middle of the past century by Ostrogradskiĭ.

We shall not try to give here a proof of the general formula of Ostrogradskiĭ, which has very wide application, but will merely illustrate it by an example of its simplest particular case.

Let us imagine, as we did before, that over a plane surface there is a horizontal flow of water that is also soaking into the ground or seeping out again from it. We mark off a region G, bounded by a curve Γ, and

---

* Since for small displacements along the curve the differential of the coordinate y is equal to $\cos(\bar{n}, \bar{x}) \, ds$ and the differential dx is equal to $-\cos(\bar{n}, \bar{y}) \, ds$, this latter integral is often written in the form

$$\int_{\Gamma} [P(x, y) \, dy - Q(x, y) \, dx].$$

assume that for each point of the region we know the components $P(x, y)$ and $Q(x, y)$ of the velocity of the water in the direction of the x-axis and of the y-axis respectively.

Let us calculate the rate at which the water is seeping from the ground at a point with coordinates (x, y). For this purpose we consider a small rectangle with sides Δx and Δy situated at the point (x, y).

As a result of the velocity $P(x, y)$ through the left vertical edge of this rectangle, there will flow approximately $P(x, y)\Delta y$ units of water per minute into the rectangle, and through the right side in the same time will flow out approximately $P(x + \Delta x, y)\Delta y$ units. In general, the net amount of water leaving a square unit of surface as a result of the flow through its left and right vertical sides will be approximately

$$\frac{[P(x + \Delta x, y) - P(x, y)] \Delta y}{\Delta x \, \Delta y} .$$

If we let Δx approach zero, we obtain in the limit

$$\frac{\partial P}{\partial x} .$$

Correspondingly, the net rate of flow of water per unit area in the direction of the y-axis will be given by

$$\frac{\partial Q}{\partial y} .$$

This means that the intensity of the seepage of ground water at the point with coordinates (x, y) will be equal to

$$\frac{\partial P}{\partial x} + \frac{\partial Q}{\partial y} .$$

But in general, as we saw earlier, the quantity of water coming out from the ground will be given by the double integral of the function expressing the intensity of the seepage of ground water at each point, namely

$$\iint_G \left(\frac{\partial P}{\partial x} + \frac{\partial Q}{\partial y}\right) dx \, dy. \tag{52}$$

But, since the water is incompressible, this entire quantity must flow out during the same time through the boundaries of the contour Γ. The quantity of water flowing out through the contour Γ is expressed, as we saw earlier, by the curvilinear integral over Γ

$$\int_\Gamma [P(x, y) \cos (\bar{n}, \bar{x}) + Q(x, y) \cos (\bar{n}, \bar{y})] \, ds. \tag{53}$$

The equality of the magnitudes (52) and (53) expresses the formula of Ostrogradskiĭ in its simplest two-dimensional case

$$\iint_G \left(\frac{\partial P}{\partial x} + \frac{\partial Q}{\partial y}\right) dx\, dy = \int_\Gamma [P(x, y) \cos(\bar{n}, \bar{x}) + Q(x, y) \cos(\bar{n}, \bar{y})]\, ds.$$

We have merely explained the meaning of this formula by a physical example, but it can be proved mathematically.

In this way the mathematical theorem of Ostrogradskiĭ reflects a widespread phenomenon in the external world, which in our example we interpreted in a readily visualized way as preservation of the volume of an incompressible fluid.

Ostrogradskiĭ established a considerably more general formula expressing the connection between an integral over a multidimensional volume and an integral over its surface. In particular, for a three-dimensional body G, bounded by the surface Γ, his formula is

$$\iiint_G \left(\frac{\partial P}{\partial x} + \frac{\partial Q}{\partial y} + \frac{\partial R}{\partial z}\right) dx\, dy\, dz$$

$$= \iint_\Gamma [P \cos(\bar{n}, \bar{x}) + Q \cos(\bar{n}, \bar{y}) + R \cos(\bar{n}, \bar{z})]\, d\sigma,$$

where $d\sigma$ is the element of surface.

It is interesting to note that the fundamental formula of the integral calculus

$$\int_a^b f(x)\, dx = F(b) - F(a) \tag{54}$$

may be considered as a one-dimensional case of the formula of Ostrogradskiĭ. The equation (54) connects the integral over an interval with the "integral" over its "null-dimensional" boundary, consisting of the two end points.

Formula (54) may be illustrated by the following analogy. Let us imagine that in a straight pipe with constant cross section $s = 1$ water is flowing with velocity $F(x)$, which is different for different cross sections (figure 36). Through the porous walls of the pipe, water is seeping into it

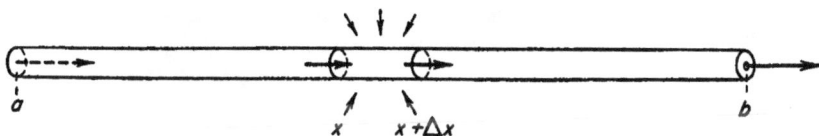

FIG. 36.

(or out of it) at a rate which is also different for different cross sections.

If we consider a segment of the pipe from x to $x + \Delta x$, the quantity of water seeping into it in unit time must be compensated by the difference $F(x + \Delta x) - F(x)$ between the quantity flowing out of this segment and the quantity flowing into it along the pipe. So the quantity seeping into the segment is equal to the difference $F(x + \Delta x) - F(x)$, and consequently the rate of seepage per unit length of pipe (the ratio of the seepage over an infinitesimal segment to the length of the segment) will be equal to

$$f(x) = \lim_{\Delta x \to 0} \frac{F(x + \Delta x) - F(x)}{\Delta x} = F'(x).$$

More generally, the quantity of water seeping into the pipe over the whole section $[a, b]$ must be equal to the amount lost by flow through the ends of the pipe. But the amount seeping through the walls is equal to $\int_a^b f(x)\, dx$ and the amount lost by flow through the ends is $F(b) - F(a)$. The equality of these two magnitudes produces formula (54).

§14. Series

Concept of a series.   A series in mathematics is an expression of the form

$$u_0 + u_1 + u_2 + \cdots.$$

The numbers u_k are called the terms of the series. There is an infinite number of them, and they are arranged in a definite order, so that to each natural number $k = 0, 1, 2, \cdots$ there corresponds a definite value u_k.

The reader must keep in mind that we have not said whether it is possible to calculate a value for such expressions or, in case it is possible, how to do it. The presence of a plus sign between the terms u_k in our expression seems to indicate that in some way all the terms should be added. But there are infinitely many of them and addition of numbers is defined only for a finite number of terms.

Let us denote by S_n the sum of the first n terms of the series; we will call it the *nth partial sum*. As a result we obtain a sequence of numbers

$$S_1 = u_0,$$
$$S_2 = u_0 + u_1,$$
$$\cdots\cdots\cdots\cdots\cdots\cdots$$
$$S_n = u_0 + u_1 + \cdots + u_{n-1},$$
$$\cdots\cdots\cdots\cdots\cdots\cdots$$

and we may speak of a variable quantity S_n, where $n = 1, 2, \cdots$.

The series is said to be *convergent* if, as $n \to \infty$, the variable S_n approaches a definite finite limit

$$\lim_{n\to\infty} S_n = S.$$

This limit is called the *sum of the series*, and in this case we write

$$S = u_0 + u_1 + u_2 + \cdots.$$

But if, as $n \to \infty$, the limit S_n does not exist, then the series is said to be *divergent* and in this case there is no sense in speaking of its sum.* But if all the u_n have the same sign, then it is customary to say that the sum of the series is equal to infinity with the corresponding sign.

As an example, let us consider the series

$$1 + x + x^2 + \cdots,$$

whose terms form a geometric progression with common ratio x.

The sum of the first n terms is equal to

$$S_n(x) = \frac{1 - x^n}{1 - x} \, (x \neq 1); \qquad (55)$$

if $|x| < 1$ this sum has a limit

$$\lim_{n\to\infty} S_n(x) = \frac{1}{1 - x},$$

and so for $|x| < 1$ we may write

$$\frac{1}{1 - x} = 1 + x + x^2 + \cdots$$

If $|x| > 1$, then obviously

$$\lim_{n\to\infty} S_n(x) = \infty,$$

and the series diverges. The same situation holds also for $x = 1$, as may be seen immediately without use of formula (55), which for $x = 1$ has no meaning. Finally, if $x = -1$ the partial sums take the values $+1$ and 0 alternately, so that this series also is divergent.

---

* Let us note that it is also possible to give generalized definitions of the sum of a series, by virtue of which it is possible to assign to certain divergent series a more or less natural concept of "generalized sum." Such series are said to be summable. Operations with generalized sums of divergent series are sometimes useful.

To each series there corresponds a definite sequence of values of its partial sums $S_1, S_2, S_3, \cdots$ such that the convergence of the series depends on the fact that the sums approach a limit. Conversely, to an arbitrary sequence of numbers $S_1, S_2, S_3, \cdots$ corresponds a series

$$S_1 + (S_2 - S_1) + (S_3 - S_2) + \cdots,$$

the partial sums of which will be the numbers of the sequence. Thus the theory of variables ranging over a sequence may be reduced to the theory of the corresponding series, and conversely. Yet each of these theories has independent significance. In some cases it is more convenient to study the variable directly and in others to consider the equivalent series.

Let us note that series have long served as an important method of representing various entities (above all, functions) and of calculating their value. Of course, the views of mathematicians concerning series have changed with the passage of time, corresponding to the changes in their ideas about infinitesimals. The above clear-cut definition of convergence and divergence of a series was formulated at the beginning of the last century at the same time as the closely associated concept of a limit.

If the series converges, then its general term approaches zero with increasing n, since

$$\lim_{n \to \infty} u_n = \lim_{n \to \infty} (S_{n+1} - S_n) = S - S = 0.$$

From examples given in the following paragraphs, it will be clear that the converse statement is in general false. But the criterion is still a useful one, since it provides a *necessary* condition for the convergence of a series. For example, the divergence of a geometric progression with common ratio $x > 1$ follows immediately from the fact its general term does not approach zero.

If the series consists of positive terms, then its partial sum S_n increases with increasing n and only two cases can exist: Either the variable S_n becomes and remains greater than any preassigned number A for sufficiently large n, in which case $\lim_{n \to \infty} S_n = \infty$, so that the series diverges; or else there exists a number A such that for all n the value of S_n does not exceed A; but then the variable S_n necessarily approaches a definite finite limit not greater than A and the series is convergent.

Convergence of a series. The question whether a given series converges or diverges may often be settled by comparing it with another series. Here it is customary to make use of the following criterion.

If we are given two series

$$u_0 + u_1 + u_2 + \cdots,$$
$$v_0 + v_1 + v_2 + \cdots$$

with positive terms such that for all values of n, beginning with a certain one, we have the inequality

$$u_n \leqslant v_n,$$

then the convergence of the second series implies the convergence of the first, and the divergence of the first implies the divergence of the second.

For example, let us consider the so-called harmonic series

$$1 + \frac{1}{2} + \frac{1}{3} + \frac{1}{4} + \frac{1}{5} + \frac{1}{6} + \frac{1}{7} + \frac{1}{8} + \cdots$$

Its terms are correspondingly not less than the terms of the series

$$1 + \frac{1}{2} + \frac{1}{4} + \frac{1}{4} + \frac{1}{8} + \frac{1}{8} + \frac{1}{8} + \frac{1}{8} + \frac{1}{16} + \cdots + \frac{1}{16} + \cdots$$
$$\underbrace{\qquad\qquad\qquad\qquad}_{8 \text{ times}}$$

in which the sum of the underlined terms in each case is equal to $\frac{1}{2}$.

It is clear that the sum S_n of the second series approaches infinity with increasing n, and consequently that the harmonic series diverges.

The series

$$1 + \frac{1}{2^\alpha} + \frac{1}{3^\alpha} + \frac{1}{4^\alpha} + \cdots, \tag{56}$$

where α is a positive number less than unity, also obviously diverges, since for arbitrary n

$$\frac{1}{n^\alpha} > \frac{1}{n} \, (0 < \alpha < 1).$$

On the other hand, it is possible to prove that series (56) for $\alpha > 1$ is convergent. We will prove this here only for the case $\alpha \geqslant 2$; for this purpose we consider the series

$$\left(1 - \frac{1}{2}\right) + \left(\frac{1}{2} - \frac{1}{3}\right) + \cdots + \left(\frac{1}{n-1} - \frac{1}{n}\right) + \cdots$$

with positive terms. It converges to unity as its sum, since its partial sums S_n are equal to

$$S_n = 1 - \frac{1}{n+1} \to 1 \, (n \to \infty).$$

On the other hand, the general term of this series satisfies the inequality

$$\frac{1}{n-1} - \frac{1}{n} = \frac{1}{(n-1)\,n} > \frac{1}{n^2},$$

from which it follows that the series

$$1 + \frac{1}{2^2} + \frac{1}{3^2} + \frac{1}{4^2} + \cdots$$

converges. All the more then will the series (56) converge with $\alpha > 2$.

Let us give here without proof another useful criterion for convergence and divergence of series with positive terms, the so-called criterion of d'Alembert.

Let us suppose that, as n approaches infinity, the ratio $(u_n + 1)/u_n$ has a limit q. Then for $q < 1$ the sequence will certainly converge, while for $q > 1$ it will diverge. But for $q = 1$ the question of its convergence remains open.

The sum of a finite number of summands does not change if we permute the summands. But in general this is no longer true for infinite series. There exist convergent series for which it is possible to permute the terms in such a way as to change their sum and even to turn them into divergent series. Series with unstable sums of this sort fail to possess one of the fundamental properties of ordinary sums, permutability of the summands. So it is important to distinguish those series which preserve this property. It turns out that they are the so-called absolutely convergent series. The series

$$u_0 + u_1 + u_2 + u_3 + \cdots$$

is said to be *absolutely convergent* if the series

$$|u_0| + |u_1| + |u_2| + |u_3| + \cdots$$

of absolute values of its terms is also convergent. It is possible to prove that an absolutely convergent series is always convergent; in other words, that its partial sums S_n approach a finite limit. It is obvious that every convergent series with terms of one sign is absolutely convergent.

The series

$$\frac{\sin x}{1^2} + \frac{\sin 2x}{2^2} + \frac{\sin 3x}{3^2} + \cdots$$

is an example of an absolutely convergent series, since the terms of the series

$$\left|\frac{\sin x}{1^2}\right| + \left|\frac{\sin 2x}{2^2}\right| + \left|\frac{\sin 3x}{3^2}\right| + \cdots$$

are not greater than the corresponding terms of the convergent series

$$1 + \frac{1}{2^2} + \frac{1}{3^2} + \cdots$$

An example of a series which is convergent, but not absolutely convergent, is the following

$$1 - \frac{1}{2} + \frac{1}{3} - \frac{1}{4} + \cdots$$

as the reader may prove for himself.

Series of functions; uniformly convergent series. In analysis we often have to do with series whose terms are functions of x. In the preceding paragraphs we have already had examples of this sort, for instance, the series $1 + x + x^2 + x^3 + \cdots$. For some values of x this series converges, but for others it diverges. Particularly important in applications are series of functions convergent for all values of x belonging to a certain interval, which may in particular be the whole of the real axis or the positive half of it and so forth. Then the necessity arises for differentiating such series term by term, integrating them, deciding whether their sum is continuous, and so forth. For the familiar case of the sum of a finite number of terms, there are simple general rules. We know that the derivative of a sum of differentiable functions is equal to the sum of their derivatives, the integral of a sum of continuous functions is the sum of their integrals, and a sum of continuous functions is itself a continuous function: All this holds for the sum of a finite number of terms.

But for infinite series these simple rules are in general no longer true. We could give many examples of convergent series of functions for which the rules of termwise integration and differentiation are false. In the same way a series of continuous functions may turn out to have a discontinuous sum. On the other hand many infinite series behave like finite sums with respect to these rules.

Profound investigations of this question have shown that these rules may still be applied if the infinite series in question are not only convergent at each separate point of the interval of definition (the domain over which x varies) but if they are *uniformly* convergent over the whole interval. In this way there was crystallized in mathematical analysis, in the middle of the 19th century, the important concept of the uniform convergence of a series.

Let us consider the series

$$S(x) = u_0(x) + u_1(x) + u_2(x) + \cdots,$$

whose terms are functions defined on the interval $[a, b]$. We suppose that for each separate value of x in the interval this series converges to a certain sum $S(x)$. The sum of the first n terms of the series

$$S_n(x) = u_0(x) + u_1(x) + \cdots + u_{n-1}(x)$$

is also a certain function of x, defined on $[a, b]$.

We now introduce a magnitude η_n, which is equal to the least upper bound of the values* $| S(x) - S_n(x) |$, as x varies on the interval $[a, b]$. This magnitude is written as follows†

$$\eta_n = \sup_{a \leqslant x \leqslant b} | S_n(x) - S(x) |.$$

In case the quantity $S(x) - S_n(x)$ attains its maximum value, which will certainly occur for example, when $S(x)$ and $S_n(x)$ are continuous, then η_n is simply the maximum of $| S(x) - S_n(x)|$ on $[a, b]$.

From the assumed convergence of our series, we have for every individual value of x in the interval $[a, b]$

$$\lim_{n \to \infty} | S(x) - S_n(x) | = 0.$$

But the magnitude η_n may approach zero or it may not. If η_n approaches zero as $n \to \infty$, then the series is said to be *uniformly convergent*, and in the opposite case nonuniformly convergent. In the same sense it is possible to speak of the uniform or nonuniform convergence of a sequence of functions $S_n(x)$ without necessarily interpreting them as partial sums of a series.

Example 1. The series of functions

$$\frac{1}{x + 1} - \frac{1}{(x + 1)(x + 2)} - \frac{1}{(x + 2)(x + 3)} - \cdots,$$

which we take to be defined only for nonnegative values of x, namely on the half line $[0, \infty)$, may be written in the form

$$\frac{1}{x + 1} + \left(\frac{1}{x + 2} - \frac{1}{x + 1}\right) + \left(\frac{1}{x + 3} - \frac{1}{x + 2}\right) + \cdots,$$

from which we see that its partial sums are equal to

$$S_n(x) = \frac{1}{x + n}$$

and

$$\lim_{n \to \infty} S_n(x) = 0.$$

---

* See Chapter XV.
† sup is an abbreviation for the Latin word *supremum* (highest).

Thus the series is convergent for all nonnegative x and has the sum $S(x) = 0$. Furthermore,

$$\eta_n = \sup_{0 \leqslant x < \infty} |S_n(x) - S(x)| = \sup_{0 \leqslant x < \infty} \frac{1}{x + n} = \frac{1}{n} \to 0 \ (n \to \infty),$$

so that the series is uniformly convergent to zero on the half axis $[0, \infty)$. Figure 37 shows the graphs of some of the partial sums $S_n(x)$.

Example 2. The series

$$x + x(x - 1) + x^2(x - 1) + \cdots$$

may be written in the form

$$x + (x^2 - x) + (x^3 - x^2) + \cdots,$$

from which

$$S_n(x) = x^n,$$

and therefore

$$\lim_{n \to \infty} S_n(x) = \begin{cases} 0, \text{ if } 0 \leqslant x < 1; \\ 1, \text{ if } x = 1. \end{cases}$$

Thus the sum of the series is discontinuous on the interval $[0, 1]$ with a discontinuity at the point $x = 1$. The quantity $|S_n(x) - S(x)|$ is less than unity for every x in $[0, 1]$ but for x close to $x = 1$ it is arbitrarily close to unity. So,

$$\eta_n = \sup_{0 \leqslant x \leqslant 1} |S_n(x) - S(x)| = 1$$

for all $n = 1, 2, \cdots$. Thus the series is nonuniformly convergent on the interval $[0, 1]$. Figure 38 shows some of the graphs of the function $S_n(x)$. The graph of the sum of the series consists of the segment $0 \leqslant x < 1$ of the x-axis omitting the right end point and of the point $(1, 1)$.

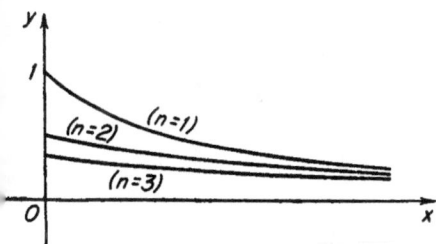

FIG. 37.

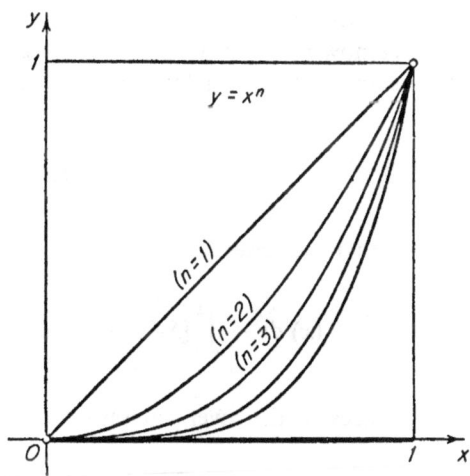

FIG. 38.

This example shows that the sum of a nonuniformly convergent series of continuous functions may in fact be a discontinuous function.

On the other hand, if we consider the series on the interval $0 \leqslant x \leqslant q$ with $q < 1$, then

$$\eta_n = \sup_{0 \leqslant x \leqslant q} |S_n(x) - S(x)| = \max_{0 \leqslant x \leqslant q} x^n = q^n \underset{n \to \infty}{\to} 0,$$

so that on this interval the series converges uniformly and its sum is continuous. The fact that the sum of a uniformly convergent series of continuous functions is itself a continuous function is a general rule, as was pointed out earlier, which can be rigorously proved.

Example 3.   The sum of the first n terms of the series $S_n(x)$ has the graph represented by the heavy broken line in figure 39. Obviously, for all n we have $S_n(0) = 0$, but if $0 < x \leqslant 1$, then for all $n \geqslant 1/x$, we will have $S_n(x) = 0$, and consequently for arbitrary x in the interval $[0, 1]$,

$$S(x) = \lim_{n \to \infty} S_n(x) = 0.$$

On the other hand,

$$\eta_n = \sup_{0 \leqslant x \leqslant 1} |S_n(x) - S(x)| = \sup |S_n(x)| = n^2.$$

So the quantity η_n does not approach zero but even approaches infinity. We now note that the series corresponding to this sequence $S_n(x)$ cannot be integrated term by term on the interval $[0, 1]$, since

$$\int_0^1 S(x)\, dx = 0, \int_0^1 S_n(x)\, dx = \frac{1}{2} n^2 \frac{1}{n} = \frac{n}{2},$$

so that the series

$$\int_0^1 S_1(x)\, dx + \int_0^1 [S_2(x) - S_1(x)]\, dx + \int_0^1 [S_3(x) - S_2(x)]\, dx + \cdots$$

reduces to the divergent series

$$\frac{1}{2} + \left(\frac{2}{2} - \frac{1}{2}\right) + \left(\frac{3}{2} - \frac{2}{2}\right) + \left(\frac{4}{2} - \frac{3}{2}\right) + \cdots$$

Let us state without proof the fundamental properties of uniformly convergent series:

1. The sum of a series of continuous functions which is uniformly convergent on the interval $[a, b]$ is a continuous function on this interval.

2. If the series of continuous functions

$$S(x) = u_0(x) + u_1(x) + u_2(x) + \cdots \tag{57}$$

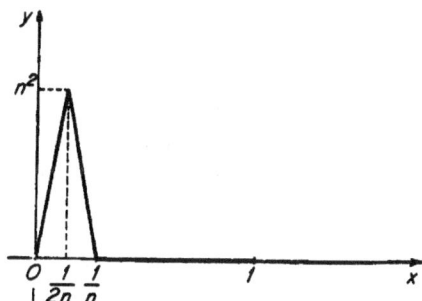

FIG. 39.

converges uniformly on the interval $[a, b]$, then it may be integrated term by term on this interval; that is, for all x_1, x_2 in $[a, b]$ we have the equality

$$\int_{x_1}^{x_2} S(t)\, dt = \int_{x_1}^{x_2} u_0(t)\, dt + \int_{x_1}^{x_2} u_1(t)\, dt + \cdots$$

3. If on the interval $[a, b]$ the series (57) converges and the functions $u_k(x)$ have continuous derivatives, then the equality

$$S'(x) = u_0'(x) + u_1'(x) + u_2'(x) + \cdots, \tag{58}$$

obtained by termwise differentiation of (57) will be valid on the interval $[a, b]$ if the series on the right in (58) converges uniformly.

Power series. In §9, a function $f(x)$ defined on an interval $[a, b]$ was called analytic, if on this interval it has derivatives of arbitrary order and if in a sufficiently small neighborhood of any point x_0 of the interval $[a, b]$ it may be expanded in a convergent Taylor series

$$f(x) = f(x_0) + \frac{f'(x_0)}{1}(x - x_0) + \frac{f''(x_0)}{2!}(x - x_0)^2 + \cdots. \tag{59}$$

If we introduce the notation

$$a_n = \frac{f^{(n)}(x_0)}{n!},$$

this series may be written in the following form

$$f(x) = a_0 + a_1(x - x_0) + a_2(x - x_0)^2 + \cdots. \tag{60}$$

A series of this sort, where the numbers a_1, a_2, $\cdots$ are constants independent of x, is called a *power series*.

As an example let us consider the power series

$$1 + x + x^2 + x^3 + \cdots, \tag{61}$$

whose terms form a geometric progression.

We know that for all values of x in the interval $-1 < x < 1$ this series converges and its sum is equal to

$$S(x) = \frac{1}{1 - x}.$$

For other values of x the series diverges.

It is also easy to see that the difference between the sum of the series and the sum of its first n terms is given by the formula

$$S(x) - S_n(x) = \frac{x^n}{1 - x}, \tag{62}$$

and if $-q \leqslant x \leqslant q$, where q is a positive number less than unity, then

$$\eta_n = \max | S(x) - S_n(x) | = \frac{q^n}{1 - q}.$$

From this it is clear that η_n approaches zero with increasing n so that the series is uniformly convergent on the interval $-q \leqslant x \leqslant q$, for all positive values of $q < 1$.

It is easy to verify that the function

$$S(x) = \frac{1}{1 - x}$$

has a derivative of nth order, which is equal to

$$S^{(n)}(x) = \frac{n!}{(1 - x)^{n+1}},$$

from which

$$S^{(n)}(0) = n!$$

and the sum of the first n terms of the Taylor series for the function $S(x)$ exactly coincides for $x_0 = 0$ with the sum of the first n terms of the series (59). Moreover, we know that the remainder term of the formula, given by the equality (62), approaches zero with increasing n, for all x

on the interval $-1 < x < 1$. Thus we have shown that the series (61) is the Taylor series of its sum $S(x)$.

Let us note one further fact. From the interval of convergence $-1 < x < 1$ of our series, let us choose an arbitrary point x_0. It is easy to see that for all x sufficiently close to x_0, namely for all x satisfying the inequality

$$\frac{|x - x_0|}{1 - x_0} < 1,$$

we have the equality

$$S(x) = \frac{1}{1 - x} = \frac{1}{1 - x_0} \frac{1}{\left(1 - \dfrac{x - x_0}{1 - x_0}\right)}$$

$$= \frac{1}{1 - x_0}\left[1 + \frac{x - x_0}{1 - x_0} + \left(\frac{x - x_0}{1 - x_0}\right)^2 + \cdots\right]$$

$$= \frac{1}{1 - x_0} + \frac{x - x_0}{(1 - x_0)^2} + \frac{(x - x_0)^2}{(1 - x_0)^3} + \cdots. \tag{63}$$

The reader may prove without difficulty that

$$\frac{S^{(n)}(x_0)}{n!} = \frac{1}{(1 - x_0)^{n+1}}.$$

Consequently, series (63) is the Taylor series of its sum $S(x)$ and converges to it in a sufficiently small neighborhood of any point x_0 belonging to the interval of convergence of (61). Since the point x_0 is arbitrary, this means that the function $S(x)$ is analytic on the interval.

All these facts that we have observed for the particular power series (61) hold for arbitrary power series.* Namely, for every power series of the form (60) where the constants a_k are chosen by any given law, there exists a certain nonnegative number R (which may also be infinite), called the *radius of convergence of the series* (60), with the following properties:

1. For all values of x from the interval $x_0 - R < x < x_0 + R$, which is called its interval of convergence, the series converges and its sum $S(x)$ is an analytic function of x in its interval. Here the convergence is uniform for every interval $[a, b]$ lying completely within the interval of convergence. The series itself is the Taylor series of its sum.

2. At the end points of the interval of convergence, the series may converge or diverge, depending on its individual character. But it will certainly diverge outside the closed interval $x_0 - R \leqslant x \leqslant x_0 + R$.

---

* For more detailed information on this point see Chapter IX.

We suggest to the reader that he consider the power series

$$1 + \frac{x}{1} + \frac{x^2}{2!} + \frac{x^3}{3!} + \cdots,$$

$$1 + x + 2!x^2 + 3!x^3 + \cdots$$

$$1 + x + \frac{x^2}{2} + \frac{x^3}{3} + \cdots$$

and convince himself that their radii of convergence are respectively infinity, zero, and unity.

By the definition given earlier every analytic function may be expanded, in a sufficiently small neighborhood of an arbitrary point where it is defined, into a power series which converges to the function. Conversely, from what has been said it follows that the sum of every power series whose radius of convergence is not zero is an analytic function in its interval of convergence.

So we see that power series are organically connected with analytic functions. We could even say that on their interval of convergence power series are the natural means of representing analytic functions, and consequently they are also the natural means of approximating analytic functions by algebraic polynomials.*

For example, from the fact that the function $1/(1 - x)$ may be expanded in the power series

$$\frac{1}{1 - x} = 1 + x + x^2 + x^3 + \cdots$$

which is convergent on the interval $-1 < x < 1$, it follows that the power series is uniformly convergent on an arbitrary interval $-a \leqslant x \leqslant a$ with $a < 1$, and this implies the possibility of approximating the function on the whole interval $[-a, a]$ by the partial sums of the series with any preassigned degree of accuracy.

Let us suppose that we are required to approximate the function $1/(1 - x)$ by polynomials on the interval $[-\frac{1}{2}, \frac{1}{2}]$ with an accuracy of 0.01. We note that for all x in this interval we have the inequality

$$\left| \frac{1}{1 - x} - 1 - x - \cdots - x^n \right| = | x^{n+1} + x^{n+2} + \cdots |$$

$$\leqslant | x |^{n+1} + | x |^{n+2} + \cdots \leqslant \frac{1}{2^{n+1}} + \frac{1}{2^{n+2}} + \cdots = \frac{1}{2^n},$$

---

* Approximations going beyond the limits of the interval of convergence of a power series require other methods. (See Chapter XII.)

and since $2^6 = 64$, and $2^7 = 128$, the desired polynomial, approximating the function on the whole interval $[-\frac{1}{2}, \frac{1}{2}]$ with an accuracy of 0.01, will have the form

$$\frac{1}{1-x} \approx 1 + x + x^2 + \cdots + x^7.$$

Let us note one further extremely valuable property of power series: They may be differentiated termwise everywhere in the interval of convergence. This property finds extremely wide application in the solution of various problems in mathematics.

For example, let it be required to find the solution of the differential equation $y' = y$ under the auxiliary condition $y(0) = 1$. We will seek the solution in the form of a power series,

$$y = a_0 + a_1x + a_2x^2 + \cdots.$$

Because of the auxiliary condition, we must set $a_0 = 1$. Assuming that this series converges, we may differentiate it termwise; as a result we obtain

$$= a_1 + 2a_2x + 3a_3x^2 + \cdots.$$

If we substitute these two series into the differential equation and equate coefficients for each of the powers of x, we obtain

$$a_k = \frac{1}{k!} \ (k = 1, 2, \cdots)$$

and the desired solution has the form

$$y = 1 + \frac{x}{1} + \frac{x^2}{2!} + \frac{x^3}{3!} + \cdots.$$

It is well known that this series converges for all values of x and that its sum is equal to $y = e^x$.

In this case we have obtained a series whose sum is a well-known elementary function. But this does not always happen; it may turn out that a convergent power series so obtained has a sum that is not an elementary function. An example is the series

$$y_p(x) = x^p \left[1 - \frac{x^2}{2(2p+2)} + \frac{x^4}{2 \cdot 4(2p+2)(2p+4)} - \cdots \right],$$

obtained as a solution of Bessel's differential equation, which is of great importance in applications. In this way power series may serve to define functions.

Suggested Reading

R. Courant, *Differential and integral calculus*, 2 vols., Interscience, New York, 1938.

A. Dresden, *Introduction to the calculus*, Henry Holt, New York, 1940.

F. Klein, *Elementary mathematics from an advanced standpoint. Arithmetic—algebra—analysis*, Dover, New York, 1953.

K. Knopp, *Infinite sequences and series*, Dover, New York, 1956.

PART 2

CHAPTER III

ANALYTIC GEOMETRY

§1. Introduction

In the first half of the 17th century a completely new branch of mathematics arose, the so-called analytic geometry, establishing a connection between curves in a plane and algebraic equations in two unknowns.

A quite rare event thereby happened in mathematics: In one or two decades there appeared a great, entirely new branch of mathematics based on a very simple concept, which until then had not received proper attention. The appearance of analytic geometry in the first half of the 17th century was not accidental. The transition in Europe to the new capitalistic methods of manufacture required the advance of a whole series of sciences. A short time before, contemporary mechanics was being created by Galileo and other scientists, experimental data were being accumulated in all regions of natural science, the means of observation were being perfected, and instead of absolute scholastic theories new ones were being created. In astronomy, among the foremost scientists the teachings of Copernicus had finally triumphed. The rapid development of long-range navigation insistently called for knowledge of astronomy and the elements of mechanics.

The art of warfare also required mechanics. Ellipses and parabolas, whose geometric properties as conic sections were already well known in detail to the ancient Greeks almost 2000 years earlier, ceased to be only part of geometry, as they were to the Greeks. After Kepler had discovered that the planets revolve around the sun in ellipses, and Galileo that a stone thrown into the air traces out a parabola, it was necessary to calculate these ellipses and to find the parabolas along which bullets fly from a gun; it was necessary to discover the law by which the atmospheric pressure, discovered by Pascal, decreases with the height; it was necessary actually to calculate the volumes of various bodies, and so forth.

All these questions almost simultaneously called to life three entirely new mathematical sciences: analytic geometry, differential calculus, and integral calculus, including the solution of the simplest differential equations.

These three new fields qualitatively changed the face of the whole of mathematics. They made it possible to solve problems never even dreamed of before.

In the first half of the 17th century, i.e., at the beginning of the 1600's, a group of the most outstanding mathematicians was already close to the idea of analytic geometry, but there were two of them, in particular, who understood clearly the possibility of creating a new branch of mathematics. These were Pierre Fermat, a counsellor of the parliament of the French city of Toulouse and a world-famous mathematician, and the famous French philosopher René Descartes. Descartes is credited with being the chief creator of analytic geometry. He was the one who, as a philosopher, raised the question of its complete generality. Descartes published the great philosophical treatise "Discourse on the method of rightly conducting the reason and seeking the truth in the sciences, with applications: dioptrics, meteorology and geometry."

The last part of this work, entitled "Geometry" and published in 1637, contains a sufficiently complete, although somewhat confusing, presentation of the mathematical theory that since then has been called analytic geometry.

§2. Descartes' Two Fundamental Concepts

Descartes wished to create a method that could equally well be applied to the solution of all problems of geometry, that is, which would provide a general method for their solution. Descartes' theory is based on two concepts: the concept of coordinates and the concept of representing by the coordinate method any algebraic equation with two unknowns in the form of a curve in the plane.

The concept of coordinates. By the *coordinates of a point* in the plane Descartes means the abscissa and ordinate of this point, i.e., the numerical values x and y of its distances (with corresponding signs) to two mutually perpendicular straight lines (coordinate axes) chosen in this plane (see Chapter II). The point of intersection of the coordinate axes, i.e., the point having coordinates (0, 0) is called the *origin*.

With the introduction of coordinates Descartes constructed, so to speak, an "arithmetization" of the plane. Instead of determining any point geometrically, it is sufficient to give a pair of numbers x, y and conversely (figure 1).

The notion of comparison of equations with two unknowns with curves in the plane.   Descartes' second concept is the following. Up to the time of Descartes, where an algebraic equation in two unknowns $F(x, y) = 0$ was given, it was said that the problem was indeterminate, since from the equation it was impossible to determine these unknowns; any value could be assigned to one of them, for example to x, and substituted in the equation; the result was an equation with only one unknown y, for which, in general, the equation could be solved. Then this *arbitrarily chosen x* together with the *so-obtained y* would satisfy the given equation. Consequently, such an "indeterminate" equation was not considered interesting.

Descartes looked at the matter differently. He proposed that in an equation with two unknowns x be regarded as the abscissa of a point and the corresponding y as its ordinate. Then if we vary the unknown x, to every value of x the corresponding y is computed from the equation, so that we obtain, in general, a set of points which form a curve (figure 2).*

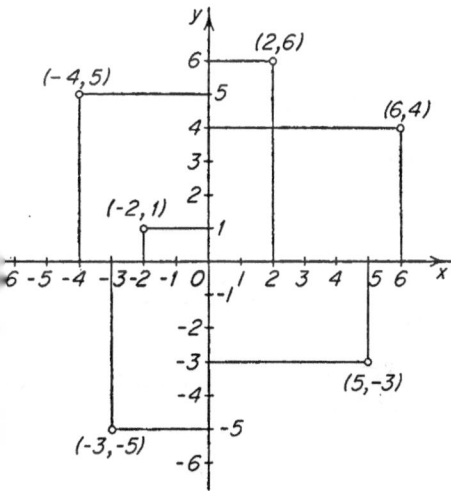

Fig. 1.

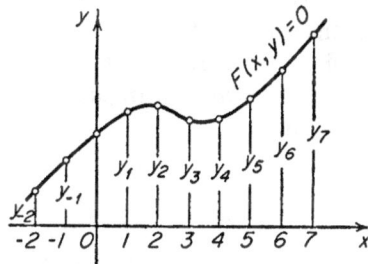

$$F(-2, y_{-2}) = 0$$
$$F(-1, y_{-1}) = 0$$
$$F(0, y_0) = 0$$
$$F(1, y_1) = 0$$
$$F(2, y_2) = 0$$

Fig. 2.

---

* Sometimes, the equation is not satisfied by any point (x, y) with real coordinates, sometimes by one or a few such points. In this case we say that the curve is imaginary or reduces to points (see §7).

Thus, to each algebraic equation with two variables, $F(x, y) = 0$, corresponds a completely determined curve of the plane, namely a curve representing the totality of all those points of the plane whose coordinates satisfy the equation $F(x, y) = 0$.

This observation of Descartes *opened up an entire new science.*

The basic problems solved by analytic geometry and the definition of analytic geometry. Analytic geometry provides the possibility: (1) of solving construction problems by computation (see for example, the division of a segment in a given ratio, see §3); (2) of finding the equation of curves defined by a geometric property (for example, of an ellipse defined by the condition that the sum of distances to two given points is constant, see §7); (3) of proving new geometric theorems algebraically (see, for example, the derivation of Newton's theory of diameters, §6); (4) conversely, of representing an algebraic equation geometrically, to clarify its algebraic properties (see, for example, the solution of third- and fourth-degree equations from the intersection of a parabola with a circle, §5).

Thus, analytic geometry is that part of mathematics which, applying the coordinate method, investigates geometric objects by algebraic means.

§3. Elementary Problems

The coordinates of a point that divide a segment in a given ratio. Given the coordinates (x_1, y_1) and (x_2, y_2) of two points M_1 and M_2, let us find the coordinates (x, y) of the point M dividing the segment M_1M_2 in the ratio m to n (figure 3). From the similarity of the shaded triangles we obtain:

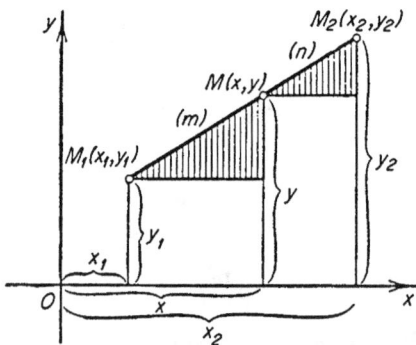

FIG. 3.

$$\frac{x - x_1}{x_2 - x} = \frac{m}{n}, \text{ from which}$$

$$x = \frac{nx_1 + mx_2}{m + n},$$

$$\frac{y - y_1}{y_2 - y} = \frac{m}{n}, \text{ from which}$$

$$y = \frac{ny_1 + my_2}{m + n}.$$

Distance between two points. Let us find the distance between the points M_1 and M_2, whose coordinates are (x_1, y_1) and

(x_2, y_2) respectively. From the shaded right triangle (figure 4), we obtain by the theorem of Pythagoras

$$d = \sqrt{(x_2 - x_1)^2 + (y_2 - y_1)^2}.$$

The area of a triangle. Let us find the area S of the triangle $M_1 M_2 M_3$ (figure 5) if the coordinates of its vertices are respectively $(x_1, y_1), (x_2, y_2), (x_3, y_3)$. Considering the area of the triangle as the sum of the areas of trapezoids with bases y_1, y_3 and y_3, y_2 minus the area of the trapezoid with bases y_1, y_2 and writing the product $- (y_1 + y_2)(x_2 - x_1)$ in the form $(y_1 + y_2)(x_1 - x_2)$, we obtain

$$S = \tfrac{1}{2} [(y_1 + y_2)(x_1 - x_2) + (y_2 + y_3)(x_2 - x_3) + (y_3 + y_1)(x_3 - x_1)].$$

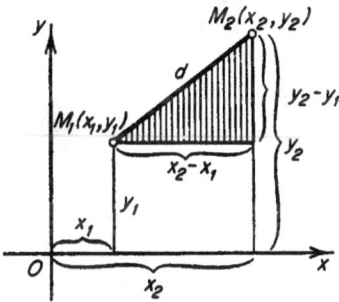

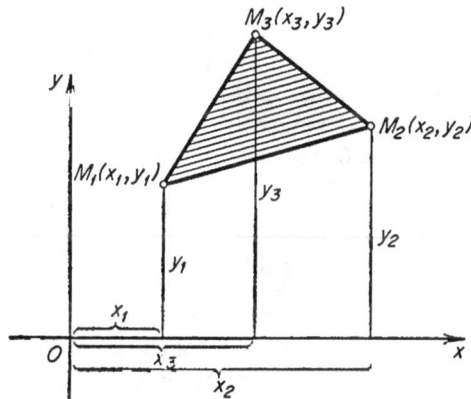

FIG. 4.   FIG. 5.

In these problems it only remains to verify that the derived formulas remain valid without any change in those cases when one or more coordinates or their differences are negative. Such verification easily follows.

Determination of the points of intersection of two curves. Relying on the second fundamental idea that the equation $F(x, y) = 0$ represents a curve, it is particularly simple to find the points of intersection of two curves. In order to find the coordinates of the points of intersection of two curves, it is obviously necessary to solve simultaneously the equations that represent them. The pair of numbers x, y obtained from the ordinary solution of these two equations will determine the point whose coordinates satisfy both of the equations, i.e., the point that lies on the first as well as on the second curve, and this is the point of their intersection.

The solution of geometric problems by the tools of analytic geometry, as we see, is very convenient for practical purposes, especially because every solution is at once obtained in the convenient form of numbers. *Such a geometry, such a science, was exactly what was lacking at that time.*

§4. Discussion of Curves Represented by First- and Second-Degree Equations

First degree equation. Making use of his second idea, Descartes first of all examined what curves correspond to an equation of the first-degree,

$$Ax + By + C = 0, \tag{1}$$

i.e., to an equation where A, B, C are numerical coefficients with A and B not both zero. It turned out that in the plane a straight line always corresponds to such an equation.

We shall prove that equation (1) always represents a straight line, and conversely, that to every line in the plane there corresponds a completely determined equation of the form (1). In fact, let us suppose, for example, that $B \neq 0$; then equation (1) can be solved for y

$$y = kx + l,$$

where $k = -\dfrac{A}{B}$; $l = -\dfrac{C}{B}$.

We examine first the equation $y = kx$. It obviously represents a straight line passing through the origin and making an angle ϕ with the x-axis whose tangent $\tan \phi$ is k (figure 6). Indeed, the equation can be

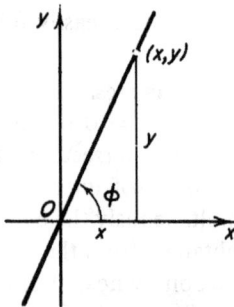

FIG. 6.

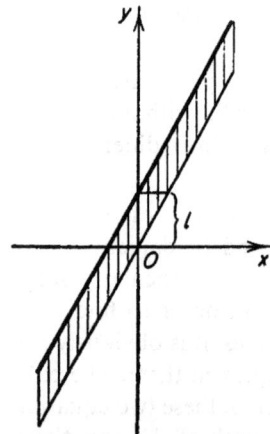

FIG. 7.

written as $y/x = k$, so that the coordinates of every point (x, y) on the straight line satisfy the equation, and the coordinates of no point $(x, \bar{y})$ not lying on the straight line satisfy the equation, since for such a point y/x will be either greater than or smaller than k. In addition, if $\tan \phi > 0$, then for this line either both x and y are positive or both negative, and if $\tan \phi < 0$ their signs are opposite.

Thus, the equation $y = kx$ represents a straight line passing through the origin O, and consequently the equation $y = kx + l$ also represents a line, namely the one which is obtained from the previous line by the parallel translation such that the ordinate of each of its points is increased by l (figure 7).

The earlier derived formulas of the coordinates of a point dividing a segment in a given ratio, the distance between two given points, and the area of a triangle as well as the information about the equation of a straight line already enable us to solve a large number of problems.

The equation of a straight line passing through one or two given points.
Let M_1 be the point with coordinates x_1, y_1 and let k be a given number. The equation $y = kx + l$ represents a straight line making with the Ox-axis an angle whose tangent is equal to k and intersecting the Oy-axis at a distance l from O. Let us choose l such that this line goes through the point (x_1, y_1). For this, the coordinates of the point M_1 must satisfy the equation, i.e., we must have $y_1 = kx_1 + l$, from which $l = y_1 - kx_1$.

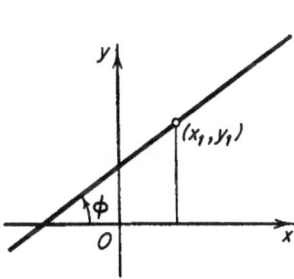

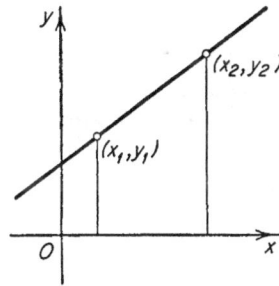

Fig. 8.   Fig. 9.

Substituting this value for l, we obtain the equation of the line that passes through the given point (x_1, y_1) and makes with the Ox-axis an angle whose tangent is equal to k (figure 8). This equation is $y = kx + y_1 - kx_1$ or

$$y - y_1 = k(x - x_1).$$

Example. Let the angle between the line and the Ox-axis be equal to 45°, and let the point M have coordinates $(3, 7)$; then the equation of the corresponding line (since $\tan 45° = 1$) will be: $y - 7 = 1 \cdot (x - 3)$ or $x - y + 4 = 0$.

If we require that the line passing through the point (x_1, y_1) also go through the point (x_2, y_2), it follows that the condition $y_2 - y_1 = k(x_2 - x_1)$ must be imposed on k. Finding k from this and substituting it in the previous equation, we obtain the equation of the line passing through two given points (figure 9):

$$\frac{x - x_1}{x_2 - x_1} = \frac{y - y_1}{y_2 - y_1}.$$

Descartes' result concerning second-degree equations. Descartes also investigated the question as to what kinds of curves in the plane are represented by the second-degree equation with two variables whose general form is

$$Ax^2 + Bxy + Cy^2 + Dx + Ey + F = 0,$$

and showed that such an equation, generally speaking, represents an ellipse, a hyperbola, or a parabola; i.e., curves very well known to the mathematicians of antiquity.

These are Descartes' most important achievements. However, his book was far from being restricted to these topics; he also investigated the equations of a number of interesting geometric loci, examined certain theorems on transformation of algebraic equations, mentioned without proof his famous *law of signs* for the number of positive roots of an equation whose roots are all real (see Chapter IV, §4) and, finally, presented a remarkable method for determining the real roots of third- and forth-degree equations from the intersection of the parabola $y = x^2$ with circles.

§5. Descartes' Method of Solving Third- and Fourth-Degree Algebraic Equations

Transformation of third- and fourth-degree equations to an equation of the fourth-degree not involving the x^3-term. We will show that the solution of an arbitrary third- or fourth-degree equation can be reduced to the solution of an equation of the form

$$x^4 + px^2 + qx + r = 0. \tag{2}$$

Let the given third-degree equation be $z^3 + az^2 + bz + c = 0$. Substituting $z = x - a/3$, we obtain

$$(x - a/3)^3 + a(x - a/3)^2 + b(x - a/3) + c = 0.$$

The x^2-terms in the expansion of the parentheses will cancel out, so that we get an equation of the form $x^3 + px + q = 0$. Multiplying this equation by x, we bring it to the form (2) with $r = 0$, which also admits a root $x_4 = 0$.

An equation of the fourth-degree $z^4 + az^3 + bz^2 + cz + d = 0$ can be reduced to the form (2) by the substitution $z = x - a/4$. Hence, the solution of all third- and fourth-degree equations can be reduced to the solution of an equation of the form (2).

The solution of third- and fourth-degree equations by the intersection of a circle with the parabola $y = x^2$. Let us first derive the equation of a circle with center (a, b) and radius R. If (x, y) is any of its points, then the square of its distance to the point (a, b) is equal to $(x - a)^2 + (y - b)^2$ (see §3). Thus, the equation of the circle in question is

$$(x - a)^2 + (y - b)^2 = R^2.$$

Now we try to find the points of intersection of this circle with the parabola $y = x^2$. In order to do this, by virtue of what was said in §3, it is necessary to solve simultaneously the equation of this circle

$$x^2 + y^2 - 2ax - 2by + a^2 + b^2 - R^2 = 0$$

and the equation of the parabola

$$y = x^2.$$

Substituting y from the second equation into the first, we obtain a fourth-degree equation in x:

$$x^2 + x^4 - 2ax - 2bx^2 + a^2 + b^2 - R^2 = 0$$

or

$$x^4 + (1 - 2b) x^2 - 2ax + a^2 + b^2 - R^2 = 0.$$

If we choose a, b and R^2 such that

$$1 - 2b = p, \ -2a = q, \ a^2 + b^2 - R^2 = r,$$

then exactly equation (2) is obtained. For this purpose we have to take

$$a = -\frac{q}{2}, b = \frac{1 - p}{2}, R^2 = \frac{q^2}{4} + \frac{(1 - p)^2}{4} - r. \tag{3}$$

In the last formula (3), generally speaking, R^2 may turn out to be negative. However, in the case when equation (2) has even one real root x_1, the following equality holds

$$x_1^4 + (1 - 2b)\, x_1^2 - 2ax_1 + a^2 + b^2 - R^2 = 0. \qquad (4)$$

Denoting x_1^2 by y_1, equation (4) can be rewritten as

$$x_1^2 + y_1^2 - 2ax_1 - 2by_1 + a^2 + b^2 - R^2 = 0$$

or as

$$(x_1 - a)^2 + (y_1 - b)^2 = R^2.$$

Hence, in the case when equation (2) has a real root, the number $R^2 = [(1 - p)^2 + q^2]/4 - r$ is *positive*, the equation

$$(x - a)^2 + (y - b)^2 = R^2$$

is the equation of a circle, and all real roots of equation (2) are the abscissas of points of intersection of the parabola $y = x^2$ with this circle. (In case $r = 0$, $R^2 = a^2 + b^2$, this circle passes through the origin.)

Thus, if the coefficients p, q, r of equation (2) are given, and it is necessary to find a, b and R^2 by formulas (3), then if $R^2 < 0$, equation (2) is known to have no real roots. But, if $R^2 \geqslant 0$ then the abscissas of the points of intersection of the circle with center (a, b) and radius R with the parabola $y = x^2$ (drawn once and for all) give all the real roots of equation (2); and also in case $R^2 < 0$, the resulting circle cannot intersect the parabola and equation (2) does not have real roots.

Example. Let the given fourth-degree equation be:

$$x^4 - 4x^2 + x + \frac{5}{2} = 0.$$

Then we have

$$a = -\frac{1}{2},\, b = \frac{5}{2} = 2\frac{1}{2},$$

$$R = \sqrt{\frac{1}{4} + \frac{25}{4} - \frac{5}{2}} = 2.$$

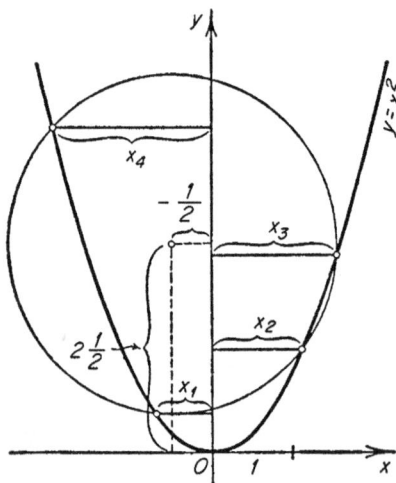

FIG. 10.

Figure 10 shows the corresponding circle and the roots x_1, x_2, x_3, x_4 of the given equation.

The 1st, 2nd, 3rd and 4th sections above contain, in an abbreviated and somewhat more modern form, the essential content of Descartes' book.

From Descartes' time up to the present, analytic geometry has undergone an immense development that has been very fruitful for many different parts of mathematics. We will attempt in the following sections of this chapter to trace the most important stages of this development.

First of all, it is necessary to say that the inventors of the infinitesimal analysis were already in possession of Descartes' method. Whether it was a question of tangents or normals (perpendiculars to the tangents at the point of tangency) to curves, or of maxima or minima of functions considered geometrically, or of the radius of curvature of a curve at a given point, etc., the equation of the curve was considered first, by the method of Descartes, and then the equations of the normal, the tangent, and so forth, were found. Thus infinitesimal analysis, namely the differential and integral calculus, would have been inconceivable without the preliminary development of analytic geometry.

§6. Newton's General Theory of Diameters

The first mathematician to take a further great step forward in analytic geometry itself was Newton. In 1704 he examined the theory of third-order curves, i.e., curves which are represented by third-degree algebraic equations in two unknowns. At the same time he found, among other things, an elegant general theorem about "diameters," which correspond to secants in a given direction. He proved the following.

Let an nth-order curve be given, i.e., a curve which is represented by an nth-degree algebraic equation in two unknowns; then an arbitrary straight line intersecting it has in general n common points with it. Let M be the point of the secant that is the "center of gravity" of these points of its intersec-

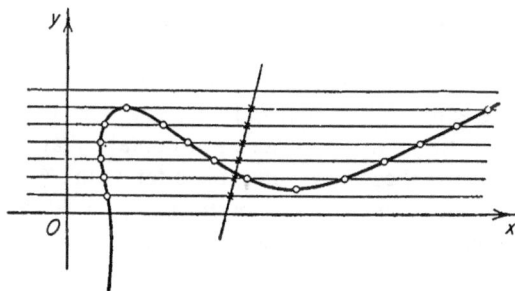

Fig. 11.

tion with the given nth-order curve, i.e., the center of gravity of a set of n equal point masses situated at these points. It turns out that if we take all possible sets of mutually parallel secants and for each of them consider these centers of mass M, then for any given set of parallel secants all the points M lie on a straight line. Newton called this line the "diameter" of the nth-order curve corresponding to the given direction of the secants. Since the proof of this theorem is quite easy with the help of analytic geometry, we will give it here.

Let an nth-order curve be given and some set of mutually parallel secants of the curve. Choose the coordinate axes so that these secants are parallel to the Ox-axis (figure 11). Then their equations will have the form $y = l$, where the constant l will be different for different secants. Let $F(x, y) = 0$ be the equation that represents the nth-order curve with respect to these coordinate axes. It is easy to show that under a transformation from one rectangular coordinate system to another, although the equation of the curve changes, its order does not change (this will be shown in §8). Therefore $F(x, y)$ will also be an nth-degree polynomial. To determine the abscissas of the points of intersection of the curve with the secant $y = l$, it is necessary to solve the simultaneous equations $F(x, y) = 0$ and $y = l$; as a result, in general, an nth-degree equation in x is obtained

$$F(x, l) = 0, \tag{5}$$

from which we find the abscissas $x_1, x_2, \cdots, x_n$. The abscissa x_c of the center of gravity of the n points of intersection is equal, by the very definition of center of gravity, to

$$x_c = \frac{x_1 + x_2 + \cdots + x_n}{n}.$$

But, as is known from the theory of algebraic equations, the sum $x_1 + x_2 + \cdots + x_n$ of the roots of an equation is equal to the coefficient of the $(n-1)$th power of the unknown x, taken with the opposite sign, divided by the coefficient of the nth power of x. But because the sum of the exponents of x and y in every term of $F(x, y)$ is equal to or less than n, the term in x^n does not contain y at all but has the form Ax^n, where A is a constant; and if the terms in x^{n-1} contain y, they do so to no higher than the first power; i.e., they have the form $x^{n-1}(By + C)$. Consequently, the coefficient of x^n is A and that of x^{n-1} is $Bl + C$, and we have for any given l

$$x_c = -\frac{Bl + C}{nA}.$$

But the secant is parallel to the Ox-axis so that for all of its points $y = l$, and hence the ordinate y of the center of gravity of the points of its intersection with the given nth-order curve is also equal to l; thus finally we obtain $nAx_c + By_c + C = 0$, i.e., the coordinates x_c, y_c of the centers of gravity for all these secants satisfy a first-degree equation, and consequently lie on a straight line.

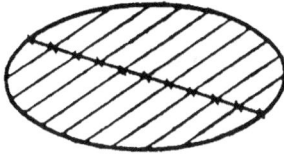

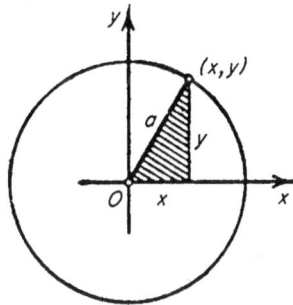

<table>
<tr><td>Fig. 12.</td><td>Fig. 13.</td></tr>
</table>

The case when $F(x, y)$ does not contain x^n can be investigated analogously.

In case the curve is of the 2nd order ($n = 2$) the center of gravity of two points is simply the midpoint between them, so that the locus of midpoints of parallel chords of a second-order curve is a straight line (figure 12), a result that for the ellipse, as well as for the hyperbola and the parabola, was already well known to the ancients. But this was proved by them, even though only for these partial cases, with quite difficult geometric arguments, and here a new general theorem, unknown to the ancients, is proved in an entirely simple way.

Such examples reveal the power of analytic geometry.

§7. Ellipse, Hyperbola, and Parabola

In this and the following sections, we consider second-order curves. Before investigating the general second-degree equation, it is useful to examine some of its simplest forms.

The equation of a circle with center at the origin. First of all, we consider the equation

$$x^2 + y^2 = a^2.$$

It evidently represents a circle with center at the origin and radius a, as follows from the theorem of Pythagoras applied to the shaded right triangle (figure 13), since whatever point (x, y) of this circle is taken, its x and y coordinates satisfy this equation, and conversely, if the coordinates x, y of a point satisfy the equation, then the point belongs to the circle; i.e., the circle is the set of all those points of the plane that satisfy the equation.

The equation of an ellipse and its focal property.   Let two points F_1 and F_2 be given, the distance between which is equal to $2c$. We will find the equation of the locus of all points M of the plane; the sum of whose distances to the points F_1 and F_2 is equal to a constant $2a$ (where, of course, a is greater than c). Such a curve is called an *ellipse* and the points F_1 and F_2 are its *foci*.

Let us choose a rectangular coordinate system such that the points F_1 and F_2 lie on the Ox-axis and the origin is halfway between them. Then the coordinates of the points F_1 and F_2 will be $(c, 0)$ and $(-c, 0)$. Let us take an arbitrary point M with coordinates (x, y), belonging to the locus in question, and let us write that the sum of its distances to the points F_1 and F_2 is equal to $2a$,

$$\sqrt{(x - c)^2 + (y - 0)^2} + \sqrt{(x + c)^2 + (y - 0)^2} = 2a. \qquad (6)$$

This equation is satisfied by the coordinates (x, y) of any point of the locus under consideration. Obviously the converse is also true, namely that any point whose coordinates satisfy equation (6) belongs to this locus. Equation (6) is therefore the equation of the locus. It remains to simplify it.

Raising both sides to the second power, we obtain

$$x^2 - 2cx + c^2 + y^2 + 2\sqrt{(x^2 - 2cx + c^2 + y^2)(x^2 + 2cx + c^2 + y^2)}$$
$$+ x^2 + 2cx + c^2 + y^2 = 4a^2,$$

or after simplification

$$x^2 + y^2 + c^2 - 2a^2 = -\sqrt{(x^2 + y^2 + c^2)^2 - 4c^2x^2}.$$

Squaring again both sides, we obtain

$$(x^2 + y^2 + c^2)^2 - 4a^2(x^2 + y^2 + c^2) + 4a^4 = (x^2 + y^2 + c^2)^2 - 4c^2x^2$$

or after simplification

$$(a^2 - c^2)x^2 + a^2y^2 = (a^2 - c^2)a^2.$$

Let us set $a^2 - c^2 = b^2$ (as may be done since $a > c$); then we obtain $b^2x^2 + a^2y^2 = a^2b^2$, and dividing by a^2b^2 we have

$$\frac{x^2}{a^2} + \frac{y^2}{b^2} = 1. \qquad (7)$$

The coordinates (x, y) of any point M of the locus thus satisfy equation (7).

It can be shown on the other hand that if the coordinates of a point satisfy equation (7) then they also satisfy equation (6). Consequently, equation (7) is the equation of this locus, i.e., the equation of the ellipse (figure 14).

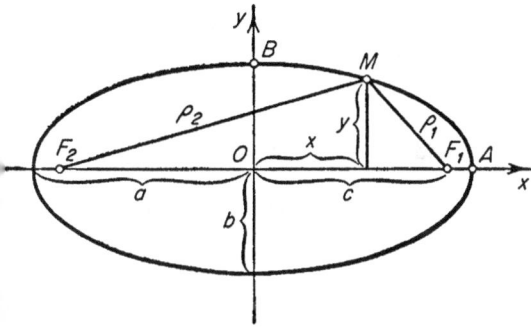
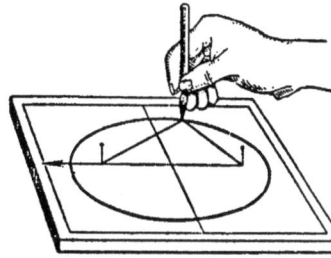

FIG. 14.   FIG. 15.

This argument is a classical example of finding the equation of a curve given by some of its geometrical properties.

The well-known method of tracing an ellipse by means of a thread (figure 15) is based on the property of the ellipse that the sum of the distances of any of its points to two given points is a constant.

Remark. In order to determine an ellipse, we could have taken, instead of the focal property considered here, any other geometric property characteristic of it, for example, that the ellipse is the result of a "uniform contraction" of a circle toward one of its diameters or any other property.

Substituting $y = 0$ in equation (7) of the ellipse, we obtain $x = \pm a$, i.e., a is the length of the segment OA (see figure 14), which is called the *major semiaxis* of the ellipse. Analogously, substituting $x = 0$, we obtain $y = \pm b$, i.e., b is the length of the segment OB, which is called the *minor semiaxis* of the ellipse.

The number $\epsilon = c/a$ is called the *eccentricity* of the ellipse, so that, since $c = \sqrt{a^2 - b^2} < a$, the eccentricity of an ellipse is less than 1. In the case of a circle, $c = 0$ and consequently $\epsilon = 0$; both foci are at one point, the center of the circle (since $OF_1 = OF_2 = 0$), but the previous method of drawing the curve with a thread is still valid.

Laws of planetary motion. In studying Tycho Brahe's long-continued observations on the motion of the planet Mars, Kepler discovered that the planets revolve around the Sun in ellipses such that the Sun occupies one focus of the ellipse (the other focus remains unoccupied and plays no part in the motion of a planet around the Sun) (figure 16). Kepler also observed that the focal radius ρ in equal times sweeps out sectors of equal area,* and Newton showed that the necessity of such a motion follows mathematically from the law of inertia (proportionality of acceleration to force) and the law of universal gravitation.

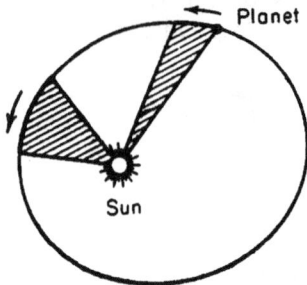

FIG. 16.

The ellipse of inertia. As an example of the application of the ellipse in a technical problem, we consider the so-called ellipse of inertia of a plate.

Let the plate be of uniform thickness and homogeneous material, for example a zinc plate of arbitrary shape. We rotate it around an axis in its plane. A body in rectilinear motion has, as is well known, an inertia with respect to this rectilinear motion that is proportional to its mass (independently of the shape of the body and the distribution of the mass). Similarly, a body rotating around an axis, for instance a flywheel, has inertia with respect to this rotation. But in the case of rotation, the inertia is not only proportional to the mass of the rotating body but

FIG. 17a.

FIG. 17b.

---

* The eccentricities of planetary orbits are not very large, so that the orbits of planets are almost circles.

also depends on the distribution of the mass of the body with respect to the axis of rotation, since the inertia with respect to rotation is greater if the mass is farther from the axis. For example, it is very easy to bring a stick at once into fast rotation around its longitudinal axis (figure 17a). But if we try to bring it at once to fast rotation around an axis perpendicular to its length, even if the axis passes through its midpoint, we will find that unless this stick is very light, we must exert considerable effort (figure 17b).

It is possible to show that the inertia of a body with respect to rotation about an axis, the so-called *moment of inertia* of the body relative to the axis, is equal to $\Sigma r_i^2 m_i$ (where by $\Sigma r_i^2 m_i$ we mean the sum $r_1^2 m_1 + r_2^2 m_2 + \cdots + r_n^2 m_n$ and think of the body as decomposed into very small elements, with m_i as the mass of the ith element and r_i the distance of the ith element from the axis of rotation, the summation being taken over all elements of the body).

Let us return to our plate. Let O (figure 18) be a point of this plate. We consider the moments of inertia J_u of the plate relative to an axis

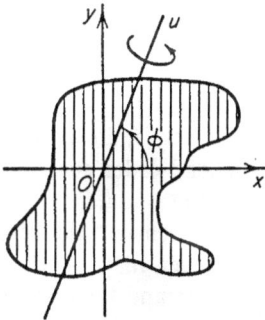

FIG. 18.

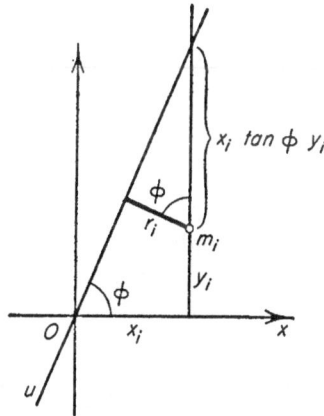

FIG. 19.

of rotation u passing through O and lying in the plane of the plate. For this purpose we take the point O as the origin of a Cartesian coordinate system and choose arbitrary axes Ox and Oy in the plane of the plate; then we will characterize the axis of rotation u by the angle ϕ which it makes with the Ox-axis. It is easy to see (figure 19) that

$$r_i = |(x_i \tan\phi - y_i) \cos\phi| = |x_i \sin\phi - y_i \cos\phi|.$$

Hence

$$\Sigma\, r_i^2 m_i = \Sigma\, (x_i^2 \sin^2\phi - 2x_i y_i \sin\phi \cos\phi + y_i^2 \cos^2\phi)\, m_i$$

$$= \sin^2\phi\, \Sigma\, x_i^2 m_i - 2\sin\phi \cos\phi\, \Sigma\, x_i y_i m_i + \cos^2\phi\, \Sigma\, y_i^2 m_i\,.$$

The quantities $\sin^2\phi$, $2\sin\phi\cos\phi$, and $\cos^2\phi$ are taken outside the summation sign, since they are constant for a given axis u. We now write

$$\Sigma\, x_i^2 m_i = A, \quad -\Sigma\, x_i y_i m_i = B, \quad \Sigma\, y_i^2 m_i = C.$$

The quantities A, B, and C do not depend on the choice of the axis u, but only on the shape of the plate, the distribution of its mass, and the fixed choice of the coordinate axes Ox and Oy. Consequently,

$$J_u = A\sin^2\phi + 2B\sin\phi\cos\phi + C\cos^2\phi.$$

We consider all possible axes u in the plane of the plate passing through the point O and lay off on each of these axes from the point O a length equal to ρ, the inverse of the square root of the moment of inertia J_u of the plate relative to that axis, i.e., $\rho = 1/\sqrt{J_u}$. Then we obtain

$$\frac{1}{\rho^2} = A\sin^2\phi + 2B\sin\phi\cos\phi + C\cos^2\phi.$$

But

$$x = \rho\cos\phi, \quad y = \rho\sin\phi,$$

so that the equation of this locus has the following form:

$$Cx^2 + 2Bxy + Ay^2 = 1.$$

A second-order curve is obtained that is evidently finite and closed; i.e., it is an ellipse (figure 20), since all other second-order curves, as we will later show, are either infinite or reduce to one point.

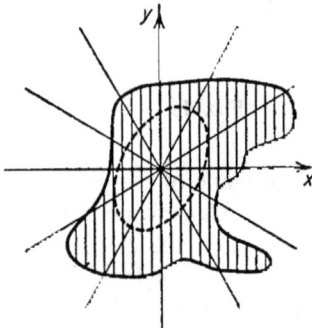

FIG. 20.

The following remarkable result is obtained: Whatever may be the form and size of a plate and the distribution of its mass, the magnitude of its moment of inertia (more precisely, of the quantity ρ inversely proportional to the square root of the moment of inertia) with respect to the various axes lying in the plane of the plate and passing through the given point O, is characterized by a certain ellipse. This ellipse is called the ellipse of inertia of the plate relative to the point O. If the point O is the center of gravity of the plate, then the ellipse is called its central ellipse of inertia.

The ellipse of inertia plays a great role in mechanics; in particular, it has an important application in the strength of materials. In the theory of strength of materials, it is proved that the resistance to bending of a beam with given cross section is proportional to the moment of inertia of its cross section relative to the axis through the center of gravity of the cross section and perpendicular to the direction of the bending force. Let us clarify this by an example. We assume that a bridge across a stream consists of a board that sags under the weight of a pedestrian passing over it. If the same board (no thicker than before) is placed "on its edge," it scarcely bends at all, i.e., a board placed on its edge is, so to speak, stronger. This follows from the fact that the moment of inertia of the cross section of the board (it has the shape of an elongated rectangle that we may think of as evenly covered with mass) is greater relative to the axis perpendicular to its long side than relative to the axis parallel to its long side. If we set the board not exactly flat nor on edge but obliquely, or even if we do not take a board at all but a rod of arbitrary cross section, for example a rail, the resistance to bending will still be proportional to the moment of inertia of its cross section relative to the corresponding axis. The resistance of a beam to bending is therefore characterized by the ellipse of inertia of its cross section.

For an ordinary rectangular beam this ellipse will have the form shown in figure 21. The rigidity of such a beam under a load in the direction of the Oz-axis is proportional to bh^3.

Steel beams often have a ʃ -shaped cross section; for such beams the cross section and the ellipse of inertia are represented in figure 22. The

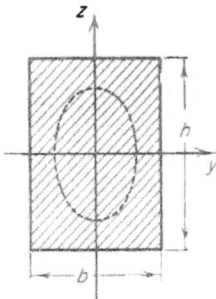

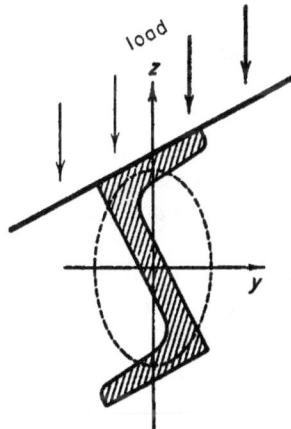

FIG. 21.   FIG. 22.

greatest resistance to bending is in the z direction. When they are used, for example as roof rafters under a load of snow and their own weights, they work directly against bending in a direction close to this most advantageous direction.

The hyperbola and its focal property. Now we consider the equation

$$\frac{x^2}{a^2} - \frac{y^2}{b^2} =$$

representing a curve which is called a *hyperbola*. If we denote by c a number such that $c^2 = a^2 + b^2$, then it is possible to show that a hyperbola is the locus of all points the difference of whose distances to the points F_1 and F_2 on the Ox-axis with abscissas c and $-c$ is a constant: $\rho_2 - \rho_1 = 2a$ (figure 23). The points F_1 and F_2 are called the *foci*.

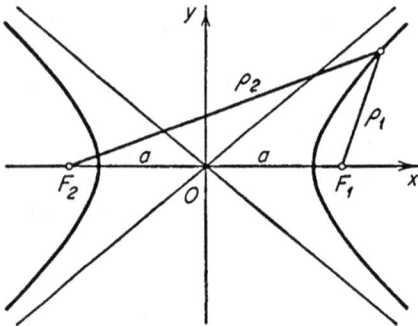

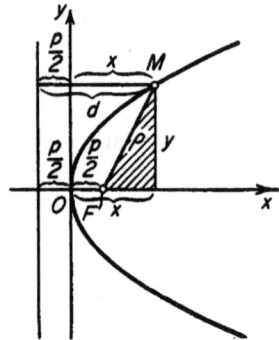

FIG. 23.   FIG. 24.

The parabola and its directrix. Finally, we consider the equation

$$y^2 = 2px$$

and call the corresponding curve a *parabola*. The point F lying on the Ox-axis with abscissa $p/2$ is called the *focus* of the parabola, and the straight line $y = -p/2$, parallel to the Oy-axis, is its *directrix*. Let M be any point of the parabola (figure 24), ρ the length of its focal radius MF, and d the length of the perpendicular dropped from it to the directrix. Let us compute ρ and d for the point M. From the shaded triangle we obtain $\rho^2 = (x - p/2)^2 + y^2$. As long as the point M lies on the parabola, we have $y^2 = 2px$, hence

$$\rho^2 = \left(x - \frac{p}{2}\right)^2 + 2px = \left(x + \frac{p}{2}\right)^2.$$

But directly from the figure it is clear that $d = x + p/2$. Therefore $\rho^2 = d^2$, i.e., $\rho = d$. The inverse argument shows that if for a given point we have $\rho = d$, then the point lies on the parabola. Thus a parabola is the locus of points equidistant from a given point F (called the focus) and a given straight line d (called the directrix).

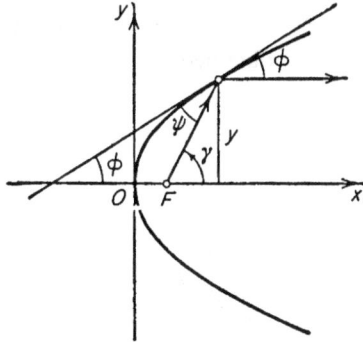

Fig. 25.

The property of the tangent to a parabola. Let us examine an important property of the tangent to a parabola and its application in optics. Since for a parabola $y^2 = 2px$ we have $2y\, dy = 2p\, dx$, it follows that the derivative, or the slope of the tangent, is equal to $dy/dx = \tan\phi = p/y$ (figure 25).

On the other hand, it follows directly from the figure that

$$\tan \gamma = \frac{y}{x - p/2}.$$

But

$$\tan 2\phi = \frac{2\,p/y}{1 - p^2/y^2} = \frac{2py}{y^2 - p^2} = \frac{2py}{2px - p^2} = \frac{y}{x - p/2},$$

i.e., $\gamma = 2\phi$, and since $\gamma = \phi + \psi$, therefore $\psi = \phi$. Consequently, by virtue of the law (angle of incidence is equal to angle of reflection) a beam of light, starting from the focus F and reflected by an element of the parabola (whose direction coincides with the direction of the tangent) is reflected parallel to the Ox-axis, i.e., parallel to the axis of symmetry of the parabola.

On this property of the parabola is based the construction of reflecting telescopes, as invented by Newton. If we manufacture a concave mirror whose surface is a so-called paraboloid of revolution, i.e., a surface obtained by the rotation of a parabola around its axis of symmetry, then all the light rays originating from any point of a heavenly body lying strictly in the direction of the "axis" of the mirror are collected by the mirror (figure 26) at one point, namely its focus. The rays originating from some other point of the heavenly body, being not exactly parallel to the axis of the mirror, are collected almost at one point in the neighborhood of the focus. Thus, in the so-called focal plane through the focus of the mirror and perpendicular to its axis, the inverse image of the star is obtained; the farther away this image is from the focus, the more diffuse it will be, since it is only the rays exactly parallel to the axis of the mirror that are collected by the mirror at one point. The image so obtained can be viewed in a special microscope, the so-called

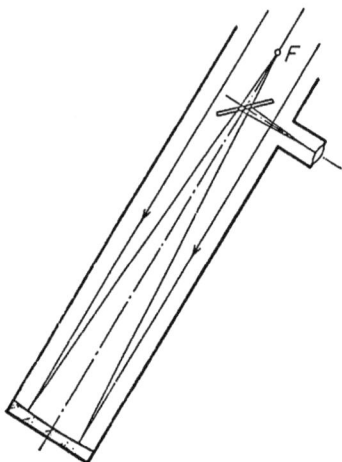

FIG. 26.   FIG. 27.

eye piece of the telescope, either directly or, in order not to cut off the light from the star with one's own head, after reflection in a small plane mirror, attached to the telescope near the focus (somewhat nearer than the focus to the concave mirror) at an angle of 45°.

The searchlight (figure 27) is based on the same property of the parabola. In it, conversely, a strong source of light is placed at the focus of a paraboloidal mirror, so that its rays are reflected from the mirror in a beam parallel to its axis. Automobile headlights are similarly constructed (figure 28).

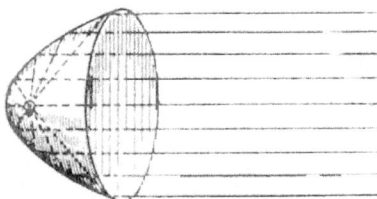
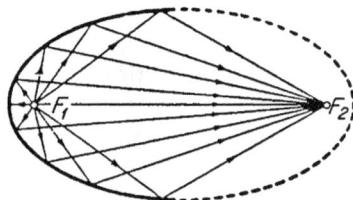

FIG. 28.   FIG. 29.

In the case of an ellipse, as it is easy to show, the rays issuing from one of its foci F_1 and reflected by the ellipse are collected at the other focus F_2 (figure 29), and in the hyperbola the rays originating from one of its foci F_1 are reflected by it as if they originated from the other focus F_2 (figure 30).

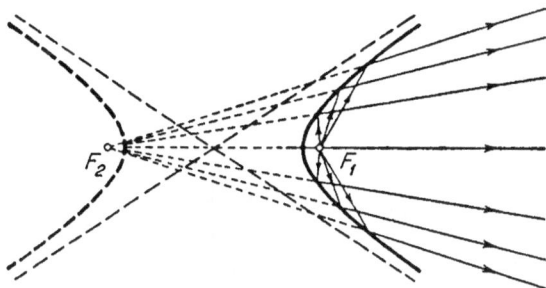

Fig. 30.

The directrices of the ellipse and the hyperbola. Like the parabola, the ellipse and the hyperbola have directrices, in this case two apiece. If we consider a focus and the directrix "on the same side with it," then for all points M of the ellipse we have $p/d = \epsilon$, where the constant ϵ is the eccentricity, which for an ellipse is always smaller than 1; and for all points of the corresponding branch of the hyperbola, we also have $p/d = \epsilon$, where ϵ is again the eccentricity, which for a hyperbola is always greater than 1.

Thus the ellipse, the parabola and one branch of the hyperbola are the loci of all those points in the plane for which the ratio of their distance p from the focus to their distance d from the directrix is constant (figures 31 and 32). For the ellipse this constant is smaller than unity, for the parabola it is equal to unity, and for the hyperbola it is greater than unity. In

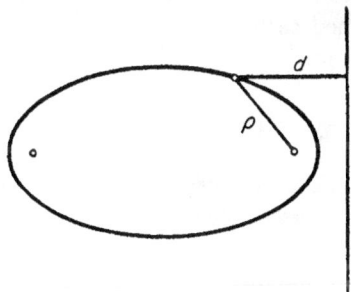

Fig. 31.

Fig. 32.

this sense the parabola is the "limiting" or "transition" case from the ellipse to the hyperbola.

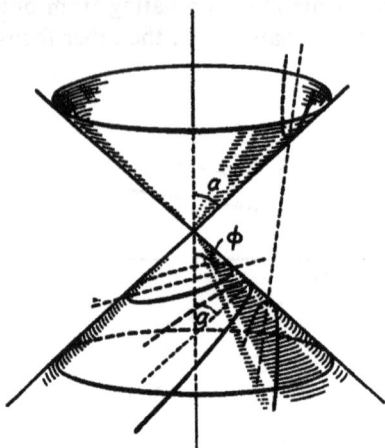

Conic sections. The ancient Greeks had already investigated in detail the curves obtained by intersecting a straight circular cone by a plane. If the intersecting plane makes with the axis of the cone an angle ϕ of 90°, i.e., is perpendicular to it, then the section obtained is a circle. It is easy to show that if the angle ϕ is smaller than 90°, but greater than the angle α which the generators of the cone make with its axis, then an ellipse is obtained. If ϕ is equal to α, a parabola results and if ϕ is smaller than α, then we obtain a hyperbola as the section (figure 33).

The parabola as the graph of quadratic proportion and the hyperbola as the graph of inverse proportion. We recall that the graph of quadratic proportion

$$y = kx^2$$

is a parabola (figure 34) and that the graph of inverse proportion

$$y = \frac{k}{x} \quad \text{or} \quad xy = k$$

is a hyperbola (figure 35), as we will easily prove later. A hyperbola was defined earlier as the curve represented by the equation

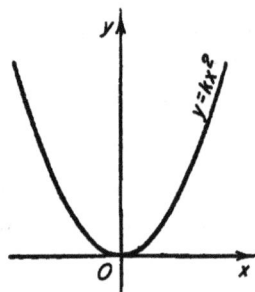

FIG. 34.

$$\frac{x^2}{a^2} - \frac{y^2}{b^2} = 1.$$

In the special case $a = b$ the so-called *rectangular hyperbola* plays the same role among hyperbolas as the circle plays among ellipses. In this case, if we rotate the coordinate axes by 45° (figure 36) the equation in the new coordinates (x', y') will have the form

$$x'y' = k.$$

We have now considered three important second-order curves: the ellipse, the hyperbola, and the parabola, and for their definitions we have taken the so-called canonical equations

$$\frac{x^2}{a^2} + \frac{y^2}{b^2} = 1, \frac{x^2}{a^2} - \frac{y^2}{b^2} = 1 \quad \text{and} \quad y^2 = 2px,$$

by which they are represented.

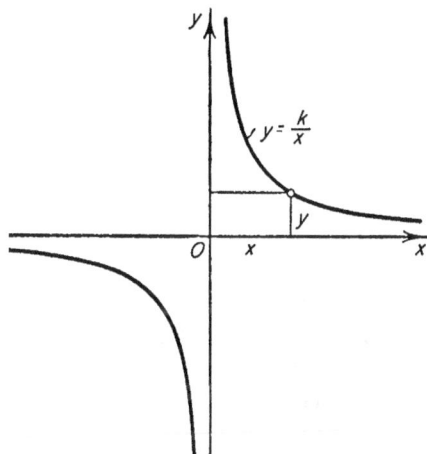

FIG. 35.

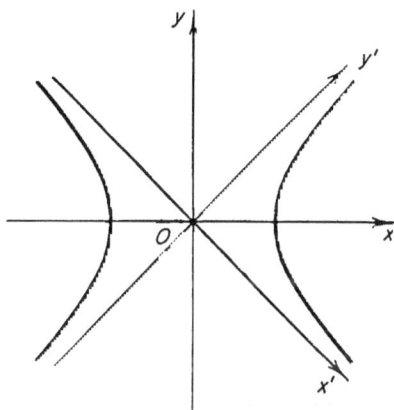

FIG. 36.

We now pass to the study of the general second-degree equation in two unknowns, namely to the question what kinds of curves are represented by this equation.

§8. The Reduction of the General Second-Degree Equation to Canonical Form

The first consistent presentation of analytic geometry by Euler. A significant step in the development of analytic geometry was the appearence in 1748 of the book "Introduction to analysis" in the second volume of which, among other things related to the theory of functions and other branches of analysis, for the first time a presentation was given of analytic geometry in the plane with a detailed investigation of second-order curves, very close to the one given in contemporary textbooks of analytic geometry, and also with an investigation of higher order curves. This was the first exposition of analytic geometry in the contemporary sense of the word.

The notion of reducing an equation to canonical form. A second-degree equation*

$$Ax^2 + 2Bxy + Cy^2 + 2Dx + 2Ey + F = 0$$

contains six terms, not three or only two as in the above canonical equations of the ellipse, hyperbola, and parabola. This is not because such an equation represents a more complicated curve but because the system of coordinates is possibly not suited to it. It turns out that if we select a suitable Cartesian coordinate system, then a second-degree equation with two variables always can be reduced to one of the following canonical forms:

1. $\dfrac{x^2}{a^2} + \dfrac{y^2}{b^2} - 1 = 0.$ ⬭ Ellipse

2. $\dfrac{x^2}{a^2} + \dfrac{y^2}{b^2} + 1 = 0.$ ⬭ Imaginary ellipse

3. $\dfrac{x^2}{a^2} + \dfrac{y^2}{b^2} = 0.$ Point (a pair of imaginary lines intersecting in a real point)

4. $\dfrac{x^2}{a^2} - \dfrac{y^2}{b^2} - 1 = 0.$) (Hyperbola

5. $\dfrac{x^2}{a^2} - \dfrac{y^2}{b^2} = 0.$ ✕ A pair of intersecting lines

6. $y^2 - 2px = 0.$ A parabola

7. $x^2 - a^2 = 0.$ | | A pair of parallel lines

8. $x^2 + a^2 = 0.$ ┊ ┊ A pair of imaginary parallel lines

9. $x^2 = 0.$ ▌ A pair of coincident straight lines

where a, b, p, are not equal to zero.

Equations 1, 4, and 6 of the enumerated canonical forms are already well known to us; these are the canonical equations of the ellipse, hyperbola, and parabola. Two of them are not satisfied by any points, namely equations 2 and 8. Indeed, the square of a real number is always positive or zero, so that on the left-hand side of equation 2 the sum of the terms $x^2/a^2 + y^2/b^2$ is never negative, and since the term $+1$ also appears, the

---

* The coefficients of xy, x, y will be denoted not by B, D, E but by $2B$, $2D$, $2E$ for simplicity of subsequent formulas.

left-hand side cannot be equal to zero; analogously in equation 8, the number x^2 is not negative, and a^2 is positive. From these considerations, it follows that only ($x = 0$, $y = 0$) satisfies equation 3, i.e., one point, the origin. Equation 5 can be written as $(x/a - y/b)(x/a + y/b) = 0$, from which we see that it is satisfied by those points and only those points for which one of the first-degree expressions $x/a - y/b$ or $x/a + y/b$ is equal to zero; so the curve it represents is this pair of intersecting lines. Equation 7 analogously gives $(x - a)(x + a) = 0$; i.e., the corresponding curve is a pair of parallel lines $x = a$ and $x = -a$. Finally, curve 9 is a special limiting case of curve 7, when $a = 0$; i.e., it is a pair of coincident lines.

Formulas of coordinate transformations. In order to obtain the indicated important result about the possible types of second-order curves, it is necessary first to deduce the formulas by which the rectangular coordinates of points vary under a change of the coordinate system.

Let x, y be the coordinates of a point M relative to the axes Oxy. Let us translate these axes parallel to themselves to the position $O'x'y'$ and let the coordinates of the new origin O' relative to the old axes

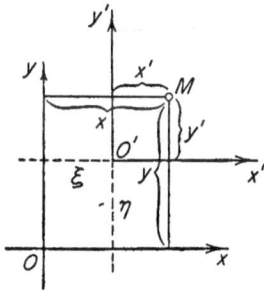

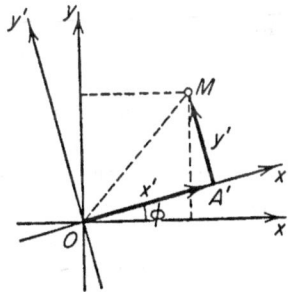

FIG. 37.   FIG. 38.

be ξ and η. It is evident (figure 37) that the new coordinates x', y' of the point M are connected with its old coordinates x, y by the formulas

$$x = x' + \xi,$$

$$y = y' + \eta,$$

which are the formulas of the so-called parallel translation of axes. If we rotate the original axes Oxy about the origin counterclockwise by an angle ϕ then, as is easy to see (figure 38), if we project the polygonal line $OA'M$ composed of the new coordinate segments x', y' on the Ox-axis and the Oy-axis, respectively, we obtain

$$x = x' \cos\phi - y' \sin\phi,$$
$$y = x' \sin\phi + y' \cos\phi,$$

which are the formulas for transformation of coordinates under rotation of a rectangular coordinate system.

If we are given an equation $F(x, y) = 0$ of a curve relative to the axes Oxy and we wish to write the transformed equation of the same curve, i.e., relative to the new axes $O'x'y'$, then we must replace x and y in the equation $F(x, y) = 0$ by their expressions in terms of x' and y', given by the formulas of the transformation. For example, under parallel translation of the axes, we obtain the transformed equation

$$F(x' + \xi, y' + \eta) = 0,$$

and under rotation of the axes the equation

$$F(x' \cos\phi - y' \sin\phi, x' \sin\phi + y' \cos\phi) = 0.$$

We note that under a transformation to new axes the degree of an equation does not change. Indeed, the degree cannot increase, since the transformation formulas are of the first-degree. But the degree cannot decrease either, since then the inverse coordinate transformation would increase it (and it is also of the first degree).

The reduction of a general second-degree equation to one of the 9 canonical forms. We now show that given any second-degree equation in two unknowns we can always first rotate the axes and then translate them parallel to themselves in such a way that the transformed equation for the final axes will have one of the forms 1, 2, ···, 9.

Indeed, let the given second-degree equation have the form

$$Ax^2 + 2Bxy + Cy^2 + 2Dx + 2Ey + F = 0. \qquad (8)$$

Let us rotate the axes through some angle ϕ, which we select in the following way. Replacing x and y in equation (8) by their expressions in terms of the new coordinates (according to the formulas for rotation), we find, after collecting similar terms, that the coefficient $2B'$ in the transformed equation

$$A'x'^2 + 2B'x'y' + C'y'^2 + 2D'x' + 2E'y' + F' = 0$$

is equal to

$$2B' = -2A \sin\phi \cos\phi + 2B(\cos^2\phi - \sin^2\phi) + 2C \sin\phi \cos\phi$$
$$= 2B \cos 2\phi - (A - C) \sin 2\phi.$$

Setting it equal to zero, we obtain $2B \cos 2\phi = (A - C) \sin 2\phi$, from which

$$\cot 2\phi = \frac{A - C}{2B}.$$

Since the cotangent varies from $-\infty$ to $+\infty$, we can always find an angle ϕ for which this equality is satisfied. By rotating the axes through this angle, we find that for the rotated axes $Ox'y'$ the equation of our curve, represented for the initial axes by equation (8), has the form

$$A'x'^2 + C'y'^2 + 2D'x' + 2E'y' + F = 0, \tag{9}$$

i.e., that it does not contain the term with the product of the coordinates (F remains unchanged, since the formulas of rotation do not contain constant terms).

Now we translate the already rotated axes $Ox'y'$ parallel to themselves to the position $O''x''y''$, and let the coordinates of the new origin O'' relative to the axes $Ox'y'$ be ξ', η'. The equation of our curve for these final axes will be

$$A'(x'' + \xi')^2 + C'(y'' + \eta')^2 + 2D'(x'' + \xi') + 2E'(y'' + \eta') + F = 0. \tag{10}$$

We now show that we can always select ξ' and η' (i.e., we can translate the axes $Ox'y'$ parallel to themselves) in such a way that the final equation for the axes $O''x''y''$ has one of the canonical forms $1, 2, \cdots, 9$.

Removing all parentheses in equation (10) and collecting similar terms, we obtain

$$A'x''^2 + C'y''^2 + 2(A'\xi' + D')x'' + 2(C'\eta' + E')y'' + F' = 0, \tag{10'}$$

where we have denoted by F' the sum of all constant terms; its value does not interest us at the moment.

We consider three possible cases.

I. A' and C' both not equal to zero. In this case, taking $\xi' = -D'/A'$, $\eta' = -E'/C'$, we eliminate the terms with the first powers of x'' and y'' and obtain an equation of the form

$$A'x''^2 + C'y''^2 + F' = 0. \tag{I}$$

II. $A' \neq 0, C' = 0$, but $E' \neq 0$. Letting $\xi' = -D'/A', \eta' = 0$, i.e., $y'' = y'$, we obtain the equation

$$A'x''^2 + 2E'y' + F' = 0,$$

or

$$A'x''^2 + 2E'\left(y' + \frac{F'}{2E'}\right) = 0.$$

Then making a parallel translation along the Oy'-axis by an amount $\eta'' = - F'/2E'$, we find that $y' = y'' - F'/2E'$, i.e., $y' + F'/2E' = y''$ so that we obtain the equation

$$A'x''^2 + 2E'y'' = 0. \tag{II}$$

If we have $A' = 0$, $C' \neq 0$, $D' \neq 0$, we can simply interchange the roles of x and y and obtain the same result.

III. $A' \neq 0$, $C' = 0$, $E' = 0$. Taking again $\xi' = - D'/A'$, $\eta' = 0$, we obtain the equation

$$A'x''^2 + F' = 0. \tag{III}$$

If we have $A' = 0$, $C' \neq 0$, $D' = 0$, we can again interchange the roles of x and y.

We have now considered all the possibilities, in view of the fact that A' and C' cannot simultaneously be zero, since then the degree of the equation would be reduced, and we have seen that under our coordinate transformations this degree does not change.

Thus, with the appropriate choice of rectangular coordinates every second-degree equation can be brought to one of the three so-called "reduced" equations (I), (II), (III).

Let the equation have the form (I) (in this case A' and C' are not equal to zero). If $F' \neq 0$, then writing equation (I) as

$$\frac{x''^2}{-F'/A'} + \frac{y''^2}{-F'/C'} - 1 = 0,$$

we arrive, depending on the signs of A', C', F', at one of the equations 1, 2, or 4. If the denominator of x''^2 is negative and that of y''^2 is positive, then we must also interchange the axes $O''x''$ and $O''y''$.

If $F' = 0$, then equation (I) can be written in the form

$$\frac{x''^2}{1/A'} + \frac{y''^2}{1/C'} = 0,$$

and we arrive at equations 3 or 5.

If the equation has the form (II) (in this case A' and E' are not both zero), then we can write it as

$$x''^2 + \frac{2E'}{A'} y'' = 0,$$

and denoting $-E'/A'$ by p and interchanging the names of the axes $O''x''$ and $O''y''$ we obtain equation 6.

Finally, if we have an equation of form (III) (where $A' \neq 0$), it can be rewritten as $x''^2 + F'/A' = 0$ and one of the equations 7, 8, or 9 is obtained.

This important theorem on the possibility of reducing every 2nd-degree equation to one of the 9 canonical forms was already examined in detail by Euler. The arguments in Euler's book differ only in form from the ones just given.

§9. The Representation of Forces, Velocities, and Accelerations by Triples of Numbers; Theory of Vectors

Following Euler an important step was taken by Lagrange. In his "Analytic mechanics," published in 1788, Lagrange arithmetized forces, velocities and accelerations in the same way as Descartes arithmetized points. This idea that Lagrange developed in his book subsequently took the form of the so-called theory of vectors and proved to be an important help in physics, mechanics, and technology.

Rectangular coordinates in space. We remark, first of all, that neither Descartes nor Newton developed analytic geometry in space. This was done later on, in the first half of the 18th century, by Laguerre and Clairaut. In order to specify a point M in space they selected three mutually perpendicular axes Ox, Oy, and Oz and considered (figure 39) the numerical values of the distances of the point M from the planes Oyz, Oxz, and Oxy, taken with the corresponding signs, the so-called *abscissa x*, *ordinate y*, and *altitude z* of the point M.

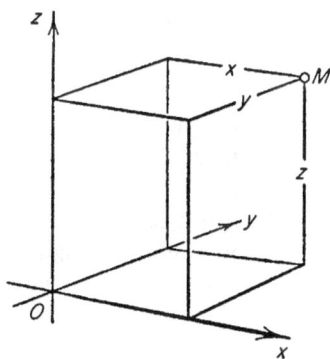
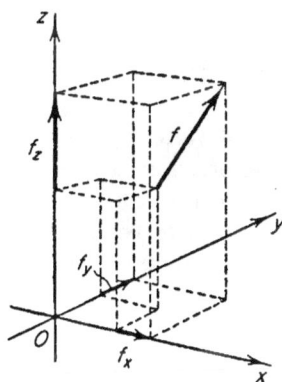

Fig. 39. Fig. 40.

Arithmetization of forces, velocities, and accelerations, introduced by Lagrange. We consider (figure 40) a force f which can be represented in conventional units by a segment with an arrow, having a specific length and direction. Lagrange points out that this force f can be decomposed into three components f_x, f_y, and f_z in the direction of the corresponding axes Ox, Oy, and Oz; these components, as directed segments on the axes, can be given simply by numbers, positive or negative depending on whether the component is directed in the positive or the opposite direction of the axis. Thus, we can consider, for example, the force (2, 3, 4) or the force (1, —2, 5), etc. In the composition of forces according to the parallelogram law, as can easily be shown (it will be shown later), their corresponding components have to be added. For example, the sum of the given forces is the force

$$(2 + 1, 3 - 2, 4 + 5) = (3, 1, 9).$$

The same can be done for velocities and accelerations. In every problem of mechanics, all the equations connecting forces, velocities, and accelerations can also be written as equations connecting their components, i.e., connecting simply numbers; then the mechanical equation will necessarily be written in the form of three equations: first for the x's, the second for the y's, and the third for the z's.

But it was only after a hundred years from the time of Lagrange that mathematicians and physicists, particularly under the influence of the developing theory of electricity, began on a wide scale to consider the general theory of such segments, having a definite length and direction. Such segments were called vectors.

The theory of vectors has a great significance in mechanics, physics, and technology, and its algebraic side, the so-called algebra of vectors (in contrast to vector analysis) appears at once as an essential constituent part of analytic geometry.

Algebra of vectors. Any directed segment (whether it represents a force, a velocity, an acceleration, or some other entity) i.e., a segment having a given length and a definite direction, is called a *vector*. Two vectors are said to be *equal*, if they have the same length and the same direction; i.e., in the very concept of "vector" only its length and its direction are taken into account. Vectors can be added. Let the vectors **a, b, ⋯ , d** be given. We lay out the vector **a** from some point, then from its end point we draw the vector **b**, etc. We obtain a so-called vector polygon **ab ⋯ d** (figure 41). The vector **m** whose initial point coincides with the initial point of the first vector **a** of this polygon, and whose end point coincides with the end point of the last vector **d**, is called the *sum* of these vectors

$$\mathbf{m} = \mathbf{a} + \mathbf{b} + \cdots + \mathbf{d}. \tag{11}$$

It is easy to show that the vector **m** does not depend on the order in which the summands **a, b,** $\cdots$, **d** are taken.

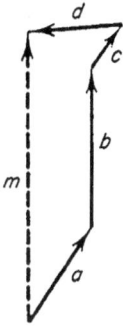

FIG. 41.

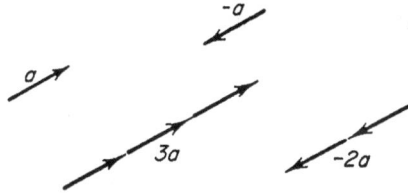

FIG. 42.

The vector equal in length to the vector **a** but opposite in direction is called its *inverse* vector and is denoted by —**a**.

Subtraction of the vector **a** is defined as addition of its inverse vector.

In vector calculus ordinary real numbers are customarily called scalars. Let a vector **a** (figure 42) and a scalar λ be given, then by the product of the vector **a** with the scalar (number) λ, i.e., λ**a**, is meant the vector whose length is equal to the product of the length |**a**| of the vector **a**

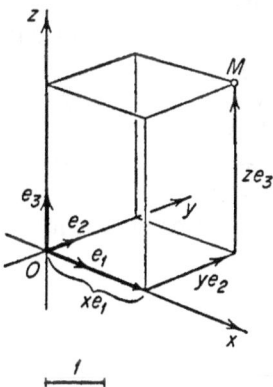

FIG. 43.

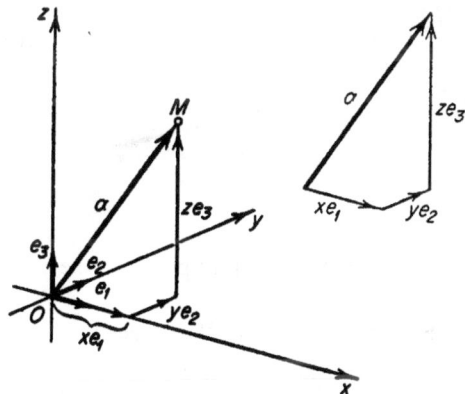

FIG. 44.

and the absolute value $|\lambda|$ of the number λ, and whose direction is the same as that of **a** if $\lambda > 0$ and the opposite if $\lambda < 0$.

Let us consider a system of rectangular Cartesian coordinates $Oxyz$ and the vectors $\mathbf{e}_1$, $\mathbf{e}_2$, $\mathbf{e}_3$ having length equal to unity and directions coinciding with the positive directions of the axes Ox, Oy, Oz, respectively. It is obvious that any given point M (figure 43) of space can be reached from the origin O by traversing a certain number of "times" (an integral, fractional or irrational, positive or negative "number of times") the vector $\mathbf{e}_1$, then so many "times" the vector $\mathbf{e}_2$, and finally so many "times" the vector $\mathbf{e}_3$. It is clear that the numbers x, y, z showing how many "times" it is necessary to traverse the vectors $\mathbf{e}_1$, $\mathbf{e}_2$, $\mathbf{e}_3$, are simply the Cartesian coordinates of the point M.

Let a vector **a** be given; if we cause a point to move from the initial point of **a** to its end point and decompose this motion into motions parallel to the axes Ox, Oy, and Oz, and if it is hereby necessary to shift the point through a distance $x\mathbf{e}_1$ parallel to the Ox-axis, through $y\mathbf{e}_2$ parallel to the Oy-axis and through $z\mathbf{e}_3$ parallel to the Oz-axis, then

$$\mathbf{a} = x\mathbf{e}_1 + y\mathbf{e}_2 + z\mathbf{e}_3 . \tag{12}$$

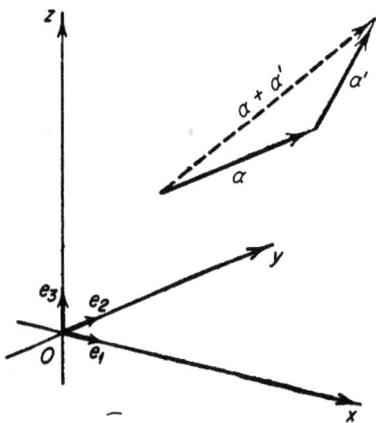

The numbers x, y, z are called the *coordinates* of the vector **a**. These are obviously just the coordinates of the end point M of this vector, if its initial point lies at the origin O of the coordinate system (figure 44). From this it follows at once that in adding vectors their corresponding coordinates are to be added, and in subtraction they are to be subtracted. If the first vector "carries" us along the Ox-axis by a distance $x\mathbf{e}_1$, and the second by $x'\mathbf{e}_1$, then clearly their sum "carries" us along the Ox-axis by a distance $(x + x')\mathbf{e}_1$, etc. (figure 45). It also follows at once that in multiplication of a vector by a number, its coordinates are multiplied by the number.

FIG. 45.

Scalar product and its properties. If we are given two vectors **a** and **b**, then the number equal to the product of their lengths by the cosine of the angle between them $|\mathbf{a}||\mathbf{b}|\cos\phi$ is called their *scalar product* and is denoted by **ab** or (**ab**). Let x, y, z be the coordinates of the vector **a**

and $\bar{x}$, $\bar{y}$, $\bar{z}$ the coordinates of the vector **b**; then the scalar product is equal to

$$\mathbf{ab} = x\bar{x} + y\bar{y} + z\bar{z}, \tag{13}$$

i.e., to the sum of the products of their corresponding coordinates.

This important result can be proved as follows. First we make the following remarks:

(1) If we multiply one of the vectors of a scalar product, for example **a**, by a number λ, this is obviously the same as multiplying their scalar product by the same number, i.e.,

$$(\lambda\mathbf{a})\mathbf{b} = \lambda(\mathbf{ab}).$$

(2) The scalar product is distributive, i.e., if $\mathbf{a} = \mathbf{a}_1 + \mathbf{a}_2$, then $\mathbf{ab} = \mathbf{a}_1\mathbf{b} + \mathbf{a}_2\mathbf{b}$.

In fact, the left-hand side of this equality is equal to the product of the length of the vector **b** by the numerical value of the projection of the vector **a** on the axis of the vector **b** (figure 46), and the right-hand side is equal to the product of the length of **b** by the sum of the numerical values

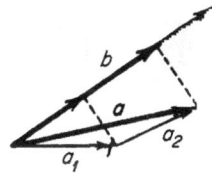

FIG. 46.

of the projections of the vectors $\mathbf{a}_1$ and $\mathbf{a}_2$ on the axis of **b**. But proj $\mathbf{a}$ = proj $\mathbf{a}_1$ + proj $\mathbf{a}_2$, which proves the equality.

Now we consider two vectors **a** and **b** whose decompositions in terms of the vectors $\mathbf{e}_1$, $\mathbf{e}_2$, $\mathbf{e}_3$ are $\mathbf{a} = x\mathbf{e}_1 + y\mathbf{e}_2 + z\mathbf{e}_3$, $\mathbf{b} = \bar{x}\mathbf{e}_1 + \bar{y}\mathbf{e}_2 + \bar{z}\mathbf{e}_3$, so that

$$\mathbf{ab} = (x\mathbf{e}_1 + y\mathbf{e}_2 + z\mathbf{e}_3)(\bar{x}\mathbf{e}_1 + \bar{y}\mathbf{e}_2 + \bar{z}\mathbf{e}_3).$$

By the distributivity (2) of the scalar product, the sums of vectors in parentheses can be multiplied as polynomials, and by (1) the scalar factors in each of the terms can be taken outside the parentheses, so that

$$\mathbf{ab} = x\bar{x}\mathbf{e}_1\mathbf{e}_1 + x\bar{y}\mathbf{e}_1\mathbf{e}_2 + x\bar{z}\mathbf{e}_1\mathbf{e}_3 + y\bar{x}\mathbf{e}_2\mathbf{e}_1 + y\bar{y}\mathbf{e}_2\mathbf{e}_2 + y\bar{z}\mathbf{e}_2\mathbf{e}_3$$
$$+ z\bar{x}\mathbf{e}_3\mathbf{e}_1 + z\bar{y}\mathbf{e}_3\mathbf{e}_2 + z\bar{z}\mathbf{e}_3\mathbf{e}_3 .$$

But

$$|\mathbf{e}_1| = |\mathbf{e}_2| = |\mathbf{e}_3| = 1, \quad \cos 0° = 1 \text{ and } \cos 90° = 0.$$

Consequently,

$$\mathbf{e}_1\mathbf{e}_1 = 1, \quad \mathbf{e}_1\mathbf{e}_2 = 0, \quad \mathbf{e}_1\mathbf{e}_3 = 0,$$
$$\mathbf{e}_2\mathbf{e}_1 = 0, \quad \mathbf{e}_2\mathbf{e}_2 = 1, \quad \mathbf{e}_2\mathbf{e}_3 = 0,$$
$$\mathbf{e}_3\mathbf{e}_1 = 0, \quad \mathbf{e}_3\mathbf{e}_2 = 0, \quad \mathbf{e}_3\mathbf{e}_3 = 1.$$

Thus,

$$\mathbf{ab} = x\bar{x} + y\bar{y} + z\bar{z}. \tag{14}$$

We remark, in particular, that if the vectors **a** and **b** are mutually perpendicular, then $\phi = 90°$ and $\cos \phi = 0$. Therefore the equality

$$x\bar{x} + y\bar{y} + z\bar{z} = 0 \qquad (15)$$

serves as an easily verifiable condition of perpendicularity of the vectors **a** and **b**.

Angle between two directions. Let us consider a direction characterized by its angles α, β, γ with the coordinate axes. We draw the line

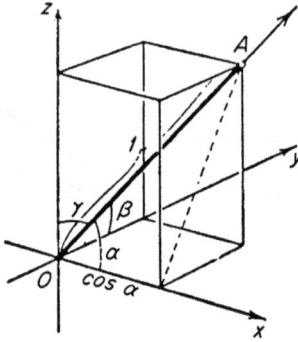
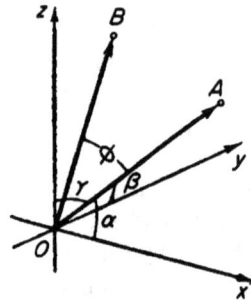

FIG. 47.   FIG. 48.

in this direction through the origin of the coordinate system and mark off on it from the origin a segment OA of unit length (figure 47). In this case the coordinates of the point A, i.e., the coordinates of the vector $\overrightarrow{OA}$ are exactly $\cos \alpha$, $\cos \beta$, and $\cos \gamma$. If we have a second direction given by the angles $\bar{\alpha}$, $\bar{\beta}$, $\bar{\gamma}$, then the analogous vector $\overrightarrow{OB}$ for this second direction has coordinates $\cos \bar{\alpha}$, $\cos \bar{\beta}$, $\cos \bar{\gamma}$ (figure 48). Let ϕ be the angle between these vectors; then their scalar product is equal to $1 \cdot 1 \cos \phi$ from which we find

$$\cos \phi = \cos \alpha \cos \bar{\alpha} + \cos \beta \cos \bar{\beta} + \cos \gamma \cos \bar{\gamma}. \qquad (16)$$

This is the very important formula for the cosine of the angle between two directions.

§10. Analytic Geometry in Space; Equations of a Surface in Space and Equations of a Curve

If an equation $z = f(x, y)$ is given and if x and y are regarded as the abscissa and ordinate and z the altitude of a point, then this equation itself represents some surface P, which can be obtained by erecting perpendiculars of length z at the points (x, y) of the Oxy-plane. The locus of the end points of these perpendiculars gives the surface P represented by this equation. If the equation connecting x, y, and z is not already solved with respect to z, then it can be solved for z and after that we can construct the surface P. In general, in analytic geometry the totality of all those points of space whose coordinates x, y, z satisfy a given equation (figure 49) in three variables x, y, z is said to be the surface represented by the equation.

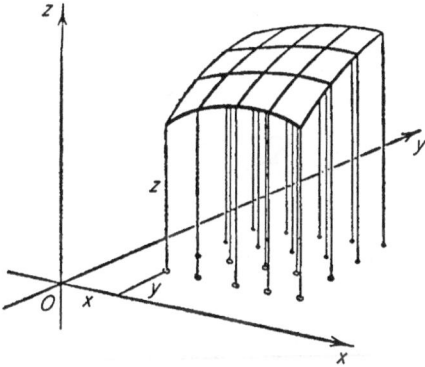
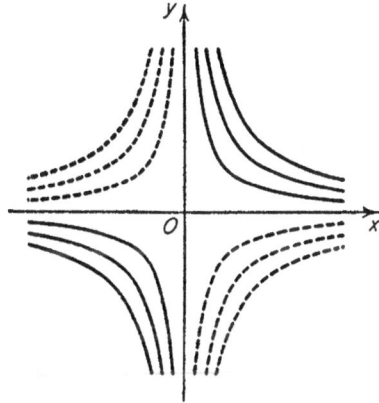

FIG. 49.   FIG. 50.

A function of two variables $f(x, y)$, as was pointed out already in Chapter II, can represent not only a surface P, but also its system of level curves, i.e., curves in the Oxy-plane on each of which the function $f(x, y)$ has a constant value. This system of curves is clearly nothing else than the topographical map of the surface P on the Oxy-plane.

Example. The equation $xy = z$ gives, for instance, the level curves: $\cdots$, $xy = -3$, $xy = -2$, $xy = -1$, $xy = 0$, $xy = 1$; $xy = 2$, $xy = 3$, $\cdots$. All of them are hyperbolas (figure 50) except $xy = 0$, which represents the two coordinate axes. What is obtained is clearly a saddlelike surface (figure 51) (the so-called hyperbolic paraboloid).

In order to define a curve in space, we can give the equations of any two surfaces P and Q which intersect along the curve. For example, the system

$$xy = z,$$
$$x^2 + y^2 = 1$$

gives a space curve (figure 52). The equation $xy = z$ determines the earlier hyperbolic paraboloid, and the equation $x^2 + y^2 = 1$ determines a circular cylinder of unit radius, whose axis is the Oz-axis. The system of equations consequently defines the curve of intersection of the paraboloid with the cylinder, which is represented in figure 52.

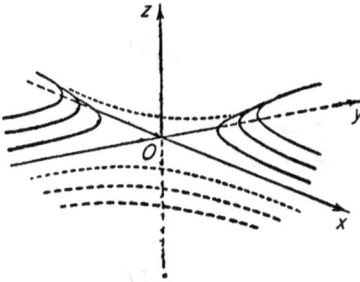

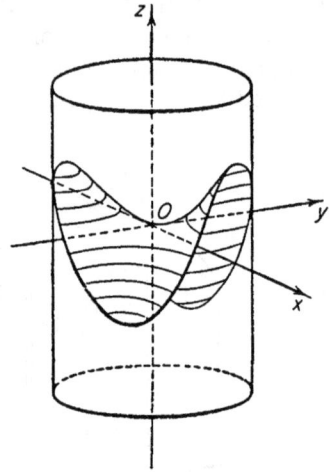

FIG. 51.   FIG. 52.

If in this system one of the unknowns, say x, is chosen arbitrarily, and then the system is solved with respect to y and z, we will obtain the coordinates x, y, z of the various points of the curve.

Equation of a plane and equations of a straight line. It can be shown that every equation of the first degree with three variables

$$Ax + By + Cz + D = 0$$

represents a plane, and conversely. By what has already been said, it is clear that a line can be given by a system of two such equations:

$$A_1x + B_1y + C_1z + D_1 = 0,$$
$$A_2x + B_2y + C_2z + D_2 = 0,$$

i.e., as the curve of intersection of two planes.

The general second-degree equation in three variables and its 17 canonical forms. A second-degree equation in three variables

$$A_1x^2 + A_2y^2 + A_3z^2 + 2B_1yz + 2B_2xz + 2B_3xy$$
$$+ 2C_1x + 2C_2y + 2C_3z + D = 0, \quad (17)$$

contains 10 terms. Analogously to what was done earlier for an equation with two variables, it can be shown that by a suitable rotation of the given coordinate system about the origin, equation (17) can be reduced to the form

$$A_1'x'^2 + A_2'y'^2 + A_3'z'^2 + 2C_1'x' + 2C_2'y' + 2C_3'z' + D = 0, \qquad (18)$$

i.e., so as to eliminate the terms with products of the variables. However, the proof here of the possibility of such a simplification of the equation is considerably more difficult than in the case of the plane. The difficulty of the proof arises from the fact that in the plane a rotation about a point is given by one angle ϕ, which we selected suitably, while in space the rotation of a body about a fixed point is given by three independent angles (Euler angles) ϕ, θ, ψ and in a quite complicated way. So the equation must be cleared of the terms with products of variables in a roundabout way (see Chapter XVI on the theory of reduction by orthogonal transformations of a quadratic form to a sum of squares). Then, as in the case of the plane, a parallel translation of the axes is made and the equation is simplified, after which equation (18) finally assumes one of the following canonical forms:

1. $\dfrac{x^2}{a^2} + \dfrac{y^2}{b^2} + \dfrac{z^2}{c^2} - 1 = 0$ Ellipsoid

2. $\dfrac{x^2}{a^2} + \dfrac{y^2}{b^2} + \dfrac{z^2}{c^2} + 1 = 0$ Imaginary ellipsoid

3. $\dfrac{x^2}{a^2} + \dfrac{y^2}{b^2} - \dfrac{z^2}{c^2} - 1 = 0$ Hyperboloid of one sheet

4. $\dfrac{x^2}{a^2} + \dfrac{y^2}{b^2} - \dfrac{z^2}{c^2} + 1 = 0$ Hyperboloid of two sheets

5. $\dfrac{x^2}{a^2} + \dfrac{y^2}{b^2} - \dfrac{z^2}{c^2} = 0$ Second-order cone

6. $\dfrac{x^2}{a^2} + \dfrac{y^2}{b^2} + \dfrac{z^2}{c^2} = 0$ Imaginary second-order cone

7. $\dfrac{x^2}{a^2} + \dfrac{y^2}{b^2} - 2cz = 0$ Elliptic paraboloid

X

8. $\dfrac{x^2}{a^2} - \dfrac{y^2}{b^2} - 2cz = 0$ Hyperbolic paraboloid

9. $\dfrac{x^2}{a^2} + \dfrac{y^2}{b^2} - 1 = 0$ Elliptic cylinder

10. $\dfrac{x^2}{a^2} + \dfrac{y^2}{b^2} + 1 = 0$ Imaginary elliptic cylinder

11. $\dfrac{x^2}{a^2} + \dfrac{y^2}{b^2} = 0$ A pair of intersecting
     imaginary planes

12. $\dfrac{x^2}{a^2} - \dfrac{y^2}{b^2} - 1 = 0$ Hyperbolic cylinder

13. $\dfrac{x^2}{a^2} - \dfrac{y^2}{b^2} = 0$ A pair of intersecting planes

14. $y^2 - 2px = 0$ Parabolic cylinder

15. $x^2 - a^2 = 0$ A pair of parallel planes

16. $x^2 + a^2 = 0$ A pair of imaginary parallel
     planes

17. $x^2 = 0$ A pair of coincident planes.

The last nine canonical equations 9-17 do not contain terms in z and represent exactly the canonical equations of second-order curves in the Oxy-plane. In space these equations represent cylinders, whose directrices are the corresponding second-order curves in the Oxy-plane and whose generators are parallel to the Oz-axis. Indeed, if one of these equations is satisfied by a point with coordinates $(x_1, y_1, 0)$, then it will also be satisfied by any point with coordinates (x_1, y_1, z) whatever z may be, since there are in any case no terms with z in the equation.

Among the equations 1-8 as can easily be seen, equation 2 is not satisfied by any point with real x, y, z and equation 6 is satisfied only by one such point $(0, 0, 0)$, i.e., the origin. It remains, therefore, to study only the six equations 1, 3, 4, 5, 7, 8.

Ellipsoid. Let us compare the surfaces represented by the equations $x^2/a^2 + y^2/b^2 + z^2/c^2 - 1 = 0$ and $x^2 + y^2 + z^2 - 1 = 0$. The second of these is obviously the equation of a sphere C with center at the origin

and with unit radius, since $x^2 + y^2 + z^2$ is the square of the distance
from the point (x, y, z) to the
origin O. If (x, y, z) is a point
lying on the sphere, i.e., satisfying the second equation, then
(ax, by, cz) is a point whose
coordinates satisfy the first
equation. The surface represented by the first equation is
thus obtained from the sphere
C if all abscissas x of points of
the sphere are replaced by ax,
y by by, and z by cz, i.e., if the

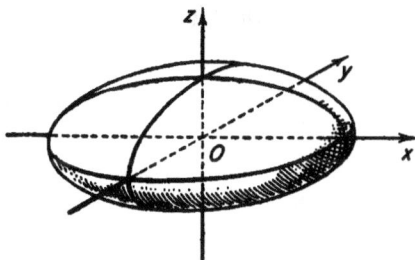

FIG. 53.

sphere C is uniformly stretched from the Oyz-, Oxz-, and Oxy-planes with
coefficients of expansion a, b and c, respectively. This surface is called
an ellipsoid (figure 53).

Hyperboloids and the second-order cone. Let us consider equations
3, 4, and 5, i.e., the equation of the form

$$\frac{x^2}{a^2} + \frac{y^2}{b^2} - \frac{z^2}{c^2} = \delta, \tag{19}$$

where $\delta = 1, -1$ or 0. Let us compare it with the equation

$$\frac{x^2}{a^2} + \frac{y^2}{a^2} - \frac{z^2}{c^2} = \delta, \tag{20}$$

in which the denominator of y^2 is also a^2 and not b^2, as in equation (19).
As before, we observe that surface (19) is obtained from surface (20) by
expansion from the Oxz-plane with coefficient b/a.

Let us now see what surface is represented by (20). We take a plane
$z = h$ perpendicular to the Oz-axis and examine its intersection with the
surface (20). Substituting $z = h$ in equation (20), we obtain

$$x^2 + y^2 = a^2 \left(\delta + \frac{h^2}{c^2} \right).$$

If $\delta + h^2/c^2$ is positive, then this equation together with $z = h$ gives
a circle, lying in the plane $z = h$ with center on the Oz-axis. If $\delta + h^2/c^2$
is negative, which can be the case only with $\delta = -1$ and h^2 small, then
the plane $z = h$ does not intersect surface (20) at all, since the sum of
squares $x^2 + y^2$ cannot be a negative number.

The whole surface (20) thus consists of circles lying in planes perpendicular to the Oz-axis and having their centers on the Oz-axis. But in this case the surface (20) is a surface of revolution about the Oz-axis.

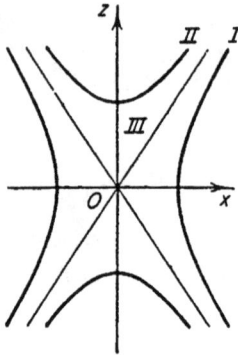

FIG. 54.

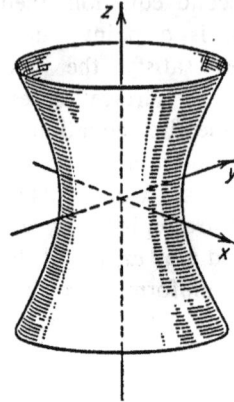

FIG. 55.

If we intersect it with a plane passing through the Oz-axis, we obtain its "meridian," i.e., a curve, lying in a plane passing through the axis, by the revolution of which the surface is generated.

If we intersect the surface (20) with the coordinate plane Oxz, i.e., the plane $y = 0$ (figure 54), by substituting $y = 0$ in equation (20), we obtain the equation of the meridian $x^2/a^2 - z^2/c^2 = \delta$. In case $\delta = 1$

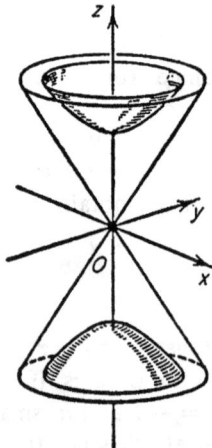

FIG. 56.

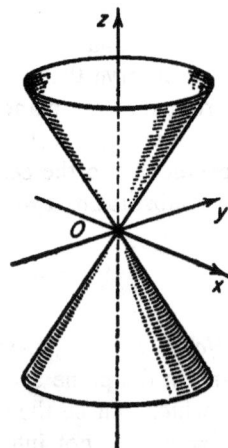

FIG. 57.

this is the hyperbola *I*, for $\delta = -1$, it is the hyperbola *II*, and for $\delta = 0$, the pair of intersecting lines *III*. By revolution around the *z*-axis these produce, respectively, a so-called hyperboloid of revolution of one sheet (figure 55), a hyperboloid of revolution of two sheets (figure 56) and a straight circular cone (figure 57).

The general hyperboloid of one sheet, hyperboloid of two sheets, and second-order cone 3, 4, and 5 are obtained from these surfaces of revolution by an expansion from the *Oxz*-plane with coefficient b/a.

Paraboloids. Only equations 7 and 8 remain. Let us compare the first of these $x^2/a^2 + y^2/b^2 = 2cz$ with the equation

$$\frac{x^2}{a^2} + \frac{y^2}{a^2} = 2cz,$$

which we investigate in the same way as before. It represents a surface obtained by revolving the parabola $x^2 = 2a^2cz$ about the *Oz*-axis,

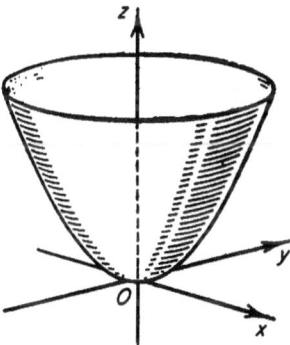

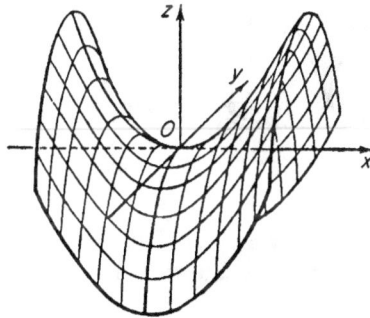

FIG. 58.   FIG. 59.

namely the so-called paraboloid of revolution (figure 58) discussed earlier in connection with parabolic mirrors. The general elliptic paraboloid 7 is obtained from the paraboloid of revolution by an expansion from the *Oxz*-plane.

The surface 8 has to be studied in a different way, namely by examining its intersections with planes $z = h$, which are hyperbolas. The contour map of the surface 8 is represented in figure 50; in a different position of the coordinate axes we considered this surface in figure 51. It is saddle-shaped, as illustrated in figure 59 and is called a hyperbolic paraboloid. Its intersections with planes parallel to the *Oxz*-plane turn out to be identical parabolas. The same result is obtained by intersections with planes parallel to the *Oyz*-plane.

Rectilinear generators of a hyperboloid of one sheet. It is a very curious and not at all obvious fact that the hyperboloid of one sheet and the hyperbolic paraboloid can be obtained, just like the cone and the cylinder, by the motion of a straight line. In case of the hyperboloid, it is sufficient to prove this fact for a hyperboloid of revolution of one sheet $x^2/a^2 + y^2/b^2 - z^2/c^2 = 1$, since the general hyperboloid of one sheet is obtained by a uniform expansion from the Oxz-plane and under such an expansion any straight line will go into a straight line. Let us

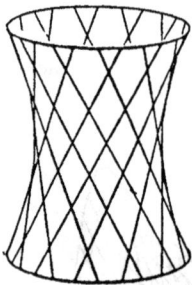

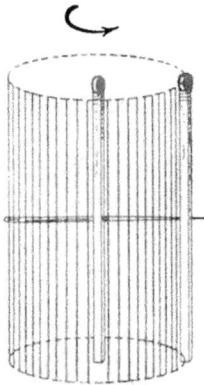

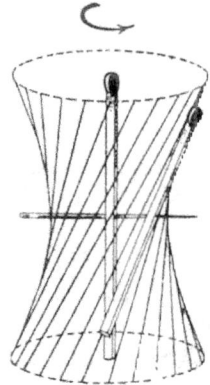

Fig. 60.   Fig. 61.   Fig. 62.

intersect the hyperboloid of revolution with the plane $y = a$ parallel to the Oxz-plane. Substituting $y = a$ we obtain

$$\frac{x^2}{a^2} + \frac{a^2}{a^2} - \frac{z^2}{c^2} = 1 \quad \text{or} \quad \frac{x^2}{a^2} - \frac{z^2}{c^2} = 0.$$

But this equation together with $y = a$ gives in the plane $y = a$ a pair of intersecting lines: $x/a - z/c = 0$ and $x/a + z/c = 0$.

Thus we have already discovered that there is a pair of intersecting lines lying on the hyperboloid. If now we revolve the hyperboloid about the Oz-axis, then each of these lines obviously traces out the entire hyperboloid (figure 60).

It is easy to show that: (1) two arbitrary straight lines of one and the same family of lines so obtained do not lie in the same plane (i.e., they are skew lines), (2) any line of one of these families intersects all the lines of the other family (except its opposite, which is parallel to it), and (3) three lines of one and the same family are not parallel to any one and the same plane.

With two matches and a needle it is easy to obtain a representation of the hyperboloid of revolution of one sheet. Let us puncture one of the matches through its middle by the needle, and on the sharp end point of the needle we pin the other match parallel to the first match. If we then revolve the whole apparatus about the first match as an axis, the second match will trace the surface of a cylinder (figure 61). But if the second match is not parallel to the first match, then during a revolution it will trace the surface of a hyperboloid of revolution of one sheet, as can easily be visualized if the rotation is rapid (figure 62).

Summary of the investigation of the second-degree equation. Although the general second-degree equation with three variables can represent essentially 17 different surfaces, it is not difficult to remember them. The last nine are cylinders over the nine possible second-order curves, while the first eight are divided into four pairs: two ellipsoids (real and imaginary), two hyperboloids (of one sheet and two sheets), two second-order cones (real and imaginary), and two paraboloids (elliptic and hyperbolic). All these surfaces play an essential role in mechanics, physics, and technology (ellipsoid of inertia, ellipsoid of elasticity, hyperboloid in the Lorentz transformation in physics, paraboloid of revolution for parabolic mirrors, etc.).

§11. Affine and Orthogonal Transformations

The next important step in the development of analytic geometry was the introduction into it, and into geometry in general, of the theory of transformations. Here it will be necessary to explain the matter in some detail.

"Contraction" of the plane toward a line. Let us consider one of the simplest transformations of the plane, namely uniform "contraction" toward a line with coefficient k. In the plane let there be given a line a and a positive coefficient k, for example, $k = 2/3$. All points of the line a are fixed, and every point M not lying on this line is sent into the point M' such that M' lies on the same side of the line as M on the perpendicular from M to a at a distance from a equal to 2/3 of the distance from M to a. If the coefficient k, as here, is smaller than unity, then we have a proper contraction of the

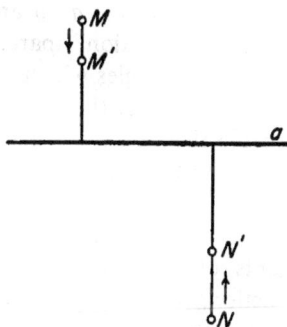

Fig. 63.

plane to the line; but if k is greater than unity, we have an expansion of the plane from the line, but for convenience we will in this and other cases talk about "contraction," except that the word "contraction" will be put in quotation marks.

The point or figure to be transformed is called the preimage and the one into which it is sent is its image. The point M', for example, is the image of the point M (figure 63).

We show that under a uniform "contraction" of a plane to a line, any line of the plane is transformed into a line. For let the plane be "contracted" to a line a lying in it with coefficient of "contraction" k. Let b be any line of the plane, O the point in which it intersects the line a, B another arbitrary point of b, and BA the perpendicular to the line a from the point B (figure 64). In the "contraction" the point B goes to

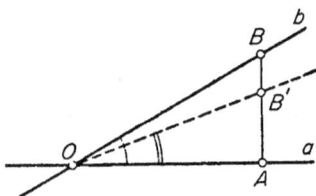

FIG. 64.

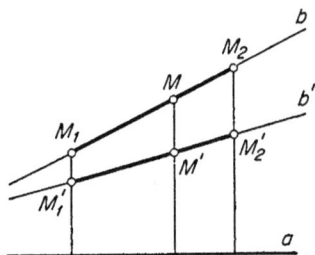

FIG. 65.

the point B' on this perpendicular such that $B'A = k \cdot BA$. Therefore, the tangent of the angle $B'OA$ will be equal to $AB'/OA = k \cdot AB/OA$, i.e., will be equal to k times the tangent of the angle which the line b makes with line a, i.e., for all points B' into which different points of the line b are transformed, it will be one and the same. All points B' consequently lie on one and the same line, passing through the point O and making with line a an angle with this tangent.

Under "contraction" parallel lines remain parallel. Indeed, if the tangents of the angles which lines b and c make with line a are the same, then the tangents of those angles which the images b' and c' make with a differ from them only by a factor k, i.e., they are still equal to each other, which means that the lines b' and c' are also parallel to each other.

Any rectilinear segment of the plane under "contraction" to a line is contracted (or expanded) uniformly (although to various degrees for segments of various directions). When we speak here of "uniform" contraction, we mean that the midpoint of the segment remains the midpoint, the third remains the third, etc., i.e., the segment shrinks uniformly along its full length. Indeed, in whatever ratio the point M

divides the segment M_1M_2, its image M' will divide $M_1'M_2'$, in the same ratio, since parallel lines (in this case perpendiculars to the line a) cut lines intersecting them (in this case b and b') in proportional parts (figure 65).

The ellipse as the result of "contraction" of a circle. We consider a circle with center at the origin and radius a. By the theorem of Pythagoras its equation is $x^2 + \bar{y}^2 = a^2$, where we have written $\bar{y}$ instead of y, since y will be needed later. Let us see what this circle is contracted into if we "contract" the plane to the Ox-axis with coefficient b/a (figure 66). After this "contraction" the x-values of all points remain the same, but the $\bar{y}$-values become equal to $y = \bar{y}\,(b/a)$, i.e., $\bar{y} = (a/b)\,y$. Substituting for $\bar{y}$ in the above equation of the circle, we will have:

$$x^2 + \frac{a^2}{b^2} y^2 = a^2 \quad \text{or} \quad \frac{x^2}{a^2} + \frac{y^2}{b^2} = 1$$

as the equation, in the same coordinate system, of the curve obtained from the given circle by contraction to the Ox-axis. As we see, we obtain an ellipse. Thus we have proved that an ellipse is the result of a "contraction" of a circle.

From the fact that an ellipse is a "contraction" of a circle, many properties of the ellipse follow directly. For example, the aforementioned property of diameters, namely that if parallel secants of an ellipse are given, then their midpoints lie on a straight line (see figure 12), can be shown in the following way. We perform the inverse expansion of the ellipse into the circle. Under this expansion parallel chords of the ellipse go into parallel chords of the circle, and their midpoints into the

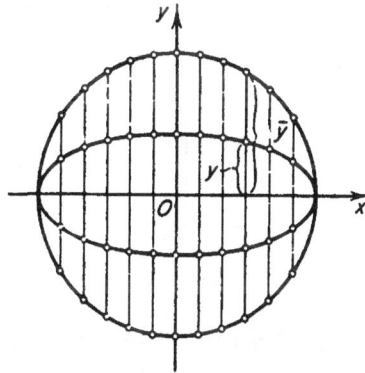

FIG. 66.

midpoints of these chords. But the midpoints of parallel chords of a circle lie on a diameter, i.e., on a straight line, and so that the midpoints of parallel chords of the ellipse also lie on a straight line. Namely, they lie on that line which is obtained from the diameter of the circle under the "contraction" which sends the circle into the ellipse.

Here is another application of the theory of "contraction." Since any vertical strip of the circle under its contraction to the Ox-axis does not change its width and its length is multiplied by b/a, the area of this strip after contraction is equal to its initial area multiplied by b/a, and since the area of the circle is equal to πa^2, the area of the corresponding ellipse is equal to $\pi a^2 (b/a) = \pi ab$.

Example of the solution of a more complicated problem. Let an ellipse be given and let it be required to find the triangle with smallest area circumscribed to this ellipse. We first solve the problem for a circle. We show that in the case of a circle, this is an equilateral triangle. Indeed, let the circumscribed triangle be nonequilateral; i.e., the smallest of its angles (denoted by B) is less than 60°, and the largest $C > 60°$. If then, without varying the angle A, we move side BC into the position B_0C_0 (figure 67) by shifting the vertex B toward A until one of the angles B_0 or C_0 becomes equal to 60°, we obtain a circumscribed triangle AB_0C_0

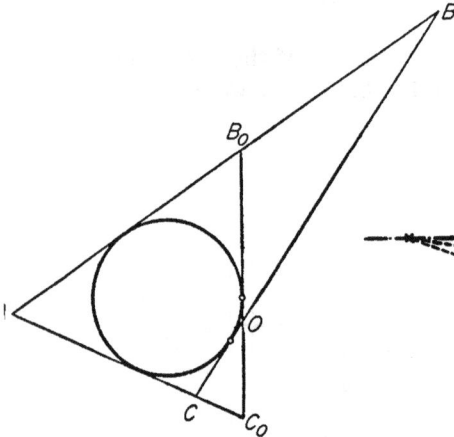 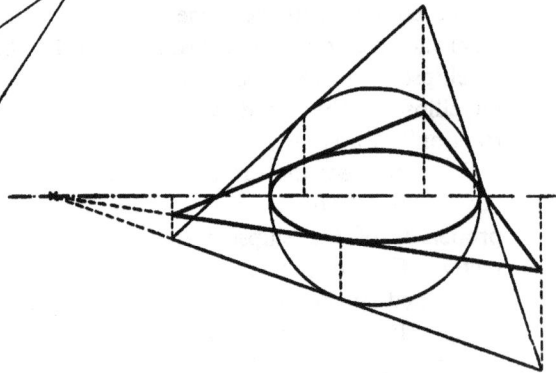

FIG. 67.   FIG. 68.

with smaller area, since here* $OC < OB$, $OC_0 \leqslant OB_0$ and therefore the discarded area OBB_0 is greater than the added one OCC_0. If the triangle so obtained is not equilateral, then by repeating the above procedure we reduce its area still further and arrive at an equilateral triangle. Hence, any nonequilateral triangle circumscribed to a given circle has a greater area than an equilateral one.

We now return to the ellipse. Let us make an expansion of it from the major axis, thereby converting it back into the circle from which it was obtained by "contraction." Under this expansion (figure 68): (1) all

---

* As can be easily shown.

triangles circumscribed to the ellipse are transformed into triangles circumscribing the resultant circle; (2) the areas of all figures, and in particular of these triangles are increased in one and the same ratio. From this we see that the triangles circumscribing the given ellipse with the smallest area will be those that are converted into equilateral triangles circumscribing the circle. There are infinitely many such triangles; each of them has its center of gravity at the center of the ellipse and the points of tangency are in the middle of its sides. Any of these triangles can easily be constructed (figure 68), starting from the aforementioned circle.

"Contractions" of the plane to a line are only a particular case of more general, so-called *affine* transformations of the plane.

General affine transformations.   A pair of vectors e_1, e_2 starting from a common origin O and not lying on the same line will be called a co-ordinate "frame" of the plane. The coordinates of a point M of the plane relative to this frame Oe_1e_2 will then be numbers x, y such that in order to reach the point M from the origin O it is necessary to lay off from the point O x-times the vector e_1 and then y-times the vector e_2. This is a general Cartesian coordinate system of the plane. Analogously, a general Cartesian coordinate system can be introduced in space. The ordinary, so-called rectangular Cartesian coordinate system that we have made use of up to now corresponds to the particular case when the coordinate vectors e_1, e_2 are mutually perpendicular and their lengths are equal to the unit of measurement.

A general affine transformation of the plane is one under which a given net of equal parallelograms is transformed into another arbitrary net of equal parallelograms. More precisely, it is a transformation of the plane under which a given coordinate frame Oe_1e_2 is transformed into a certain other frame (generally speaking, with another "metric," i.e., with different lengths for the vectors e_1' and e_2' and a different angle between them) and an arbitrary point M is sent into the point M' having the same coordinates relative to the new frame as M had relative to the old (figure 69).

"Contraction" to the Ox-axis with coefficient k is a special case in which the rectangular frame $Oe_1'e_2'$ passes into the frame $Oe_1'ke_2'$.

It can easily be shown that under an affine transformation every straight line is sent into a straight line, parallel lines are mapped into parallel lines, and if a point divides a segment in a given ratio, then its image divides the image of this segment in the same ratio. Moreover, we can prove the remarkable theorem that any affine transformation of the plane can be obtained by performing a certain rigid motion of the plane onto itself, and then, in general, two "contractions" with different coefficients k_1 and k_2 to two mutually perpendicular lines.

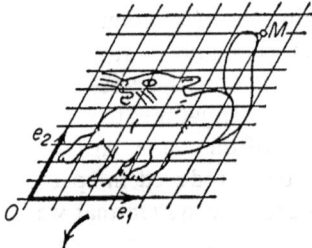

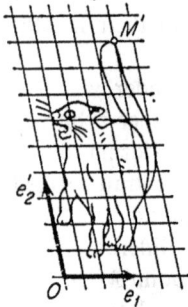

For the proof of this assertion, we consider all radii of some circle of the plane (figure 70). Let radius OA be the one which, after the transformation, turns out to be the shortest, and let it be mapped into $O'A'$. The perpendicular AB to OA is then transformed into $A'B'$, which must be perpendicular to $O'A'$, since if the perpendicular $O'C'$ were different from $O'A'$, then it would be the image of the oblique OC, and the image $O'D'$ of the radius OD would be a part of the perpendicular $O'C'$, i.e., shorter than the oblique $O'A'$, contrary to assumption.

The mutually perpendicular lines OA and AB are therefore mapped into mutually perpendicular lines $O'A'$ and $A'B'$. Consequently, the square net constructed on OA and AB is transformed into a net of equal rectangles (figure 71) and uniform "contractions" take place along the straight lines of this square net.

In a completely analogous way a general affine transformation of space can be defined as one under which a space coordinate frame $Oe_1e_2e_3$ is transformed into some other frame $O'e_1'e_2'e_3'$, generally speaking, with a different "metric," i.e., with unit segments of different lengths and with different angles between them, and a point M is sent into point M' having the same coordinates relative to the new frame as those of the point M relative to the old frame.

FIG. 69.

All the properties enumerated here also hold for affine transformations of space, except that in the last theorem there will be a rigid motion of space and then three "contractions" to three mutually perpendicular planes with certain coefficients k_1, k_2, k_3.

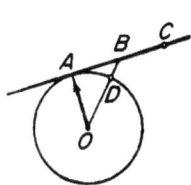 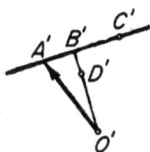 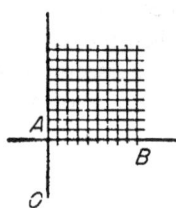 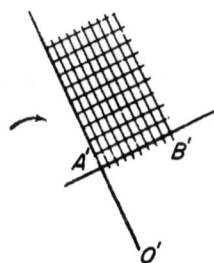

FIG. 70.   FIG. 71.

Applications of affine transformations.   The most important applications of affine transformations are:

1. In the first place there is the application in geometry to solving problems concerning affine properties of figures, i.e., properties that are preserved under affine transformations. The theorem about the diameters of an ellipse and the problem of circumscribed triangles were examples. To solve such problems we make an affine transformation of the figure to some simpler one, for which we prove the desired property and then return to the original figure.

2. Second, there is the application in analytic geometry to the classification of second-order curves and surfaces. The main point is, as can be shown, that different ellipses are related to one another in the sense that one can be obtained from another by an affine transformation (the Latin word *affinis* means "related"). Also all hyperbolas are affine to one another, and so are all parabolas. But we cannot convert an ellipse into a parabola, or a hyperbola into a parabola, by an affine transformation, i.e., they are not affinely related to one another. It is natural to divide up all second-order curves into affine classes of curves, affinely related to one another. It turns out that the reduction of an equation to canonical form gives exactly this classification; i.e., there are nine affine classes of second-order curves. (We will not go into detail why imaginary ellipses and pairs of imaginary parallel lines belong to different affine classes. Properly speaking, neither in one case nor in the other are there any curves on the plane at all. The question here is really about algebraic properties of the equation itself.)

Similarly, the classification of second-order surfaces according to their canonical equations into 17 forms is the same as the affine classification.

Let us give a simple example of the application of the affine classification of second-order surfaces. We show that if we arbitrarily select in space three lines *a*, *b*, *c* such that (1) any two of them do not lie in the same plane (i.e., they are skew to each other) and (2) they are not all parallel to one and the same plane, then the set of all straight lines d of space, each of which simultaneously intersects all three given lines a, b, c (figure 72) constitutes the entire surface of a hyperboloid of one sheet.

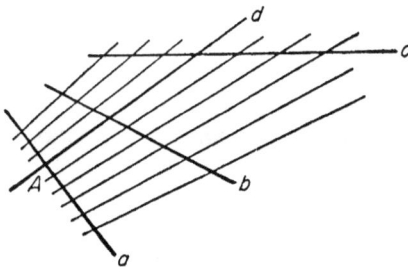

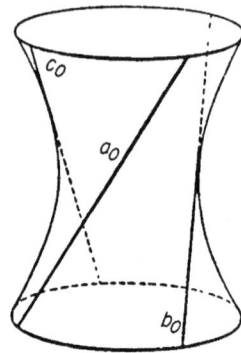

FIG. 72.   FIG. 73.

Let us explain more fully the set of lines d we are discussing here. Through an arbitrary point A of line a, we can pass a plane P containing the line b and a plane Q containing the line c. These planes P and Q intersect in a unique line d, which passes through the point A of line a and intersects lines b and c. Drawing all such lines d through arbitrary points of line a, we obtain the set of all those lines d of space each of which intersects all three given lines a, b, and c. This collection of lines determines a surface. We note that any given hyperboloid of one sheet can be obtained in this way, since we only need to take for the lines a, b, and c three distinct straight lines a_0, b_0, c_0 of one family (figure 73) and for the lines d all the straight lines of the other family. Conversely, let there be given three arbitrary pairwise skew lines of space a, b, c, not all parallel to one and the same plane. Then, as can be shown, these lines always form the three edges (without common points) of some parallelepiped (figure 74). After constructing such parallelepipeds for the given lines a, b, c and for three lines a_0, b_0, c_0 of one and the same family of an arbitrary hyperboloid of one sheet, we make an affine transformation of space that sends the parallelepiped a_0, b_0, c_0 into the parallelepiped a, b, c; obviously, this transformation maps the hyperboloid onto the surface in question. But according to the affine classification of second-order surfaces, the affine image of a hyperboloid of one sheet is again a hyperboloid of one sheet.

3. Third, there is the application to the theory of continuous transformations of continuous media, for example, in the theory of elasticity,

in the theory of electric or magnetic fields, etc. Very small elements of the given continuous medium transform "almost" affinely. We say, "in the small the transformation is linear" (we call a first-degree expression linear, and in the following section we see that in analytic geometry the formulas of affine transformations are of the first degree). This is evident in figure 75. On the lines of the large square net, their distortion or

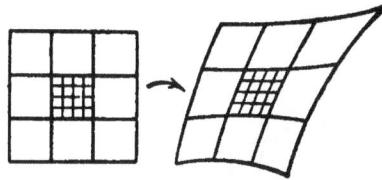

FIG. 75.

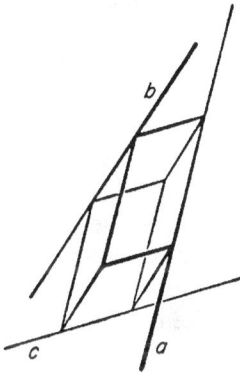

FIG. 74.

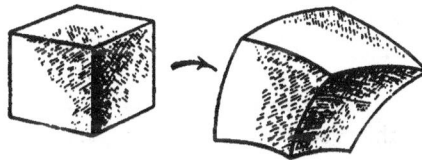

FIG. 76.

"fanning out" is clearly noticeable. But for a small piece of the very dense square net, all this shows itself very little, and the square net transforms "almost" into a net of equal parallelograms. A similar picture is obtained in space also (figure 76). By the fact that any affine transformation of space reduces to a motion and three mutually perpendicular "contractions," it follows that an element of a body under an elastic deformation first moves as a rigid body and then undergoes three mutually perpendicular "contractions."

Formulas of affine transformations. If the frame Oe_1e_2 is affinely transformed and $O'e_1'e_2'$ is its image, while the coordinates of the new origin O' relative to the old frame are ξ, η and the coordinates of the vectors e_1' and e_2' relative to the old frame are a_1, a_2 and b_1, b_2, respectively, then the formulas of the affine transformation, as can be easily seen from figure 77, are

$$x' = a_1x + b_1y + \xi,$$
$$y' = a_2x + b_2y + \eta$$

in the sense that if x, y are the coordinates of any point M relative to the old frame Oe_1e_2, then x', y' given by these formulas, are the coordinates relative to the same frame of the image M' of this point.

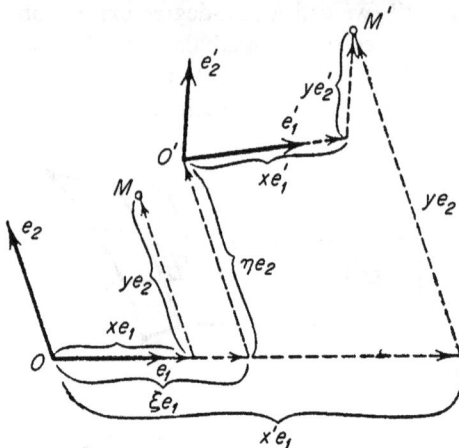

FIG. 77.

Indeed, let Oe_1e_2 be a frame before transformation, and $O'e_1'e_2'$ its image, while M is an arbitrary point of the plane and M' is its image. Then by the very definition of an affine transformation, if the coordinates of the point M relative to the frame Oe_1e_2 are x, y, then the coordinates of its image M' relative to the image $O'e_1'e_2'$ of this frame are exactly the same x, y.

Now consider a vector m' joining the origin O of the old frame to the image M' of the point M. Then $m' = x'e_1 + y'e_2$. But this vector is equal to a certain vector sum

$$m' = \xi e_1 + \eta e_2 + x e_1' + y e_2',$$

and the vectors e_1' and e_2' are

$$e_1' = a_1 e_1 + a_2 e_2, \quad e_2' = b_1 e_1 + b_2 e_2$$

so that

$$m' = \xi e_1 + \eta e_2 + a_1 x e_1 + a_2 x e_2 + b_1 y e_1 + b_2 y e_2$$

or

$$m' = (a_1 x + b_1 y + \xi)e_1 + (a_2 x + b_2 y + \eta)e_2.$$

Comparing this expression with the first expression for m' we obtain

$$\begin{aligned} x' &= a_1 x + b_1 y + \xi, \\ y' &= a_2 x + b_2 y + \eta. \end{aligned} \tag{21}$$

The determinant

$$\Delta = \begin{vmatrix} a_1 & b_1 \\ a_2 & b_2 \end{vmatrix} = a_1 b_2 - a_2 b_1,$$

as can be shown, is not zero and is equal to the ratio of area of the parallelogram constructed on the vectors of the new frame to the area of the same parallelogram constructed on the vectors of the old frame. Analogous formulas are obtained for space

$$\left.\begin{array}{l} x' = a_1x + b_1y + c_1z + \xi, \\ y' = a_2x + b_2y + c_2z + \eta, \\ z' = a_3x + b_3y + c_3z + \zeta, \end{array}\right\} \qquad (22)$$

where (ξ, η, ζ) are the coordinates of the origin O' of the transformed frame $O'e_1'e_2'e_3'$ and $(a_1, a_2, a_3), (b_1, b_2, b_3), (c_1, c_2, c_3)$ are the coordinates of its vectors e_1', e_2', e_3' relative to the old frame $Oe_1e_2e_3$.

The determinant*

$$\Delta = \begin{vmatrix} a_1 & b_1 & c_1 \\ a_2 & b_2 & c_2 \\ a_3 & b_3 & c_3 \end{vmatrix} = a_1b_2c_3 + a_2b_3c_1 + a_3b_1c_2 - a_1b_3c_2 - a_2b_1c_3 - a_3b_2c_1$$

is not zero and is equal to the ratio of the volume of the parallelepiped formed by the vectors of the new frame to the volume of the parallelepiped formed by the vectors of the old frame.

Orthogonal transformations. Rigid motions of the plane onto itself or such motions plus a reflection about a line lying in the plane, are called *orthogonal transformations of the plane*, and rigid motions of space, or such motions plus a reflection of the space about one of its planes, are called *orthogonal transformations of space*. It is clear that orthogonal transformations are affine transformations under which the "metric" of the frame does not change, since it only undergoes a rigid motion, or else such a motion plus a reflection.

We will investigate orthogonal transformations by means of rectangular coordinates, i.e., when the vectors of the original frame are mutually perpendicular and have lengths equal to the unit of measurement. After an orthogonal transformation the vectors of the frame remain mutually perpendicular, i.e., their scalar product remains equal to zero and their lengths remain equal to unity. Therefore (see formula (14), this chapter) in the case of the plane, we have

$$a_1b_1 + a_2b_2 = 0, \quad a_1^2 + a_2^2 = 1, \quad b_1^2 + b_2^2 = 1, \qquad (21')$$

and in the case of space

$$\begin{aligned} a_1b_1 + a_2b_2 + a_3b_3 &= 0, & a_1^2 + a_2^2 + a_3^2 &= 1, \\ a_1c_1 + a_2c_2 + a_3c_3 &= 0, & b_1^2 + b_2^2 + b_3^2 &= 1, \qquad (22') \\ b_1c_1 + b_2c_2 + b_3c_3 &= 0, & c_1^2 + c_2^2 + c_3^2 &= 1. \end{aligned}$$

* On determinants see Chapter XVI.

Hence, if the initial frame is taken to be rectangular, then formulas (21) give an orthogonal transformation if and only if the conditions (21') of orthogonality are fulfilled, and formulas (22) give an orthogonal transformation of space if the conditions (22') of orthogonality are satisfied. It can be shown that if $\Delta > 0$, we have a rigid motion, and if $\Delta < 0$ a rigid motion plus a reflection.

§12. Theory of Invariants

The concept of invariant.* Invariants of a second-degree equation with two variables. In the second half of the last century still another important new concept was introduced, that of invariant.

Consider, for example, a second-degree polynomial in two variables

$$Ax^2 + 2Bxy + Cy^2 + 2Dx + 2Ey + F. \qquad (23)$$

If we regard x, y as rectangular coordinates and make a transformation to new rectangular axes, then after replacing x, y in (23) by their expressions in terms of the new coordinates x', y', removing parentheses, and reducing similar terms, we obtain a new transformed polynomial with different coefficients

$$A'x'^2 + 2B'x'y' + C'y'^2 + 2D'x' + 2E'y' + F'. \qquad (24)$$

It turns out that there exist expressions formed from the coefficients which under this transformation do not change their numerical value, although the coefficients themselves change. Such an expression in A', B', C', D', E', F' has exactly the same numerical value as when it is formed with the A, B, C, D, E, F.

Expressions of this kind are called *invariants* of the polynomial (23) with respect to the group of orthogonal transformations (i.e., relative to transformations from one set of rectangular coordinates x, y to any other rectangular coordinates x', y').

Invariants of this sort, as it turns out, are

$$I_1 = A + C,$$

$$I_2 = \begin{vmatrix} A & B \\ B & C \end{vmatrix} = AC - B^2,$$

$$I_3 = \begin{vmatrix} A & B & L \\ B & C & E \\ D & E & F \end{vmatrix} = ACF + 2BDE - AE^2 - CD^2 - FB^2,$$

---

* *Invarians* in Latin means "unchanged."

i.e.,

$$A + C = A' + C', \quad AC - B^2 = A'C' - B'^2,$$

$$ACF + 2BDE - AE^2 - CD^2 - FB^2$$
$$= A'C'F' + 2B'D'E' - A'E'^2 - C'D'^2 - F'B'^2.$$

It is possible to prove the important theorem that any orthogonal invariant of the polynomial (23) can be expressed in terms of these three basic invariants.

If we equate the polynomial (23) to zero, we obtain an equation of some second-order curve. Any quantity, connected with this curve but not with its location in the plane, will clearly not depend on what coordinates its equation is written in, and therefore, when expressed in terms of the coefficients, it will be an orthogonal invariant of the polynomial (23), and thus it will be expressible in terms of the three basic invariants. Moreover, since under multiplication of all six coefficients of the equation by any given number t (different from zero) the curve represented by the equation remains the same, an expression of any property of the curve in terms of the I_1, I_2, I_3 must certainly be such that if the A, B, C, D, E, F in it are multiplied by t, the number t cancels out. The expression in question must be, as they say, homogeneous of degree zero relative to A, B, C, D, E, F.

Let us verify this by an example. For instance, let the equation

$$Ax^2 + 2Bxy + Cy^2 + 2Dx + 2Ey + F = 0$$

represent an ellipse. Since the equation completely determines this ellipse, we can calculate from it (i.e., from its coefficients) all the basic quantities connected with the ellipse. For example, we can calculate its semiaxes a and b, i.e., we can express the semiaxes in terms of the coefficients. The expressions for these semiaxes will be invariants and therefore, expressible in terms of I_1, I_2, I_3. By reduction of the equation to canonical form and some subsequent calculation, the following rather complicated expressions for the semiaxes are obtained in terms of I_1, I_2, I_3:

$$\sqrt{\frac{2\,|I_3|}{|I_2|\,|\,I_1 \pm \sqrt{I_1^2 - 4I_2}\,|}},$$

which are homogeneous relative to A, B, C, D, E, F.

From this it is clear that the invariants I_1, I_2, I_3, themselves, being homogeneous but not of degree zero, do not have straightforward geometric meanings; they are algebraic entities.

It can be shown that the expression

$$K_1 = \begin{vmatrix} A & D \\ D & F \end{vmatrix} + \begin{vmatrix} C & E \\ E & F \end{vmatrix} = AF - D^2 + CF - E^2$$

can be varied by parallel translation but not by pure rotation of the given rectangular axes, and it is therefore called a *semi-invariant*.

As an example of an application of invariants and semi-invariants, we give Table 1, which if we calculate I_1, I_2, I_3 and K_1, allows us to determine directly from its equation the affine class of a second-order curve.

In Table 1 the necessary and sufficient conditions are given that an

Table 1

Criterion of the class	Name	Reduction equation	Canonical equation
$I_2 > 0,\ I_1 I_3 < 0$	ellipse		$\dfrac{x^2}{a^2} + \dfrac{y^2}{b^2} = 1$
$I_2 > 0,\ I_1 I_3 > 0$	imaginary ellipse		$\dfrac{x^2}{a^2} + \dfrac{y^2}{b^2} = -1$
$I_2 > 0,\ I_3 = 0$	point	$\lambda_1 x^2 + \lambda_2 y^2 + \dfrac{I_3}{I_2} = 0$	$\dfrac{x^2}{a^2} + \dfrac{y^2}{b^2} = 0$
$I_2 < 0,\ I_3 \neq 0$	hyperbola		$\dfrac{x^2}{a^2} - \dfrac{y^2}{b^2} = 1$
$I_2 < 0,\ I_3 = 0$	pair of intersecting lines		$\dfrac{x^2}{a^2} - \dfrac{y^2}{b^2} = 0$
$I_2 = 0,\ I_3 \neq 0$	parabola	$I_1 x^2 + 2\sqrt{-\dfrac{I_3}{I_1}}\, y = 0$	$x^2 = 2py$
$I_1 = 0, I_3 = 0,$ $K_1 < 0$	pair of parallel lines		$x^2 = a^2$
$I_2 = 0,\ I_3 = 0,$ $K_1 > 0$	pair of imaginary parallel lines	$I_1 x^2 + \dfrac{K_1}{I_1} = 0$	$x^2 = -a^2$
$I_2 = 0,\ I_3 = 0,$ $K_1 = 0$	pair of coincident lines		$x^2 = 0$

equation of a second-order curve be reducible to one or another of the nine canonical forms ($I_1 I_3$ designates the product of I_1 and I_3).

Consider, for example, the equation $x^2 - 6xy + 5y^2 - 2x + 4y + 3 = 0$. We have $A = 1$, $B = -3$, $C = 5$, $D = -1$, $E = 2$, $F = 3$, so that $I_1 = 6$, $I_2 = -4$, $I_3 = -9$. The conditions of the 4th line of the table are satisfied: $I_2 < 0$, $I_3 \neq 0$, i.e. this is a hyperbola. Its semiaxes are equal to

$$\sqrt{\frac{2 \cdot 9}{4 \cdot |6 \pm \sqrt{36 + 16}|}} \approx 0.57 \quad \text{and} \quad 1.93.$$

The coefficients of the reduced equation (I), (II) and (III) are given in terms of invariants and semi-invariants as follows:

$$\lambda_1 x''^2 + \lambda_2 y''^2 + \frac{I_3}{I_2} = 0, \tag{I}$$

$$I_1 x''^2 + 2\sqrt{-\frac{I_3}{I_1}}\, y'' = 0, \tag{II}$$

$$I_1 x''^2 + \frac{K_1}{I_1} = 0, \tag{III}$$

where λ and λ_2 are the roots of the so-called characteristic quadratic equation

$$\lambda^2 - I_1 \lambda + I_2 = 0.$$

Formulas (I-III) allow a quick calculation of the semiaxes a and b of an ellipse and a hyperbola, the parameter p of an ellipse and the distance $2a$ between parallel lines. The formulas for semiaxes were given earlier. The parameter p is equal to

$$p = \sqrt{-\frac{I_3}{I_1^3}} \quad \text{and the distance} \quad 2a = 2\sqrt{-\frac{K_1}{I_1^2}}.$$

A completely analogous theory of invariants and semi-invariants, with a corresponding table for the determination of the affine class and the formulas of the coefficients of reduced equations, can be given for second-order surfaces in three-dimensional space.

It should be pointed out that so far we have been discussing only those invariants that are considered in analytic geometry for curves and surfaces of the second order. The concept of invariant, however, has a far broader meaning.

By an invariant of some object under study, relative to certain of its transformations, we mean any quantity numerical, vectorial, etc. connected with this object that does not vary under these transformations. In the previous problem the object is a second-degree polynomial with two variables (i.e., more precisely, the set of its coefficients), and the transformations are those of the polynomial obtained by the transition from one rectangular coordinate system to another.

Another example: The object is a given mass of a given gas under a given temperature. The transformations are changes in volume or pressure of this mass of gas. The invariant, according to the Boyle-Mariotte law, is the product of the volume by the pressure. We can speak of lengths of segments in space or the size of angles as invariants of the group of motions of space, of ratios in which a point divides a segment, or of ratios of areas, as the invariants of the group of affine transformations of space, etc.

Various invariants are particularly important in physics.

§13. Projective Geometry

Perspective projections. Artists began long ago to study the laws of perspectivity. This was necessary because a human being sees objects in perspective projection on the retina of the eye, in such a way that the

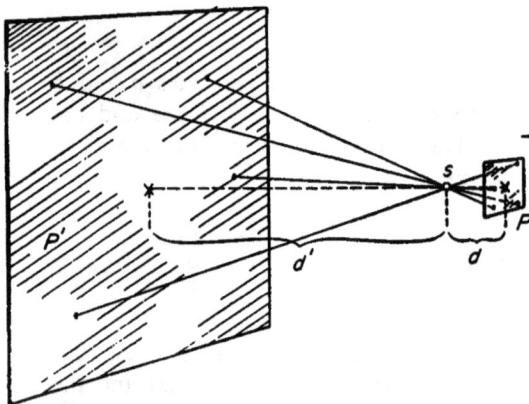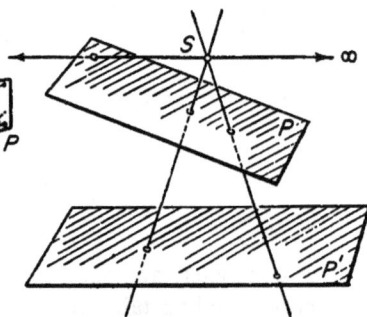

FIG. 78.   FIG. 79.

form and mutual location of objects are distorted in a characteristic manner. For example, telegraph poles in the distance look smaller and closer together, parallel tracks of a railway seem to converge, etc. We will not consider here space perspectivities, i.e., properties of perspective projections of objects in space onto a plane but only the properties of perspective projections of a plane onto a plane.

Let us consider a photograph (for example, one frame of a moving picture film) P, a screen P', and between them a lens S (figure 78). Then, if the photograph is transparent and is illuminated from behind (if it is nontransparent, let it be illuminated from the front, i.e., from the side where the lens is), then the illuminated points of the photograph radiate beams of light, which are collected by the lens in such a way that they appear again on the screen P' in the form of points. We will assume that this projection takes place as if the points of the photograph P were projected on the screen P' on straight lines passing through the optical center S of the lens.

The situation will be a very simple one if the planes P and P' are parallel. In this case we will obviously obtain on the plane P' an undistorted image of everything that is on the plane P. This image will be smaller or larger than the original depending on whether the ratio $d':d$, where d and d' are the distances from the center of the lens to the planes P and P' respectively, is smaller or larger than 1.

The situation will be considerably more difficult if the planes P and P' are not parallel (figure 79). In this case, under projection through the point S not only the size of the figure changes but also its form is distorted. Parallel lines under such projection may become convergent, the ratio in which a point divides a segment may change, etc. In general, some of the relations that remain invariant under arbitrary affine transformation may change here.

This sort of projection takes place, for example, in aerial photography. The airplane oscillates in flight and therefore the photographic apparatus (figure 80a) rigidly attached to it is, in general, not oriented altogether vertically but at the moment of exposure is usually in an oblique position, i.e., we obtain a distorted image of the locality (which we assume to be plane).

How are we to correct this image? For this it is necessary to study the properties of projection of a plane P onto another plane II (in general, the two planes are not parallel) by lines passing through a point S which is not on plane P nor on plane II. Such projections are called *perspective projections*.

We will prove later the following important theorem.

Theorem. *If we have two perspective projections of a plane P on plane* II *such that under both projections the points A, B, C, D of a quadruple of points of "general position" on the plane P (i.e., a quadruple in which no three of the points lie on one line), are projected into the same points A',*

B', C', D' respectively of plane II, *then all points of plane P are also projected under both projections into the same points of plane* II.

In other words, the result of a perspective projection is completely determined if it is known into which points this projection sends the points of an arbitrary quadruple of points of general position in the figure to be projected.

This is the so-called uniqueness theorem of the theory of projective transformations or the fundamental theorem of plane perspectivity.

Application of the fundamental theorem of plane perspectivity in aerial photography. Let us show how this theorem provides a suitable method for correcting this image in photography.

If at the moment of aerial exposure, we imagine a horizontal screen II placed at a distance h below the center S of the lens (figure 80a), then the projection onto this screen through the center S of the image recorded on the photographic plate P will obviously not be distorted but will be similar to the horizontal locality with a scale $h:H$, where H is the height of the airplane at the moment of exposure. In order to correct the image received on the photograph P so as to convert it into an undistorted image, we treat it as follows. The developed photograph P is placed in a projecting apparatus resting on a special tripod on which, by means of adjustable screws, the apparatus can be moved closer to the screen II or farther from it and can be rotated in every way.

To the screen II (figure 80b) we attach a topographical map of the locality made by measurements on the surface of the Earth (not a detailed map, since the details of interest to us are to be provided by the aerial photograph). On this map attached to the screen II we select four points A', B', C', D' that can be found easily on the photograph also (for example, an intersection of roads, a corner of a house, etc.), and at the corresponding points A, B, C, D of the picture P we pierce the film with a needle. We then place a projection lamp behind the plate P in such a position that the picture is projected onto the screen II through a lens S of the supporting apparatus. By using the adjustable screws we arrange that the light beams from the pinholes fall on the corresponding points A', B', C', D' of the map attached to the screen. After this has been done, we replace the topographical map by a plateholder with a photographic plate and then, without changing the settings of the screws, we photograph the image projected on the screen II of the picture P taken from the airplane.

By the theorem stated previously, we thereby obtain a true (i.e., similar to the locality) and not a distorted map of the photographed region.

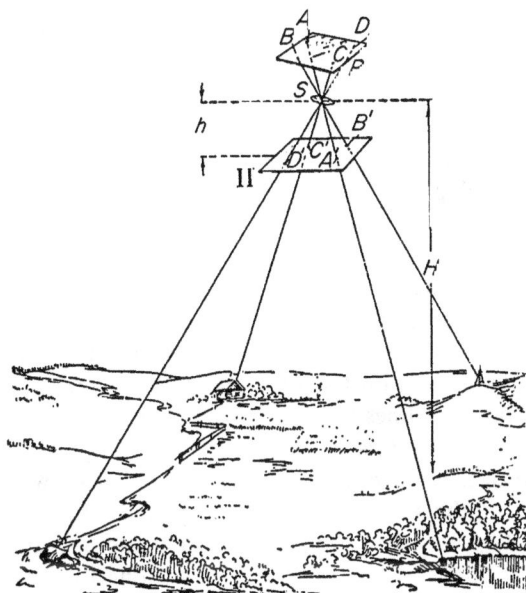

(a)

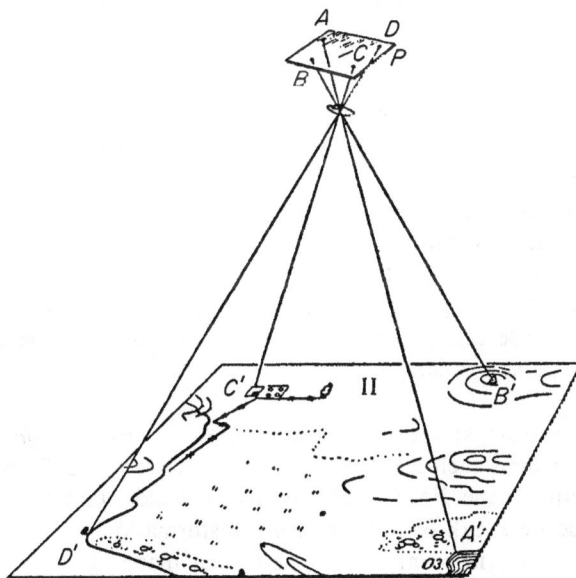

(b)

FIG. 80.

We now pass to the presentation of the theory necessary for proving the fundamental theorem.

The projective plane. The totality of *all* lines and planes of space passing through a given point S of the space is called the projecting bundle of lines and planes with center S. If this bundle is intersected by a plane P, not passing through the center, then to every point of the plane P will correspond a line of the bundle intersecting the plane P in this point, and to each line of the plane P will correspond that plane of the bundle which intersects the plane P along this line. However, we do not in this way establish a one-to-one mapping from the set of lines and planes of the bundle on the set of points and lines of the plane P. As a matter of fact, the lines and planes of the bundle which are parallel to the plane P do not in this sense correspond to any points or lines of the plane P, since they do not intersect it. Nevertheless, we *agree* to say that these lines of the bundle intersect the plane P but in its ideal (or infinitely distant) points, lying in the corresponding directions, and that such a plane of the bundle intersects the plane P along an ideal (or infinitely distant) line. The plane P, complemented by these ideal points and ideal line, is called a *complemented* or *projective plane*. We will denote it by P^*. The sets of lines and planes of the bundle S are then mapped one-to-one onto the sets of points (real and ideal) and lines (real and ideal) of this projective plane P^*.

Hence, we *agree* to say that a point (real or ideal) lies on a line (real or ideal) of the projective plane P^* if the corresponding line of the bundle lies in the corresponding plane of the bundle. From this point of view, any two lines of the projective plane intersect (in a real or ideal point), since any two planes of the bundle intersect along some line of the bundle. It follows from this, among other things, that the ideal line consists simply of the set of all ideal points.

In essence, the complementation of the plane by its ideal elements means that we use this plane as a cross section to study the bundle of all lines and planes passing through one point.

Projective mappings; the fundamental theorem. By a *projective mapping* we understand such a mapping of a projective plane P^* onto some other projective plane $P^{*\prime}$ (which can also coincide with the plane P^*, in which case we speak of a projective transformation of the plane P^*), which, first of all, is pointwise a one-to-one mapping, and second, is such that collinear sets of points of the plane P^* go into collinear sets of points of the plane $P^{*\prime}$, and conversely. (Here, by points and lines we always understand real as well as ideal points and lines.)

It is clear that two arbitrary perspective projections of one and the same plane P^* onto a plane II^* may be obtained from each other by projective transformations.

In fact, 1, their points (real or ideal) are in one-to-one correspondence with the points (real or ideal) of the protective plane P^* and consequently with each other, and 2, collinear points of the first projection correspond to collinear points of the plane P^* and consequently also of the second projection, and conversely. Therefore, the aforementioned theorem of the theory of perspectivities is a direct consequence of the following theorem about projective transformations: if under a projective transformation of the plane II^*, four of its points A, B, C, D, forming a quadruple of general position remain fixed, then all of its points remain fixed.

Let us outline the idea of the proof of this theorem by means of the so-called Möbius net.

We note that (1) if under a projective transformation two points remain fixed, then the line that passes through them is mapped into itself, and (2) if two lines are mapped into themselves, then the point of their intersection remains fixed. Therefore, from the fact that the points A, B, C, D of the plane II^* remain fixed, it follows in turn that also the points E, F, G, H, K, L, etc. remain fixed (figure 81). The con-

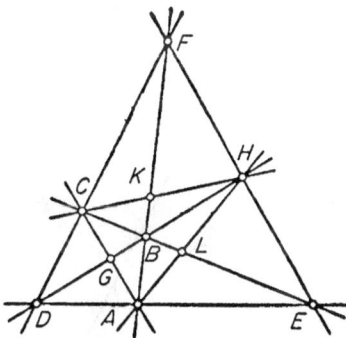

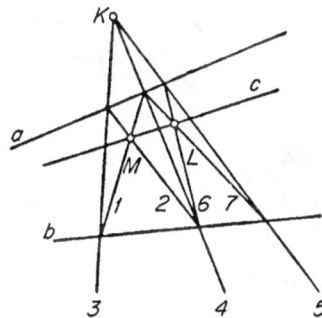

FIG. 81. FIG. 82.

struction of such points can be continued by joining the points already obtained. This is the so-called Möbius net. By continuing its construction, we can find points as densely placed as we like. It can be shown that the set of these nodes everywhere densely covers the whole plane. Therefore, if we further assume the continuity of a projective transformation (which in fact, already follows from its definition, although the proof of this fact is not easy), the result is that if under a projective transformation of the plane II* the points *A*, *B*, *C*, *D* remain fixed, then all the points of the plane II* remain fixed.

Projective geometry.   By *two-dimensional projective geometry* we mean the totality of theorems about those properties of figures in the projective plane, i.e., the ordinary plane complemented by ideal elements, which do not change under arbitrary projective transformations.

Here is an example of a problem of projective geometry. Given two lines *a* and *b* and a point *M* (figure 82), the problem is to construct the line *c* passing through the point *M* and through the point of intersection of lines *a* and *b*, not using this point of intersection (as may be necessary, if this point is very distant). If through the point *M* we draw the two secants 1 and 2 and then the lines 3 and 4 through the points of their intersection with lines *a* and *b*, we obtain the point *K*. Let us draw through it line 5 and secants 6 and 7; then it can be shown that the line *c* passing through point *L* of intersection of lines 6 and 7 and point *M*, is the desired line.

From the theory of conic sections, it follows (figure 83) that the ellipse, hyperbola and parabola are perspective projections of one another, and moreover all of them are perspective projections of the circle.

If we regard perspective projections as projective transformations of projective planes *P** and *P**′ one onto the other, then by superposing these planes we obtain the result that all ellipses, hyperbolas, and parabolas are projective transformations of the circle. The difference in them

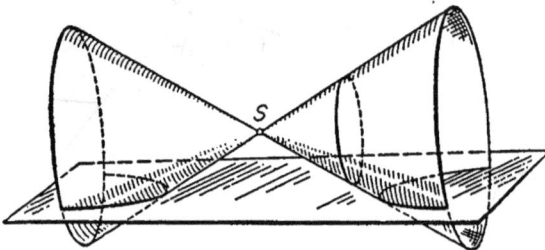

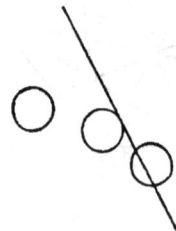

FIG. 83.   FIG. 84.

is that projective images of the circle under transformations in which a line not intersecting the circle is mapped into the infinitely distant line are ellipses; on the other hand, if a line tangent to the circle is mapped into the infinitely distant line, then a parabola is obtained, and if a secant, then a hyperbola (figure 84).

The notation of projective transformations in formulas. If on the plane P^* we take an ordinary Cartesian coordinate system, then, as can be shown, the formulas for projective transformations of the plane are as follows

$$x' = \frac{a_1 x + b_1 y + c_1}{a_3 x + b_3 y + c_3}, \quad y' = \frac{a_2 x + b_2 y + c_2}{a_3 x + b_3 y + c_3},$$

where the determinant

$$\begin{vmatrix} a_1 & b_1 & c_1 \\ a_2 & b_2 & c_2 \\ a_3 & b_3 & c_3 \end{vmatrix} \neq 0,$$

and conversely.

If for some point (x, y) the denominators are equal to zero, this means that its image (x', y') is an ideal (infinitely distant) point. The equation

$$a_3 x + b_3 y + c_3 = 0$$

represents the line which under the given projective transformation goes into the ideal (infinitely distant) line.

§14. Lorentz Transformations

The derivation of the formulas of the Lorentz transformation for motion on a straight line and in the plane from the condition of the constancy of the speed of light. At the very end of the 19th century a fundamental contradiction was discovered in physics. Michelson's well-known experiment, in which the speed of light (which is about 300,000 km/sec) was measured in the direction of motion of the Earth along its orbit around the Sun (the speed of the Earth is about 30 km/sec) and perpendicular to this direction, showed irrefutably that all moving bodies in nature, even if they are moving in a vacuum, are contracted in the direction of motion. The theory of this contraction was investigated in detail by the Dutch physicist, Lorentz. He showed that this contraction is greater as the speed of the moving body gets closer to the speed of light in a vacuum, and at a speed equal to the speed of light the contraction becomes infinite. Lorentz derived the formulas for this contraction. But shortly afterwards the physicist Einstein introduced into this problem a completely different point of view, to which Poincaré was already close. Einstein argued as follows. If we assume that for the propagation of light, as for ordinary motion of a material body, Galileo's law of composition of velocities is valid, then the speed of light is $c' = c + v$, where v is the speed of the observer moving toward the source of the light, and c is the speed of light for a stationary observer. From Michelson's experiment it follows that $c' = c$. The law $c' = c + v$ is based on the transformation

$$x' = x + v_x t,$$
$$t' = t, \qquad \qquad (25)$$

connecting the coordinate x of a point relative to a coordinate system I with its coordinate x' relative to a coordinate system II which has its axes parallel to the axes of system I and which moves parallel to the Ox-axis with velocity v_x relative to system I. Clearly, these are the formulas, as Einstein says, that must be changed.

It can be shown, as was recently done, for example, by A. D. Aleksandrov, that from the equality of the speed of light in both coordinate systems x, y, z, t and x', y', z', t' it already follows that the formulas of transformation from coordinates x, y, z, t to coordinates x', y', z', t' are linear and homogeneous, i.e., have the form

$$x' = a_1 x + b_1 y + c_1 z + d_1 t,$$
$$y' = a_2 x + b_2 y + c_2 z + d_2 t,$$
$$z' = a_3 x + b_3 y + c_3 z + d_3 t, \qquad \qquad (26)$$
$$t' = a_4 x + b_4 y + c_4 z + d_4 t.$$

From other considerations one can show that their determinant* is equal to unity.

If a point in system I moves rectilinearly and uniformly in an arbitrary given direction with the speed of light c, then $x = v_x t$, $y = v_y t$, $z = v_z t$ and $v_x^2 + v_y^2 + v_z^2 = c^2$, from which

$$x^2 + y^2 + z^2 - c^2 t^2 = 0. \qquad \qquad (27)$$

But according to Michelson's experiment this point in system II also necessarily moves with the same speed of light c, so that it is also necessary that

$$x'^2 + y'^2 + z'^2 - c^2 t'^2 = 0.$$

Consequently the formulas (26) are not just arbitrary transformations which are linear, homogeneous, and with determinant equal to 1, but must at the same time satisfy the condition that if the coordinates x, y, z, t are such that

$$x^2 + y^2 + z^2 - c^2 t^2 = 0,$$

then the transformed coordinates x', y', z', t' must also satisfy this equation. Such transformations (26) are called *Lorentz transformations*.

---

* See Chapter XVI.

Let us first consider the simplest case, when the point moves along the Ox-axis. In this case formulas (26) have the form

$$x' = a_1 x + d_1 t,$$
$$t' = a_2 x + d_2 t, \qquad (26')$$

and equation (27)

$$x^2 - c^2 t^2 = 0. \qquad (27')$$

Let us introduce the notation $ct = u$, when formulas (26') and equation (27') take the form

$$x' = a_1 x + \frac{d_1}{c} u,$$
$$u' = a_2 cx + \frac{d_2 c}{c} u \qquad (26_1)$$

and

$$x^2 - u^2 = 0.$$

Let us find the explicit forms of formulas (26_1). Consider x and u as a Cartesian rectangular system in the plane, i.e., consider the problem geometrically; then we may regard formulas (26_1) as those of an affine transformation of the plane Oxu (whose determinant, as was shown is equal to 1). We will denote this transformation by L. If, as we assume, $x^2 - u^2 = 0$ implies $x'^2 + u'^2 = 0$, then this transformation translates the intersecting straight lines

$$x^2 - u^2 = 0$$

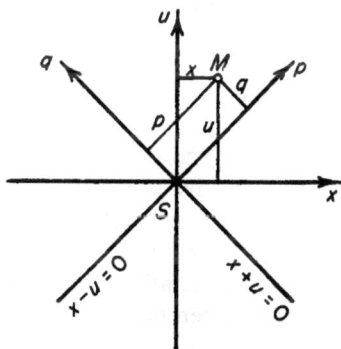

FIG. 85.

into themselves. The transformation L is therefore a combination of a contraction and expansion with identical coefficients τ along these lines.

From figure 85 we obtain

$$x = \frac{p}{\sqrt{2}} - \frac{q}{\sqrt{2}},$$
$$u = \frac{p}{\sqrt{2}} + \frac{q}{\sqrt{2}}.$$

But after the transformation L, the numbers p and q will go into $p' = p/\tau$ and $q' = q\tau$, so that

$$x' \sqrt{2} = \frac{p}{\tau} - q\tau,$$

$$u' \sqrt{2} = \frac{p}{\tau} + q\tau.$$

Expressing p and q in terms of x and u from the first pair of equations, substituting into the second and simplifying, we obtain

$$x' = \frac{x - \dfrac{\tau^2 - 1}{\tau^2 + 1} ct}{\dfrac{2\tau}{\tau^2 + 1}}, \quad t' = \frac{t - \dfrac{1}{c}\dfrac{\tau^2 - 1}{\tau^2 + 1} x}{\dfrac{2\tau}{\tau^2 + 1}},$$

or, setting $(\tau^2 - 1)/(\tau^2 + 1)c = v$, we have

$$x' = \frac{x - vt}{\sqrt{1 - (v/c)^2}}, \quad t = \frac{t - (vx/c^2)}{\sqrt{1 - (v/c)^2}},$$

which are the famous *Lorentz formulas*.

In particular, if we take $x = 0$, i.e., if we consider the motion of the origin of coordinate system I, we obtain

$$x' = \frac{-vt}{\sqrt{1 - (v/c)^2}}, \quad t' = \frac{t}{\sqrt{1 - (v/c)^2}},$$

or $x' = -vt'$, from which obviously v is the speed of motion of coordinate system II relative to system I.

Suppose, for example, that we are given two points on the Ox-axis with coordinates x_1 and x_2 relative to system I, so that the distance between them relative to system I is $r = |x_1 - x_2|$. Let us see what the distance between them is for an observer attached to system II. We have

$$x_1' = \frac{x_1 - vt}{\sqrt{1 - (v/c)^2}}, \quad x_2' = \frac{x_2 - vt}{\sqrt{1 - (v/c)^2}},$$

from which

$$r' = |x_1' - x_2'| = \frac{|x_1 - x_2|}{\sqrt{1 - (v/c)^2}}.$$

The factor $\sqrt{1 - (v/c)^2}$ is exactly the coefficient of the Lorentz contraction. Since c is very large, this coefficient is very close to 1 for moderately large v, and therefore the contraction is not significant. But such elementary particles as electrons or positrons often move with velocities comparable to the speed of light, and therefore in studying their motion it is necessary to take this contraction into account, or, as they say, to consider the relativistic effect.

We pass now to the case next in complexity, namely, when the point moves in the Oxy-plane. For this case the transformations (26) will have the form

$$\begin{aligned} x' &= a_1x + b_1y + d_1t, \\ y' &= a_2x + b_2y + d_2t, \\ t' &= a_3x + b_3y + d_3t, \end{aligned} \tag{26''}$$

where

$$\begin{vmatrix} a_1 & b_1 & d_1 \\ a_2 & b_2 & d_2 \\ a_3 & b_3 & d_3 \end{vmatrix} = 1,$$

and equation (27) will be

$$x^2 + y^2 - c^2t^2 = 0. \tag{27''}$$

These are the Lorentz formulas for motion in the Oxy-plane.

Again we put $ct = u$. Then transformations (26″) can be rewritten as

$$x' = a_1x + b_1y + \frac{d_1}{c}u,$$

$$y' = a_2x + b_2y + \frac{d_2}{c}u, \tag{26_2}$$

$$u' = a_3cx + b_3cy + \frac{d_3c}{c}u,$$

where the determinant will again be equal to one, and equation (27″) will assume the simpler form

$$x^2 + y^2 - u^2 = 0. \tag{27_2}$$

We will regard x, y, u as the Cartesian rectangular coordinates of a point in ordinary three-dimensional space and will consider formulas (26_2) as those of affine transformations of this space. Equation (27_2) represents a straight circular cone K with an angle of 90° at the vertex (figure 86).

From the point of view of this geometric interpretation (we call it geometric because here we regard $u = ct$ simply as a space coordinate) of a Lorentz transformation, the set of motions in the plane is identical with the set of all equi-affine (i.e., affine and volume-preserving) transformations of the space which map the cone K onto itself.

Let us consider some special Lorentz transformations.

1. It is clear that any simple rigid rotation about the axis of the cone K through an angle ω is an equi-affine transformation of space, mapping the cone K into itself, i.e., it is a special Lorentz transformation. We will denote it by ω.

2. Reflections of the space in an arbitrary plane π passing through the axis of the cone K are clearly also Lorentz transformations. We will denote them by π.

3. Finally, let us consider the following transformation (figure 87). Let v and w be any pair of opposite generators of the cone, and let P and Q be the planes tangent to the cone along these generators. These

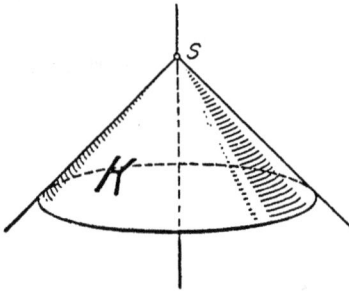

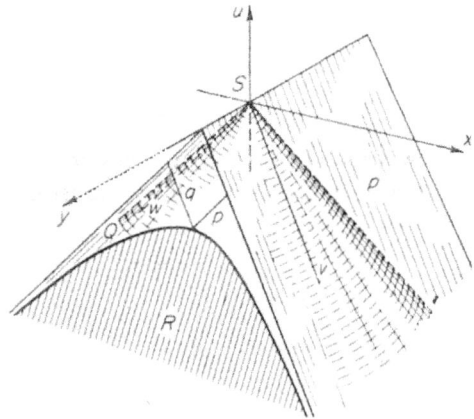

FIG. 86.   FIG. 87.

planes are mutually perpendicular. Let us make a contraction of the space to the plane P and an expansion of it with the same coefficient from the plane Q, or conversely. For example, we contract the space by a factor of three to the plane P and expand it also by a factor of three from the plane Q. Such a transformation of space is clearly also affine and preserves all volumes. We will denote it by L. We show that this transformation maps the cone K into itself. Because the cone K has the axis u as its axis of revolution, any figure can be rotated in such a way that the generators v and w lie, for example, in the plane Sxu. Therefore it is sufficient to carry out the proof for this case.

For the proof we intersect the cone K by an arbitrary plane R parallel to the Sxu plane. The equation of this plane is $y = b$, where b is a constant. Substituting this value in the equation of the cone K, we obtain

$$x^2 - u^2 = -b^2.$$

This is the equation of a hyperbola for which the lines of intersection of the plane R with the planes P and Q are exactly the asymptotes. But since for a point of such a hyperbola it is characteristic that the product of distances p and q to the asymptotes, i.e., to the planes P and Q, is constant, under transformation L all points of this hyperbola remain on the same hyperbola, and the hyperbola is mapped onto itself. But the whole surface of the cone K consists of such hyperbolas, and therefore under the transformation L of the space the cone K is sent into itself. This transformation L is therefore also a Lorentz transformation.

Since under affine transformations straight lines go into straight lines, and intersecting lines go into intersecting lines, therefore a bundle S of straight lines under any Lorentz transformation is mapped one-to-one onto itself. Moreover, under affine transformations of space all planes go into planes, so that under these transformations of the bundle S onto itself a projective transformation of the bundle is obtained. If we intersect this bundle by a plane II perpendicular to the axis of the cone K, which as a whole is not altered by the given Lorentz transformation of space, and extend this plane to the projective plane II* and then trace the points of intersection of the lines of the bundle S with the plane II*, we have the result that the Lorentz transformations of the bundle will simultaneously produce projective transformations Λ of the plane II* and these latter will transform the circle α, in which the plane II* intersects the interior part of the cone K, into itself. To analyze the properties of Lorentz transformations, it is easier to examine these projective transformations Λ of the circle α into itself.

Projective transformations of a circle into itself. A point, a ray or half line issuing from it, and one of the half planes cut off by the entire line will be called a "frame" of the plane II* (not to be confused with a coordinate frame, §11). We show (figure 88) that if we take two arbitrary frames M and M' containing interior points of the circle α, then by means of the transformations L, ω, π we can send one of these frames into the other. For this it is sufficient to make the transformations $\Lambda = L_1 \cdot \omega \cdot L_2^{-1}$ (or else $\Lambda = L_1 \cdot \omega \cdot \pi \cdot L_2^{-1}$). The transformation L_1 sends the first frame M to the center O of the circle α, the transformation ω rotates it as necessary, and finally the transformation L_2^{-1} brings it into coincidence with the second frame M'.

Let us show, in addition, that there is only one transformation Λ which translates a given frame M into a given frame M'. In order to do this we observe first that if there were two transformations Λ_1 and Λ_2 sending frame M into frame M', then the transformation $\Lambda = \Lambda_1\Lambda_2^{-1}$ would not be the identity transformation Λ and would send frame M into itself. Therefore, it is sufficient to show that if a transformation Λ sends frame M into itself, then it is the identity, i.e., leaves all points of the plane of circle α fixed.

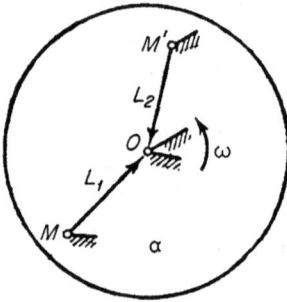

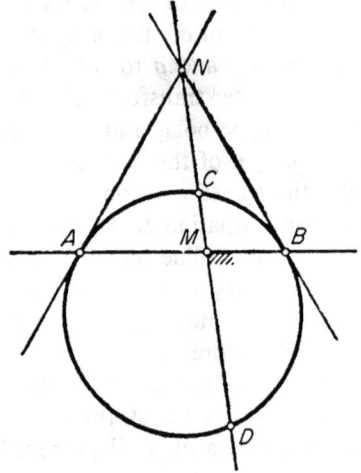

FIG. 88.   FIG. 89.

Let us show this. Suppose that the transformation Λ sends frame M into itself (figure 89). Then it maps the line AB of this frame into itself, but since it sends the circumference of the circle α into itself, it therefore leaves points A and B fixed, or else interchanges them. The latter, however, is impossible, since the half line of the frame is mapped into itself. Let us draw the tangents at points A and B to the circle α. They are mapped into themselves, since if such a tangent were mapped into a secant $A\bar{A}$, then the inverse transformation would send the different points A and $\bar{A}$ of the circle α into the one point A. But the Λ are projective transformations, and consequently one-to-one. Since under the transformations Λ these tangents go into themselves, therefore the point N of their intersection remains fixed, and consequently the line MN is mapped into itself. From the fact that the half line of the frame M is mapped into itself, we conclude as above that points C and D are not interchanged, but remain fixed. Hence, under the given projective transformation Λ of the projective plane II* four of its points A, B, C, D, no three of which lie on the same line, remain fixed. According to the uniqueness theorem of projective transformations this is the identity transformation.

Later, in §5 of Chapter XVII it will be shown that by using these properties of the Lorentz group, it is easy to construct a model of

Lobačevskiĭ's plane geometry, and if we consider Lorentz transformations for the general case of motion of a point in space, then we can do the same for Lobačevskiĭ's space geometry, and thereby prove its consistency.

We see that the theory of Lorentz transformations, projective geometry and the theory of perspectivity and non-Euclidean geometry are closely related to one another. It turns out that there is still another theory that is also closely related to them, namely, the so-called conformal transformations in the theory of functions of a complex variable, which solve such important problems of mathematical physics as the distribution of temperature in a heated plate, the flow of air around the wing of an airplane, the distribution of charge in a plane electrostatic field, the problems of elasticity in the plane, and many others.

Conclusion

Analytic geometry is an absolutely indispensable method for the investigation of other branches of mathematics, physics, and other natural sciences. Therefore it is studied not only at universities but in all technical higher institutions of learning, and also in some vocational schools. It is also a question of whether we should not include a fairly detailed treatment of the elements of analytic geometry in high school courses.

Various coördinates. The essential elements of the concept of analytic geometry, as we have seen, are the coordinate method and the investigation of equations connecting these coördinates. Besides Cartesian coordinates, other different ones can be considered. For example, in the plane, we can choose a point P (the so-called pole) and a ray originating from it (the polar axis) and determine the position of a point M by the length ρ of the polar radius from the pole to the point and the value ω of the angle made by this radius with the polar axis (figure 90).

In particular, the ellipse, hyperbola, or parabola, if for the pole we take

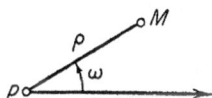

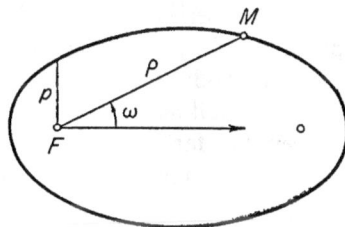

FIG. 90.   FIG. 91.

a focus, and for the polar axis the ray passing from the focus along the axis of symmetry to the side opposite the nearer vertex (figure 91), have one and the same equation

$$\rho = \frac{p}{1 - \epsilon \cos \omega},$$

where ϵ is the eccentricity of the curve, and p is its so-called parameter. This equation is of a great importance in astronomy. For it was with its help that the result was derived, from the law of inertia and the law of universal gravitation, that the planets revolve about the Sun in ellipses.

The geographical coordinates, latitude and longitude, by which the position of a point is given on a sphere, are well known.

Analogously, we can take a coordinate network on an arbitrary surface, as is done in differential geometry (see Chapter VII), etc.

Many-dimensional and infinite-dimensional analytic geometry; algebraic geometry. It would seem that in the 19th century analytic geometry underwent such an immense development, described earlier in a general way, and produced so many ideas, that it would have necessarily exhausted itself, but this is not so. In very recent times, two new, extensive branches of mathematics have been rapidly developed and have extended the concepts of analytic geometry, namely so-called *functional analysis* and general *algebraic geometry*. It is true that both of these only halfway represent a straightforward continuation of classical analytic geometry: Much of functional analysis is analysis, and in algebraic geometry there is more than a little of the theory of functions and of topology.

Let us explain what we mean. In the middle of the last century mathematicians had already begun to consider four-dimensional and general n-dimensional analytic geometry, i.e., to study those questions of algebra that are straightforward generalizations of algebraic questions of the kind involved in two- and three-dimensional analytic geometry, to the case when there are four or n unknowns. At the very end of the 19th century a series of outstanding analysts came to the idea that for the purposes of analysis and mathematical physics it is significant to consider infinite-dimensional analytic geometry.

At first glance it may seem that n-dimensional or even four-dimensional spaces seem like farfetched mathematical fictions, then the same can also be said about an infinite-dimensional space. But it is not really so. The arguments concerning an infinite-dimensional space are not at all difficult. They now constitute an important branch of mathematics, functional analysis (see Chapter XIX).

It is curious that infinite-dimensional analytic geometry has most important practical applications and plays a fundamental role in contemporary physics.

As to algebraic geometry, it is a more immediate continuation of ordinary analytic geometry, which is itself only a part of algebraic geometry. Algebraic geometry can be regarded as that part of mathematics which is occupied with curves, surfaces, and hypersurfaces, represented in Cartesian coordinates by algebraic equations of not only first and second degree, but also of higher degrees. It turns out that in these investigations it is advantageous to consider not only real but also complex coordinates, i.e., to consider everything in a so-called complex space. The most important results in this domain were obtained in the last century by Riemann. As a brilliant example of theorems about higher order curves, we point out a remarkably general result of I. G. Petrovskiĭ about the number of ovals into which an nth-order curve can be decomposed. Petrovskiĭ showed that if p is the number of such ovals which do not lie at all in other ovals, or lie in an even number of ovals, and m is the number of those ovals which lie in an odd number of ovals, and if we consider only curves whose component ovals neither intersect themselves nor each other (figure 92), then

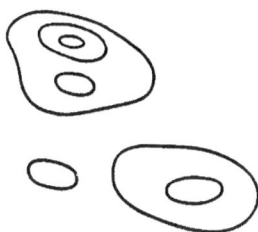

FIG. 92.

$$p - m \leqslant \frac{3n^2 - 6n}{8} + 1,$$

where n is the order of the curve, i.e., the degree of the equation by which the curve is represented.

This result is the more important as up to then almost nothing had been known about the general form of a higher order curve. It is no doubt one of the most important recent general theorems in analytic geometry.

Suggested Reading

A. A. Albert,. *Solid analytic geometry*, McGraw-Hill, New York, 1949.

J. L. Coolidge, *A history of the conic sections and quadratic surfaces*, Oxford University Press, Oxford, 1945.

H. S. M. Coxeter, *The real projective plane*, 2nd ed., Cambridge University Press, New York, 1960.

L. P. Eisenhart, *Coordinate geometry*, Dover, New York, 1960.

A. Jaeger, *Introduction to analytic geometry and linear algebra*, Holt, Rinehart and Winston, New York, 1960.

D. J. Struik, *Lectures on analytic and projective geometry*, Addison-Wesley, Cambridge, Mass., 1953.

CHAPTER IV

ALGEBRA: THEORY
OF ALGEBRAIC EQUATIONS

§1. Introduction

The characteristic features of algebra are well known to everyone, since the elementary but fundamental information about it is already given in high school. Algebra is characterized, first of all, by its method, involving the use of letters, and expressions in letters, on which we perform operations according to definite laws. In elementary algebra the letters denote ordinary numbers, so that the laws of operations on expressions in letters are based on the general laws of operations on numbers. For example, the sum does not depend on the order of the summands, a fact which in algebra is written as: $a + b = b + a$; in multiplying the sum of two numbers, we can multiply each one of the numbers individually and then add the products so obtained: $(a + b)c = ac + bc$, etc.

If we trace the proof of an algebraic theorem, it is easy to see that it depends only on these laws for operations on numbers and not at all on what the letters represent.

The algebraic method, i.e., the method of calculations with letters, penetrates all of mathematics. In fact, a substantial part of the solution of a mathematical problem often turns out to be nothing but a more or less complicated algebraic computation. Besides, in mathematics we employ various symbolic calculations in which the letters no longer denote numbers but some other entities, where the laws for operations on these entities may be different from the laws of elementary algebra. For example, in geometry, mechanics, and physics we make use of vectors, and as is well known, the laws for operations on vectors are in part the same as for numbers and in part essentially different.

261

The significance of the algebraic method in modern mathematics and the range of its applications have greatly increased in recent decades.

First of all, the growing demands of technology force us to reduce to numerical results the solutions of difficult problems of mathematical analysis, and this usually proves to be feasible only after the algebraization of these problems, a process which in turn creates new and sometimes difficult problems in algebra itself.

Second, certain problems of analysis became clear and understandable only after they were attacked by algebraic methods based on a profound generalization (to the case of infinitely many unknowns) of the theory of systems of equations of the first degree.

Finally, the more advanced parts of algebra have found application in contemporary physics: In fact, the fundamental concepts of quantum mechanics are expressed in terms of complicated and nonelementary algebraic entities.

The basic features of the history of algebra are as follows.

First of all we must point out that our ideas regarding what algebra is and what its fundamental problem consists of have changed twice: once in the first half of the past century, and the second time at the beginning of our century. Thus, algebra has meant at different times three quite different things. In this respect the history of algebra differs from the history of the three famous branches of mathematics: analytic geometry, differential calculus, and integral calculus, which were forged into shape at the hands of their creators, Fermat, Descartes, Newton, Leibnitz, and others and were later rapidly developed and amplified, sometimes by the addition of great new sections, but were comparatively little changed in their fundamental character.

In ancient times any law that was discovered for the solution of a class of mathematical problems was recorded simply in words, since symbolic calculations had not yet been invented. The word "algebra" itself was created from the name of the important work of the Kharizmian scientist of the 9th century, Mohammed Al-Kharizmi (see Chapter I), in whose works the first general law for the solution of first- and second-degree equations was deduced. However, the introduction of the symbolic notation itself is usually associated with the name of Viète, who not only began to denote the unknowns by letters but also the given quantities. Descartes also did a great deal for the development of symbolic notation, and he too, of course, took the letters to mean ordinary numbers. It is at this moment that algebra really begins as the science of symbolic calculations, of transformations of formulas composed of letters, of algebraic equations, and so forth, in contrast to arithmetic, which always operates on concrete numbers. Only now did complicated mathematical concepts become perspicuous and accessible to investigation, since by taking a look at a formula in letters, it is in most cases possible for us to see its general arrangement or law of formation and to subject it to suitable transformations. At that time everything in mathematics which was neither geometry nor infinitesimal analysis was called algebra. This is the first, the so to say Viète point of view, concerning algebra. It was very clearly expressed in the well-known book "Introduction to Algebra" by a member of the Russian Academy of Sciences, the famous L. Euler, written in the 1760's, i.e., 200 years ago.

Euler defined algebra as the theory of calculations with quantities of various kinds. The first part of his book contains the theory of calculation with integral rational numbers, ordinary fractions, square and cube roots, the theory of logarithms, progressions, the theory of calculations with polynomials, and the theory of Newton's binomial series and its applications. The second part consists of the theory of first-degree equations and of systems of such equations, the theory of quadratic equations and of solutions of third- and fourth-degree equations by radicals, and also an extensive section on methods of solutions of various indeterminate equations in integers. For example, it was shown that Fermat's equation $x^3 + y^3 = z^3$ cannot be solved in integers x, y, z.

At the end of the 18th and the beginning of the 19th century, one of the problems of algebra gradually began to occupy the central place, namely the theory of solution of algebraic equations, in which the fundamental difficulty is the solution of an nth-degree algebraic equation with one unknown

$$x^n + a_1 x^{n-1} + a_2 x^{n-2} + \cdots + a_{n-1} x + a_n = 0.$$

This happened as a natural consequence of the importance of the problem for the whole of pure and applied mathematics, and also because of the difficulty and depth of the majority of the theorems connected with it.

The general formula for the solution of a quadratic equation,

$$x = -\frac{p}{2} \pm \sqrt{\frac{p^2}{4} - q},$$

was known to everybody. Italian algebraists of the 16th century found analogous, though more complicated, general rules for the solution of arbitrary third- and fourth-degree equations. Further investigations in this direction for higher degree equations, however, met with insurmountable difficulties. The greatest mathematicians of the 16th, 17th, 18th and the beginning of the 19th century (Tartaglia, Cardan, Descartes, Newton, d'Alembert, Tschirnhausen, Bézout, Lagrange, Gauss, Abel, Galois,

Lobačevskiĭ, Sturm, and others) created an impressive edfice of theorems and methods connected with this problem. The two-volume algebra of Serret (an epoch-making work of its time, since it presented for the first time the high point of the theory of algebraic equations, namely the theory of Galois), appeared in the middle of the 19th century, exactly 100 years after Euler's text; in it algebra was already defined as the theory of algebraic equations. This is the second point of view concerning what algebra is.

In the second half of the past century there occured, on the basis of the ideas of Galois about the theory of algebraic equations, a profound development of group theory* and the theory of algebraic numbers (in the creation of which a great part was played by the Russian mathematician E. I. Zolotarev).

In this second period also, in connection with the same problems of solution of an algebraic equation, and with the theory of algebraic varieties of higher order (which were then being studied in analytic geometry) the algebraic apparatus was developed in different directions, e.g., the theory of determinants and matrices, the algebraic theory of quadratic forms and linear transformations, and, in particular, the theory of invariants. During almost the entire second half of the 19th century, the theory of invariants was a central theme in algebra. In turn, the development of group theory and the theory of invariants exerted in this period a great influence on the development of geometry.†)

A new, third point of view as to what algebra is came into existence chiefly in the following connection. In the second half of the last century, in mechanics, physics, and mathematics itself, scientists began more and more often to investigate objects for which it was natural to consider operations of addition and subtraction, and sometimes multiplication and division, but for which these operations were subjected to altogether different laws from those for rational numbers.

We have already spoken of vectors. Other sorts of mathematical objects with different laws of operation can only be mentioned here: e.g., matrices, tensors, spinors, hypercomplex numbers. All these quantities are denoted by letters, but their laws of operation differ from one another. If for some set of objects (denoted by letters) certain operations are defined together with the laws or rules that they must satisfy, then we say that an algebraic system is defined. The third point of view on what algebra is consists of regarding the whole of algebra as the study of various algebraic systems. This is the so-called axiomatic or abstract algebra. It is abstract because

---

* See Chapter XX.
† See Chapter XVII.

at a given step in the calculation we are not all concerned with what the letters in the algebraic system denote, the only important thing is the axioms or laws satisfied by the operations; and it is called axiomatic, because it is constructed exclusively from the axioms stated at the beginning. It is as though we have returned, but on a higher level, to the first or Viète point of view on algebra, that algebra is the theory of symbolic calculations. Although it makes no difference what the letters denote and only the rules of operation are important, it is still true, of course, that only those algebraic systems are interesting which have great significance either in mathematics itself or in its applications.

The great amount of algebraic material collected in the previous period served as the actual basis for the construction of contemporary abstract algebra.

The early 1930's saw the appearance of van der Waerden's well-known book "Modern Algebra," which has played a great role in the propagation of this third point of view as to what algebra is. The text of A. G. Kuroš on algebra is oriented in the same direction.

In the present century algebra has found deep applications to geometry (topology and the theory of Lie groups) and, as mentioned earlier, to contemporary physics, especially to functional analysis and quantum mechanics.

Particularly important at the present time are the problems of mechanization of algebraic calculations by means of various mathematical computing machines, especially high-speed electronic machines. The questions connected with this type of computational mathematics raise new distinctive problems in algebra.

In the present work, there are two chapters (not counting the present one) that are devoted to algebra: linear algebra (Chapter XVI) and the theory of groups and other algebraic systems (Chapter XX).

§2. Algebraic Solution of an Equation

An algebraic equation of the nth-degree with one unknown is an equation of the form

$$x^n + a_1 x^{n-1} + a_2 x^{n-2} + \cdots + a_{n-1} x + a_n = 0,$$

where a_1, a_2, $\cdots$, a_n are given coefficients.*

---

* We assume that all terms of the equation are transferred to the left-hand side and that the equation is divided by the coefficient of the highest power of the unknown.

Equations of the first- and second-degree. If the equation is of the the first-degree, then it has the form

$$x + a = 0$$

and is solved at once

$$x = -a.$$

The second-degree equation

$$x^2 + px + q = 0$$

was solved in early antiquity. Its solution is very simple: If we transfer q with the opposite sign to the right-hand side and then add $p^2/4$ to both sides we have

$$x^2 + px + \frac{p^2}{4} = \frac{p^2}{4} - q.$$

But

$$x^2 + px + \frac{p^2}{4} = \left(x + \frac{p}{2}\right)^2$$

hence

$$x + \frac{p}{2} = \pm\sqrt{\frac{p^2}{4} - q},$$

from which we obtain the well-known formula for the solutions of a quadratic equation

$$x = -\frac{p}{2} \pm \sqrt{\frac{p^2}{4} - q}.$$

Third-degree equation. It was completely different with equations of degree higher than 2. Already the general equation of the third-degree required quite profound considerations and resisted all the efforts of the mathematicians of antiquity. It was only solved at the beginning of the 1500's, in the era of the Renaissance in Italy, by the Italian mathematician Scipio del Ferro. Del Ferro, following the custom of his time, did not publish his own discoveries but communicated them to one of his pupils. After the death of del Ferro this pupil challenged to competition one of the great Italian mathematicians Tartaglia and proposed to him for solution a series of third-degree equations. Tartaglia (1500–1557) accepted the challenge and eight days before the end of the competition found a method of solving any cubic equation of the form $x^3 + px + q = 0$.

In two hours he solved all problems of his opponent. A professor of physics and mathematics in Milan, Cardan (1501–1576), learning of Tartaglia's discoveries, began to entreat Tartaglia to inform him of his secret. Tartaglia finally agreed, but with the condition that Cardan keep his method a deep secret. Cardan violated his promise and published Tartaglia's result in his work "The great art" (*Ars Magna*).

The formula for the solution of a cubic equation has since then been called Cardan's formula, although it would be correct to call it Tartaglia's formula.

Cardan's formula is derived as follows.

In the first place, the solution of the general cubic equation

$$y^3 + ay^2 + by + c = 0 \tag{1}$$

can easily be reduced to the solution of the cubic equation of the form

$$x^3 + px + q = 0, \tag{2}$$

not containing a term with the square of the unknown. To do this it is sufficient to set $y = x - a/3$. Indeed, substituting this expression into equation (1) and removing the parentheses, we obtain

$$\left(x - \frac{a}{3}\right)^3 + a\left(x - \frac{a}{3}\right)^2 + b\left(x - \frac{a}{3}\right) + c = x^3 - 3x^2\frac{a}{3} + \cdots + ax^2 + \cdots,$$

where the dots indicate those terms in which x is raised to first power or does not appear at all. We see that the terms containing x^2 cancel each other out.

Let us now consider the following equation

$$x^3 + px + q = 0.$$

We set $x = u + v$, i.e., in place of one unknown we put two, u and v, and thereby turn the whole problem into a problem with two unknowns. We have

$$(u + v)^3 + p(u + v) + q = 0,$$

or

$$u^3 + v^3 + q + (3uv + p)(u + v) = 0.$$

Whatever is the sum of the two numbers $u + v$, it is always possible to require that their product uv be equal to some quantity given beforehand. If $u + v = A$, and we require $uv = B$, then since $v = A - u$, we obtain

$$u(A - u) = B,$$

so that it is sufficient that u be a solution of the quadratic equation

$$u^2 - Au + B = 0,$$

and we know that every quadratic equation has real or complex roots, given by the well-known formula. In our case, $u + v$ is equal to the desired root x of our cubic equation and we require that

$$uv = -\frac{p}{3},$$

i.e., that $3uv + p = 0$. With this choice of u and v we obtain

$$u^3 + v^3 + q = 0,$$
$$3uv + p = 0. \tag{3}$$

Consequently, if we find the numbers u and v, satisfying this system of equations then the number $x = u + v$ will be the root of our equation.

From system (3) it is easy to form a quadratic equation whose roots will be u^3 and v^3. Indeed, it gives

$$u^3 + v^3 = -q,$$
$$u^3v^3 = -\frac{p^3}{27},$$

and, consequently by a theorem already used earlier u^3 and v^3 are the roots of the quadratic equation

$$z^2 + qz - \frac{p^3}{27} = 0.$$

Solving it by the usual formula, we obtain

$$u^3 = -\frac{q}{2} + \sqrt{\frac{q^2}{4} + \frac{p^3}{27}}, \quad v^3 = -\frac{q}{2} - \sqrt{\frac{q^2}{4} + \frac{p^3}{27}},$$

and, consequently,

$$x = \sqrt[3]{-\frac{q}{2} + \sqrt{\frac{q^2}{4} + \frac{p^3}{27}}} + \sqrt[3]{-\frac{q}{2} - \sqrt{\frac{q^2}{4} + \frac{p^3}{27}}},$$

this is the formula of Cardan.

Fourth-degree equation. Soon after the solution of the cubic equation the general fourth-degree equation was solved by Ferrari (1522-1565). For the solution of the third-degree equation we have seen that the preliminary solution of the auxiliary quadratic equation,

$$z^2 + qz - \frac{p^3}{27} = 0,$$

was necessary, where $z = u^3$ or v^3; analogously, the solution of a fourth-degree equation can be based on the preliminary solution of an auxiliary cubic equation.

Ferrari's method consists of the following. Let the general fourth-degree equation be given

$$x^4 + ax^3 + bx^2 + cx + d = 0.$$

Let us rewrite it as:

$$x^4 + ax^3 = -bx^2 - cx - d$$

and add to both sides $a^2x^2/4$; then on the left we obtain a perfect square

$$\left(x^2 + \frac{ax}{2}\right)^2 = \left(\frac{a^2}{4} - b\right)x^2 - cx - d.$$

Adding now to both sides of the equation the terms

$$\left(x^2 + \frac{ax}{2}\right)y + \frac{y^2}{4},$$

where y is a new variable, on which we later impose a necessary condition, on the left we obtain a perfect square

$$\left(x^2 + \frac{ax}{2} + \frac{y}{2}\right)^2 = \left(\frac{a^2}{4} - b + y\right)x^2 + \left(\frac{ay}{2} - c\right)x + \left(\frac{y^2}{4} - d\right). \quad (4)$$

Thus we have reduced the problem to one with two unknowns.

On the right of equation (4) we have a quadratic trinomial in x, whose coefficients depend on y. We select y such that this trinomial will be the square of the first-degree binomial $\alpha x + \beta$.

In order that the quadratic trinomial $Ax^2 + Bx + C$ be the square of the binomial $\alpha x + \beta$ it is sufficient that

$$B^2 - 4AC = 0.$$

Indeed, if $B^2 - 4AC = 0$, then

$$Ax^2 + Bx + C = (\sqrt{A}x + \sqrt{C})^2,$$

i.e.,

$$Ax^2 + Bx + C = (\alpha x + \beta)^2,$$

where

$$\alpha = \sqrt{A}, \quad \beta = \sqrt{C}.$$

Consequently, if we select y such that

$$\left(\frac{ay}{2} - c\right)^2 - 4\left(\frac{a^2}{4} - b + y\right)\left(\frac{y^2}{4} - d\right) = 0,$$

then the first part of equation (4) will be the complete square $(\alpha x + \beta)^2$. Removing the parentheses, we obtain a cubic equation in y

$$y^3 - by^2 + (ac - 4d)y - [d(a^2 - 4b) + c^2] = 0.$$

Solving this auxiliary cubic equation (for example, by Cardan's formula) we find α and β in terms of its solution y_0, namely

$$\left(x^2 + \frac{ax}{2} + \frac{y_0}{2}\right)^2 = (\alpha x + \beta)^2,$$

from which

$$x^2 + \frac{ax}{2} + \frac{y_0}{2} = \alpha x + \beta \quad \text{or} \quad x^2 + \frac{ax}{2} + \frac{y_0}{2} = -\alpha x - \beta.$$

From these two quadratic equations we can find all four roots of the given fourth-degree equation.

This is how third- and fourth-degree algebraic equations were solved by Italian mathematicians in the 1500's.

The success of the Italian mathematicians produced a very great effect. It was the first instance when modern science had exceeded the achievements of the ancients. Until then, in the whole course of the Middle Ages, the aim had always been only to understand the work of the ancients, and now, finally, certain questions were solved which the ancients had not succeeded in conquering. And this happened in the 1500's, i.e., in the century before the invention of new branches of mathematics: analytic geometry, differential calculus, and integral calculus, which finally affirmed the superiority of the new science over the old. After this there was no important mathematician who did not attempt to extend the achievements of the Italians and to solve equations of fifth, sixth, and higher degree in an analogous way by means of radicals.

The prominent algebraist of the 17th century, Tschirnhausen (1651–1708) even believed that he had finally found a general method of solution. His method was based on the transformation of an equation to a simpler one, but this very transformation required the solution of some auxiliary equations. Subsequently, by a deeper analysis it was shown that Tschirnhausen's method of transformation indeed gives the solution of second-third-, and fourth-degree equations, but already for a fifth-degree equation it requires the preliminary solution of an auxiliary equation of the sixth-degree, whose solution in turn was not known.

Factorization of a polynomial and Viète's formulas. If we accept without proof the so-called fundamental theorem of algebra* that every equation

$$f(x) = 0,$$

where $$f(x) = x^n + a_1 x^{n-1} + \cdots + a_n$$

is a polynomial in x of given degree n and the coefficients $a_1, a_2, \cdots, a_n$ are given real or complex numbers, has at least one real or complex root, and take into consideration that all computations with complex numbers are carried out by the same rules as with rational numbers, then it is easy to show that the polynomial $f(x)$ can be represented (and in only one way) as a product of first-degree factors

$$f(x) = (x - a)(x - b) \cdots (x - l),$$

where $a, b, \cdots, l$ are real or complex numbers.

Indeed, let a be a root of $f(x)$; we divide $f(x)$ by $x - a$; since the divisor is of the first-degree, the remainder will be a constant number R, i.e., we will have the identity

$$f(x) = (x - a) f_1(x) + R,$$

where $f_1(x)$ is a polynomial of degree $n - 1$ and R is a constant. Substituting here in place of x the number a, we obtain

$$f(a) = (a - a) f_1(a) + R = R.$$

But since a is a root of $f(x)$, we have $f(a) = 0$, and hence $R = 0$, i.e., a polynomial can always be divided by $(x - a)$ without remainder, where a is a root of this polynomial. Thus

$$f(x) = (x - a) f_1(x).$$

But if the fundamental theorem of algebra is true, then in turn the polynomial $f_1(x)$ has a root b, and we obtain analogously

$$f_1(x) = (x - b) f_2(x),$$

where the polynomial $f_2(x)$ is already of degree $(n - 2)$, etc. This factorization, as can easily be shown, is unique.

Every nth-degree polynomial $f(x)$ has in this sense n and only n roots $a, b, \cdots, l$. These roots may be all distinct but it can happen that some among them are identical. Then we say that the corresponding root of

---

* The proof of the fundamental theorem of algebra is difficult and was given considerably later. We devote §3 to it. But its validity was assumed long before it was rigorously proved.

the polynomial $f(x)$ is a multiple root with such and such a multiplicity. Multiplying out the expression

$$(x - a)(x - b)(x - c) \cdots (x - l)$$

and comparing the coefficients of the same powers of x, we see immediately that

$$-a_1 = a + b + c + \cdots + l,$$
$$a_2 = ab + ac + \cdots + kl,$$
$$-a_3 = abc + abd + \cdots,$$
$$\cdots\cdots\cdots\cdots\cdots\cdots\cdots\cdots\cdots\cdots$$
$$\pm a_n = abc \cdots l$$

which are Viète's formulas.

A theorem on symmetric polynomials.   Viète's formulas are polynomials in the n letters a, b, $\cdots$, l which do not vary under any permutation of these letters. Indeed, $a + b + \cdots + k + l = b + a + \cdots + k + l$, etc. In general, any such polynomials in n letters, which do not change under any permutations of these letters, are called symmetric polynomials in n letters. For example, $5x^2 + 5y^2 - 7xy$ is a symmetric polynomial in x and y. It is possible to prove the theorem that every integral symmetric polynomial in n letters with arbitrary coefficients A, B, $\cdots$ can be expressed integral rationally, i.e., with the operations of addition, subtraction, and multiplication, in terms of the coefficients A, B, $\cdots$ and of Viète's polynomials in the letters. If a, b, $\cdots$, l are the roots of an nth-degree equation $x^n + a_1 x^{n-1} + \cdots + a_n = 0$, then every symmetric polynomial in a, b, $\cdots$, l with arbitrary coefficients A, B, $\cdots$ can thus be expressed integral rationally in terms of these coefficients A, B, $\cdots$ and the coefficients $a_1, a_2, \cdots, a_n$ of the equation. This is the so-called fundamental theorem of symmetric polynomials.

Lagrange's contributions.   The famous French mathematician Lagrange in his great work "Reflections on the solution of algebraic equations" published in 1770–1771 (with more than 200 pages), critically examined all the solutions of second-, third- and fourth-degree equations that were known up to his time and showed that their success was always based on properties which did not hold for equations of degree 5 or higher. From del Ferro's time until this work of Lagrange more than two and a half centuries had passed by and nobody during this long interval had doubted the possibility of solving equations of degree 5 and higher by radicals, i.e., of finding formulas involving only the operations of addition, subtraction, multiplication, division, and radicals with integral positive exponents, which would express the solution of an equation in terms of its coefficients, that is, formulas similar to those by which the quadratic equation had been solved in antiquity and the third- and fourth-degree equations in the 1500's by the Italians. They regarded this situation as being due only to their own inability to find a valid but apparently deeply hidden solution.

Lagrange says in his memoir: "The problem of solving (by radicals) equations whose degree is higher than four is one of those problems which have not been solved although nothing proves the impossibility of solving them" and two pages later he supplements this: "From our reasoning we see that it is very doubtful that the methods which we have considered could give a complete solution of equations of the fifth-degree."

In his investigations, Lagrange introduced the expression

$$a \dot{-} \epsilon b + \epsilon^2 c \dot{\pm} \cdots \dot{+} \epsilon^{n-1} l$$

in the roots a, b, $\cdots$, l of an equation, where ϵ is an nth root of unity,* having established that such expressions are closely connected with the solution of equations by radicals. These expressions are now called "Lagrange resolvents."

In addition, Lagrange observed that the theory of permutations of roots of an equation is of great importance in the theory of solution of equations. He even expressed the thought that the theory of permutations is the "true philosophy of the whole question," in which he was completely right, as was shown in the later investigations of Galois.

Lagrange's method of solution of second-, third- and fourth-degree equations were not the same as those of the Italians, which in every case were based on special transformations of a complicated and so to speak accidental kind. Lagrange's methods were altogether orderly and developed from one general idea involving the theory of symmetric polynomials, the theory of permutations, and the theory of resolvents.

---

* I.e., a complex number which raised to the nth power is equal to one. For example, the cube roots of unity can have the values

$$1, \quad -\frac{1}{2} + \frac{\sqrt{3}}{2} i, \quad -\frac{1}{2} - \frac{\sqrt{3}}{2} i,$$

where $i = \sqrt{-1}$ (see §3). Indeed,

$$\left(-\frac{1}{2} + \frac{\sqrt{3}}{2} i\right)^3 = -\frac{1}{8} - \frac{3}{8} \sqrt{3} i + \frac{9}{8} + \frac{3\sqrt{3}}{8} i = 1,$$

and analogously

$$\left(-\frac{1}{2} - \frac{\sqrt{3}}{2} i\right)^3 = 1.$$

Let us consider, for example, the solution by Lagrange's method of the general fourth-degree equation

$$x^4 + mx^3 + nx^2 + px + q = 0.$$

Let the roots of this equation be a, b, c, d. Consider the resolvent

$$a + b - c - d,$$

i.e.,

$$a + \epsilon c + \epsilon^2 b + \epsilon^3 d,$$

where $\epsilon = -1$. If we permute a, b, c, d in all $1 \cdot 2 \cdot 3 \cdot 4 = 24$ different ways, we obtain altogether six different expressions

$$\begin{aligned}
&a + b - c - d, \\
&a + c - b - d, \\
&a + d - c - b, \\
&c + d - a - b, \\
&b + d - a - c, \\
&b + c - a - d.
\end{aligned} \tag{5}$$

An equation of the sixth-degree, whose roots are these six expressions, will thus have coefficients that do not vary with all 24 permutations of a, b, c, d, since any of the 24 permutations can only permute these expressions among themselves and the coefficients of the sixth-degree equation do not depend on the order in which we take its roots. Thus, these coefficients are symmetric polynomials in a, b, c, d. But then, by virtue of the fundamental theorem on symmetric polynomials, these coefficients are expressed integral rationally in terms of the coefficients m, n, p, q of the equation. In addition, since expressions (5) are pairwise of opposite signs, this sixth-degree equation will contain only terms of even powers. Indeed, if expressions (5) are denoted by α, β, γ, $-\alpha$, $-\beta$, $-\gamma$ respectively, then the left-hand side of the sixth-degree equation will be equal to

$$(y - \alpha)(y + \alpha)(y - \beta)(y + \beta)(y - \gamma)(y + \gamma)$$
$$= (y^2 - \alpha^2)(y^2 - \beta)(y_2 - \gamma^2).$$

Direct computation gives the sixth-degree equation

$$y^6 - (3m^2 - 8n)y^4 + 3(m^4 - 16m^2n - 16n^2 + 16mp - 64q)y^2$$
$$- (m^2 - 4m + 8p)^2 = 0.$$

Letting $y^2 = t$, we obtain a cubic equation in t, and if t', t'', t''' are its roots, then

$$\begin{aligned}
a + b - c - d &= \sqrt{t'}, \\
a + c - b - d &= \sqrt{t''}, \\
a + d - b - c &= \sqrt{t'''}.
\end{aligned}$$

We also have

$$a + b + c + d = -m.$$

Adding these equations after multiplication by 1, 1, 1, 1 or 1, -1, -1, 1, or -1, 1, -1, 1, or -1, -1, 1, 1, we obtain

$$a = \frac{1}{4}(-m + \sqrt{t'} + \sqrt{t''} + \sqrt{t'''}),$$

$$b = \frac{1}{4}(-m + \sqrt{t'} - \sqrt{t''} - \sqrt{t'''}),$$

$$c = \frac{1}{4}(-m - \sqrt{t'} + \sqrt{t''} - \sqrt{t'''}),$$

$$d = \frac{1}{4}(-m - \sqrt{t'} - \sqrt{t''} + \sqrt{t'''}).$$

Thus, the solution of a fourth-degree equation is reduced to the solution of a cubic equation; and third- and second-degree equations are solved analogously.

Lagrange achieved a great deal in the theory of algebraic equations. However, even after his persistent efforts the problem of solution in radicals of algebraic equations with degree higher than 4 remained to be settled. This problem, on which mathematicians had worked in vain for almost three centuries, constituted, in the expression of Lagrange, "a challenge to the human mind."

Abel's discovery. Consequently it was a great surprise to all mathematicians when in 1824 the work of a young Norwegian genius Abel (1802–1829) came to light, in which a proof was given that if the coefficients of an equation $a_1, a_2, \cdots, a_n$ are regarded simply as letters, then there does not exist any radical expression in these coefficients that is a root of the corresponding equation, if its degree $n \geqslant 5$. Thus, for three centuries the efforts of the greatest mathematicians of all countries to solve equations of degree 5 or higher in radicals did not lead to success for the simple reason that this problem simply does not have a solution.

Such a formula is known for second-degree equations, and as we saw analogous formulas exist for third- and fourth-degree equations, but for equations of degree 5 or greater there are no such formulas.

Abel's proof is difficult and we will not give it here.

Galois theory. But this was not yet all. A very remarkable result in the theory of algebraic equations still remained to come. The point is that there are arbitrarily many special forms of equations of any degree that are solvable in radicals, and many of them are exactly those equations that are important in the applications. Such, for instance, are the binomial equations $x^n = A$. Abel found another very broad class of such equations, the so-called cyclic equations and still more general "Abelian" equations. In connection with the problem of construction by ruler and compass of regular polygons, Gauss explicitly considered the so-called cyclotomic equations, i.e., equations of the form

$$x^{p-1} + x^{p-2} + \cdots + x + 1 = 0,$$

where p is a prime number, and showed that they can always be reduced to a chain of equations of lower degree; moreover, he found necessary and sufficient conditions that such an equation can be solved in square roots. The necessity of these conditions was rigorously proved only by Galois.

Thus, after Abel's work the situation was the following: Although, as was shown by Abel, the general equation of degree higher than 4 cannot be solved by radicals, there are arbitrarily many different special equations of arbitrary degree, all of which can be solved by radicals. The whole question of solving equations in radicals was placed by these discoveries on completely new ground. It became clear that the task now was to determine exactly which equations can be solved by radicals, or in other words, what are the necessary and sufficient conditions for the solvability of an equation in radicals. This problem, the answer to which gave in some sense the final elucidation of the whole problem, was solved by the ingenious French mathematician Evariste Galois.

Galois (1811–1832) perished at the age of 20 in a duel. In the last two years of his life he could not devote much time to mathematics, since he was carried away by the stormy whirl of political life at the time of the 1830 Revolution and languished in jail for his speech against the reactionary regime of Louis Philippe. Nevertheless, in his short life, Galois made discoveries far ahead of his time in various parts of mathematics and in particular produced some very remarkable results in the theory of algebraic equations. In a small publication "Memoir on the conditions of solvability of equations in radicals" which remained in manuscript form after his death and was first published in 1846 by Liouville, Galois started from some very simple but profound concepts and finally untangled the whole complex of difficulties surrounding the solution of equations in radicals, difficulties with which the most outstanding mathematicians had struggled unsuccessfully up to his time. The success of Galois was based on the fact that for the first time he introduced into the theory of equations a series of exceedingly important new general concepts, which subsequently played a great role in mathematics as a whole.

Let us consider the Galois theory for a special case, namely when the coefficients a_1, a_2, $\cdots$, a_n of the given nth-degree equation

$$x^n + a_1 x^{n-1} + \cdots + a_{n-1} x + a_n = 0 \tag{6}$$

are rational numbers. This case is particularly interesting and already involves essentially all the difficulties of the general Galois theory. We will also assume that the roots a, b, c, $\cdots$ of this equation are distinct.

Galois begins, like Lagrange, with considering a first-degree expression in a, b, c, $\cdots$

$$V = Aa + Bb + Cc + \cdots,$$

although he does not require that the coefficients A, B, C, $\cdots$ of this expression should be the roots of unity, but takes for A, B, C, $\cdots$ any integral rational numbers such as to give numerically distinct values for all the $n! = 1 \cdot 2 \cdot 3 \cdots n$ quantities V, V', V'', $\cdots$, $V^{(n!-1)}$ obtained from V by permuting the roots a, b, c, $\cdots$ in all $n!$ possible ways. This can always be done. Then Galois constructs the equation of degree $n!$ whose roots are V, V', V'', $\cdots$, $V^{(n!-1)}$. The theorem on symmetric polynomials shows that the coefficients of this equation $\Phi(x) = 0$ of degree $n!$ will be rational numbers.

Up to now everything is quite similar to what Lagrange did.

Next Galois introduced the first important new concept, the concept of irreducibility of a polynomial in a given field of numbers. If a polynomial in x is given, whose coefficients, for example, are rational numbers, then the polynomial is called reducible in the field of rational numbers if it can be represented in the form of a product of polynomials of lower degrees with rational coefficients. If not, then the polynomial is called irreducible in the field of rational numbers. The polynomial $x^3 - x^2 - 4x - 6$ is reducible in the rational number field, since it is equal to $(x^2 + 2x + 2)(x - 3)$, but for instance, the polynomial $x^3 + 3x^2 + 3x - 5$ is irreducible, as can be shown, in the field of rational numbers.

There exist methods, admittedly requiring long computations, of factoring any given polynomial with rational coefficients into irreducible polynomials in the field of rational numbers.

Galois then factors the above polynomial $\Phi(x)$ into irreducible factors in the field of rational numbers.

Let $F(x)$ be one of these irreducible polynomials (which one of them is immaterial for what follows) and let it be of degree m.

The polynomial $F(x)$ will then be the product of m of the $n!$ first-degree factors $x - V$, $x - V'$, $\cdots$, $x - V^{(n!-1)}$, into which the $n!$th-degree polynomial $\Phi(x)$ was decomposed. Let these m factors be $x - V$, $x - V'$, $\cdots$, $x - V^{(m-1)}$. We enumerate in any order the roots, a, b, c, $\cdots$, l

of the given nth-degree equation (6) by giving them the indices, $1, 2, \cdots, n$. Then the quantities $V, V', \cdots, V^{(n!-1)}$ correspond to all possible $n!$ permutations of the numbers $1, 2, \cdots, n$, corresponding to permutations of the roots, and the $V, V', \cdots, V^{(m-1)}$ correspond to only m of these permutations. The set G of these m permutations of the numbers $1, 2, \cdots, n$ is called the Galois group of the given equation (6).*

Then Galois introduces some new concepts and develops simple but truly remarkable arguments by which he proves that a necessary and sufficient condition for the solvability of equation (6) in radicals is that the group G of permutations of the numbers $1, 2, \cdots, n$ satisfies a certain definite condition.

Thus, Lagrange's prophecy that at the basis of the whole problem lay the theory of permutations proved to be true.

In particular, Abel's theorem on the nonsolvability of a general fifth-degree equation in radicals can now be proved as follows. It can be shown that there exist arbitrarily many fifth-degree equations, even with integral rational coefficients for which the corresponding 120th-degree polynomial $\Phi(x)$ is irreducible, i.e., whose Galois group is the group of all $5! = 120$ permutations of the indices $1, 2, 3, 4, 5$ of its roots. But this group, as can be shown, does not satisfy the Galois criterion, and therefore these fifth-degree equations cannot be solved in the radicals.

For instance, it can be shown that the equation $x^5 + x - a = 0$, where a is a positive whole number, in most cases cannot be solved by radicals. For example, it is not solvable in radicals for $a = 3, 4, 5, 7, 8, 9, 10, 11, \cdots$

The application of Galois theory to the problem of solvability of geometric problems by ruler and compass. One of the most remarkable special applications of Galois theory is the following. Many problems of plane geometry can be solved by constructions with ruler and compass alone. For example, we can construct with ruler and compass a regular triangle, square, pentagon, hexagon, octagon, decagon, etc., but it is impossible to construct a regular polygon of seven, nine, or eleven sides. Which problems can be solved by ruler and compass, and which not? Before Galois it was an unsolved problem. From the Galois theory we obtain the following answer.

The simultaneous solution of equations of two lines, a line and a circle, or two circles can be reduced to the solution of equations of first- or second-degree. For a line and a circle it is clear, and in the case of two circles $(x - a_1)^2 + (y - b_1)^2 = r_1^2$ and $(x - a_2)^2 + (y - b_2)^2 = r_2^2$ if we

---

* More will be said about Galois groups in §5, Chapter XX.

substract one equation from the other, the x^2 and y^2 cancel out, and we obtain a first-degree equation, which is to be solved simultaneously with the equation of one of the circles, so that again we have a quadratic equation. Therefore, every step of the problem to be solved by ruler and compass is reduced to an equation of first- or second-degree, and consequently, all problems solvable with ruler and compass are reduced to an algebraic equation with one unknown, whose solution involves the extraction of a chain of square roots. Conversely, if the solution of a geometric problem is reduced to such an algebraic equation, then it can be solved by ruler and compass, since square roots, as is well known, can be constructed by ruler and compass.

If a geometric problem is given, we must first set up an algebraic equation equivalent to the given problem. If it is impossible to set up such an equation, the problem is obviously not solvable by ruler and compass. If the equation has been set up, then we must select that one of its irreducible factors that is connected with the solution of the problem, and determine whether this irreducible equation can be solved in square roots. As the Galois theory shows, for this it is necessary and sufficient that the number m of permutations that constitute its Galois group be a power of 2.

With this test we can prove the theorem stated by Gauss that a regular polygon with a prime number p of sides can be constructed by ruler and compass if and only if the prime number p has the form $2^{2n} + 1$, i.e., for $p = 3, 5, 17, 257$ but not for $p = 7, 11, 13, 19, 23, 29, 31, \cdots$, etc. Gauss proved only the "if" part of this assertion.

By the same method we can prove that it is impossible to divide an arbitrary angle into three equal parts by ruler and compass, or to duplicate the cube, i.e., from the edge of a given cube to find the edge of a cube with twice as great a volume, and so forth.

The impossibility of squaring the circle, i.e., of constructing with ruler and compass the side of a square equal in area to a circle with given radius, is proved in a different way. Namely, it can be shown that the side of such a square is not connected with the radius by any algebraic equation, i.e., it is so to speak transcendental relative to the radius, and consequently it is *a fortiori* not expressible in terms of the radius by a chain of square roots. This proof is difficult and it does not follow from Galois theory.

Two fundamental unsolved problems connected with Galois theory. In Galois theory there remain two further basic problems which have not yet been solved in their general form although many excellent mathematicians have been working on them almost uninterruptedly.

The first of these is the problem of the so-called Hilbert–Čebotarev resolvents (not to be confused with the Lagrange resolvents) which is a direct generalization of the problem of solution of equations in radicals. The idea is this: Saying that an equation is solvable in radicals is exactly the same as saying that its solution is reduced to a chain of successive binomial equations, since the radical $\sqrt[n]{A}$ is a root of the binomial equation $x^n = A$. But it may happen that although the equation cannot be reduced to a chain of such simple equations as the binomial ones, it can nevertheless be reduced to a chain of certain other very simple equations. Back at the end of the 18th century, it had been shown that the general fifth-degree equation can be reduced to a chain of binomial equations together with one further equation of the form $x^5 + x + A = 0$, which, although not binomial, has like the binomial equations, only one parameter A.

Later on it was proved that even a sixth-degree equation cannot be reduced to a chain of one-parameter equations. For equations of any degree we require to solve the problem: what kind of simpler equations, i.e., with a minimum number of parameters, make up the chain to which our equation can be reduced.

If the given equation is reduced to a chain of one-parameter equations of a definite type, then for each of these one-parameter equations we can compute a table, giving its roots as a function of its parameter. Then the solution of the given equation is reduced to the use of a chain of such tables.

Second, a still deeper problem consists of the converse of Galois theory. Galois proved that the properties of the solutions of an equation depend on its group. But conversely, can any group of permutations be the Galois group of some equation and can we set up all the equations whose Galois group is a given group?

As to the first of these two questions only partial results are known, although such outstanding mathematicians as Klein and Hilbert worked on it persistently; the first general theorems were given by the remarkable Soviet algebraist H. G. Čebotarev.

The second question for the so-called solvable groups, i.e., groups satisfying Galois' criterion was solved in the affirmative in recent years by the Soviet mathematician I. R. Šafarevič.

§3. The Fundamental Theorem of Algebra

In the previous section we considered the attempts, lasting three centuries, to solve by radicals an nth-degree equation. The problem turned out to be very deep and difficult and led to the creation of new concepts, important not only for algebra but also for mathematics as a whole.

As for the practical solution of equations, the result of all this work was the following. It became clear that solution by radicals is far from being available for all algebraic equations, and even when it is available, it is of little practical value because of its complexity, except in the case of the quadratic equation.

In view of this, mathematicians long ago began to work on the theory of algebraic equations in three completely different directions, namely: (1) on the problem of the existence of a root; (2) on the problem of how can we learn from the coefficients of the equation something about its roots without solving it; for example, does it have real roots and how many; and finally, (3) on the approximate calculation of the roots of an equation.

First of all, it was necessary to prove that in general any nth-degree algebraic equation with real or complex coefficients always has at least one real or complex root.*

This theorem, which is one of the most important in the whole of mathematics, remained for a long time without rigorous proof. In view of its importance and difficulty, it is generally called the "fundamental theorem of algebra," although the majority of the methods by which it has been proved are as closely related to infinitesimal analysis as to algebra. The first proof was given by d'Alembert. One point in d'Alembert's proof, as was later made clear, turned out to be defective. Namely, d'Alembert assumed as trivial the general proposition of analysis that a continuous function, given on a bounded and closed set of points, has somewhere on the set a minimum. This is true but it had to be proved. A rigorous proof of this property was obtained only in the second half of the 19th century, i.e., a hundred years after d'Alembert's investigations.

It is generally considered that the first rigorous proof of the fundamental theorem of algebra were given by Gauss; however, some of his proofs require for full rigor no lesser additions than those required for d'Alembert's proof. Today a number of different completely rigorous proofs of this theorem are known.

In the present section we consider the proof of the fundamental theorem of algebra based on the so-called lemma of d'Alembert, and we also give a complete proof of the aforementioned proposition from analysis.

The theory of complex numbers. Before considering the proof of the fundamental theorem of algebra, we must first of all recall the theory of complex numbers as studied in high school. The difficulties which led to the creation of the theory of complex numbers are first encountered in

---

* The point is that there exist nonalgebraic equations, for example, $a^x = 0$, which definitely do not have roots, either real or complex.

solving quadratic equations. What should we do, if the number $p^2/4 - q$ under the square root in the formula for the solution of the quadratic equation is negative? There exists no real number, positive or negative, which is the square root of a negative number, since the square of any real number is either positive or zero.

After long doubts, lasting more than a century, mathematicians arrived at the conclusion that it is necessary to introduce a new form of numbers, the so-called complex numbers, with the following laws of operations on them.

Conventionally, a number of new character is introduced: $i = \sqrt{-1}$ such that $i^2 = -1$, and numbers of the form $a + bi$ are considered, where a and b are ordinary real numbers. The numbers $a + bi$ are called complex. Two such numbers $a + bi$ and $c + di$ are regarded as equal, if $a = c$, $b = d$. The sum of two such numbers is defined to be the number $(a + c) + (b + d)i$, and their difference is the number $(a - c) + (b - d)i$. In multiplication we agree to multiply these numbers as if they were binomials but to take into consideration that $i^2 = -1$, i.e.,

$$(a + bi)(c + di) = ac + bci + adi + bdi^2 = (ac - bd) + (bc + ad)i.$$

If a and b are regarded as rectangular coordinates of a point, and the point is associated with the complex number $a + bi$, then the addition and subtraction of complex numbers corresponds to the addition and subtraction of vectors, i.e., of directed segments from the origin to the points with coordinates (a, b) and (c, d), since in addition of vectors their corresponding coordinates are added. As to the geometrical meaning of a product in the so-called plane of complex numbers, we can see it more easily if we consider the length ρ of the vector from the origin of the coordinate system to the point (x, y) (this length is called the *modulus* of the complex number $z = x + iy$) and the angle ϕ which the vector makes with the Ox-axis (this angle is called the *argument* of the complex number $z = x + iy$); in other words, if we consider not the Cartesian coordinates x and y but the so-called polar coordinates ρ and ϕ (figure 1). Then $x = \rho \cos \phi$, $y = \rho \sin \phi$ and consequently the complex number itself can be written as

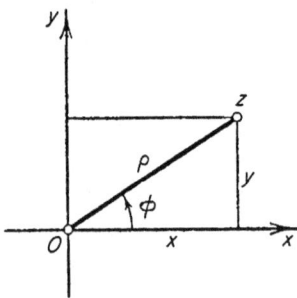

FIG. 1.

$$x + iy = \rho(\cos \phi + i \sin \phi).$$

If

$$a + bi = \rho_1(\cos \phi_1 + i \sin \phi_1), \quad c + di = \rho_2(\cos \phi_2 + i \sin \phi_2),$$

then

$$ac - bd = \rho_1\rho_2(\cos \phi_1 \cos \phi_2 - \sin \phi_1 \sin \phi_2) = \rho_1\rho_2 \cos (\phi_1 + \phi_2),$$
$$bc + ad = \rho_1\rho_2(\sin \phi_1 \cos \phi_2 + \cos \phi_1 \sin \phi_2) = \rho_1\rho_2 \sin (\phi_1 + \phi_2),$$

from this we see that in multiplication of two complex numbers their moduli ρ_1 and ρ_2 are multiplied, and the arguments ϕ_1 and ϕ_2 are added. In division, since it is the inverse operation of multiplication, one modulus is divided by the other, and the arguments are subtracted

$$\rho_1(\cos \phi_1 + i \sin\phi_1) \, \rho_2(\cos \phi_2 + i \sin \phi_2)$$
$$= \rho_1\rho_2[\cos(\phi_1 + \phi_2) + i \sin (\phi_1 + \phi_2)]$$

and

$$\frac{\rho_1(\cos \phi_1 + i \sin \phi_1)}{\rho_2(\cos \phi_2 + i \sin \phi_2)} = \frac{\rho_1}{\rho_2} [\cos (\phi_1 - \phi_2) + i \sin (\phi_1 - \phi_2)].$$

In raising to a power with positive integral exponent n, consequently, the modulus is raised to the same nth power, and the argument is multiplied by n

$$[\rho(\cos \phi + i \sin \phi)]^n = \rho^n(\cos n\phi + i \sin n\phi).$$

Conversely, taking roots

$$\sqrt[n]{\rho(\cos \phi + i \sin \phi)} = \sqrt[n]{\rho} \left(\cos \frac{\phi}{n} + i \sin \frac{\phi}{n}\right).$$

However, in taking roots a special situation arises. Let n be a positive integral exponent. Then

$$\sqrt[n]{\rho(\cos \phi + i \sin \phi)}$$

is equal to the number

$$\sqrt[n]{\rho} \left(\cos \frac{\phi}{n} + i \sin \frac{\phi}{n}\right)$$

since raising this number to the nth power gives the radicand.

But this is only one value of the root. The point is that the complex number

$$\sqrt[n]{\rho} \left[\cos \left(\frac{\phi}{n} + \frac{2k\pi}{n}\right) + i \sin \left(\frac{\phi}{n} + \frac{2k\pi}{n}\right)\right],$$

where k is any of the numbers $1, 2, \cdots, n-1$, will also be an nth root of the number

$$\rho(\cos\phi + i \sin \phi).$$

Indeed, according to the rule for raising to a power, if we raise this number to the nth power, we obtain the number

$$(\sqrt[n]{\rho})^n \left[\cos n \left(\frac{\phi}{n} + \frac{2k\pi}{n}\right) + i \sin n \left(\frac{\phi}{n} + \frac{2k\pi}{n}\right)\right]$$
$$= \rho[\cos (\phi + 2k\pi) + i \sin (\phi + 2k\pi)],$$

where the addend $2k\pi$, because of the properties of sines and cosines, can be neglected, since it changes neither sine nor cosine. Thus the nth power of this number is also

$$\rho(\cos \phi + i \sin \phi),$$

i.e., this number is

$$\sqrt[n]{\rho(\cos \phi + i \sin \phi)}.$$

It is easy to see that no other complex number, besides these n numbers for $k = 0, 1, 2, \cdots, n-1$ is an nth root of

$$\rho(\cos \phi + i \sin \phi).$$

Geometrically, the extraction of nth roots can be described as follows. The points of the complex plane corresponding to the values of the $\sqrt[n]{\ }$ of the number $\rho(\cos \phi\, i \sin \phi)$ lie at the vertices of the regular n-sided polygon inscribed in a circle drawn about the origin with radius $\sqrt[n]{\rho}$ and so rotated that one of the vertices of this n-sided polygon has argument ϕ/n (figure 2).

We make the following observation. If

$$f(z) = z^n + c_1 z^{n-1} + \cdots + c_{n-1} z + c_n$$

is a polynomial in z with given real or complex coefficients $c_1, c_2, \cdots, c_n$ and we change z continuously, i.e., continuously shift the point $z = x + iy$ in the complex plane, then the complex point $Z = X + iY = f(z)$ will also move continuously in the complex plane. This is clear from the fact that if we substitute in $f(z)$ the value of $z = x + iy$, $c_1 = a_1 + b_1 i$, $c_2 = a_2 + b_2 i$, $\cdots$, $c_n = a_n + b_n i$, and perform all computations, we find that

$$f(z) = X + iY,$$

where

$$X = P(x, y), \quad Y = Q(x, y)$$

are nth-degree polynomials in x and y with real coefficients expressed in terms of a_i and b_i. Under continuous change of x and y, these polynomials will also change continously.

We also note that, since the modulus $\rho = |f(z)|$ is equal to $\sqrt{X^2 + Y^2}$, during a continuous shift of the point z in the complex plane, the modulus $|f(z)|$ will also change continuously. In other words, if the point z is sufficiently close to the point α then the difference $|f(z)| - |f(\alpha)|$ of absolute values is smaller than any preassigned positive number.

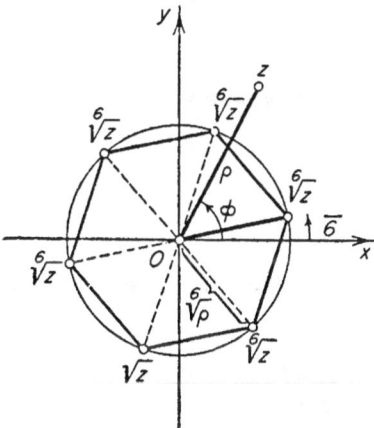

Fig. 2.

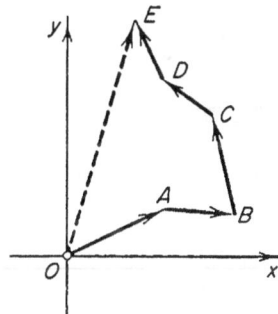

Fig. 3.

Let us also remark that the modulus of a sum of several complex numbers is always smaller than or equal to the sum of the moduli of these numbers, which is equivalent to saying that the rectilinear segment OE (figure 3) is shorter than or equal to the polygonal line $OABCDE$, being equal to it if and only if all of its segments lie on one line and in one direction.

We recall finally that to say "a complex number is equal to zero" is the same as to say that "its modulus is equal to zero," since the modulus ρ of a complex number is the distance from the origin to the corresponding point.

We now apply the theory of complex numbers to the proof of the fundamental theorem of algebra; though it must be remarked that the significance of the theory of complex numbers goes far beyond the limits of algebra. In many parts of mathematics other than algebra, we cannot get along without them. In many applications, for example in the theory of alternating currents, numerous problems are most simply solved by means of complex numbers. But what is most important is the application of complex numbers, or more precisely the theory of functions of a complex variable, to the theory of certain special functions of two real variables which are called *harmonic*. By means of these functions, important problems in the theory of airplane flight, of heat conduction in a plate, of plane electric fields, and of elasticity can be solved. A famous theorem on the lifting force on an airplane wing was obtained by the founder of contemporary aerodynamics, N. E. Žukovskiĭ, through investigations of functions of a complex variable.*

We now pass to the proof of the fundamental theorem of algebra.

Theorem. *Any polynomial*

$$f(z) = a_0 z^n + a_1 z^{n-1} + \cdots + a_{n-1} z + a_n ,$$

whose coefficients

$$a_0, a_1, \cdots, a_{n-1}, a_n$$

are any given real or complex numbers, has at least one real or complex root.

We will assume that the given polynomial is of degree n, i.e., that $a_0 \neq 0$.

The surface of the modulus of a polynomial. We consider the whole problem geometrically. Above each point z of the complex plane, we erect a perpendicular altitude t, equal in length to the modulus $|f(z)|$ of the polynomial $f(z)$ at this point z. The ends of these altitudes define a surface M, which can be called the modulus surface of the polynomial $f(z)$. We see that this surface: (1) nowhere drops below the complex plane, since the modulus of any complex number (in this case, the number $f(z)$) is nonnegative; (2) for any given point z of the complex plane, the surface has one and only one point which either lies vertically above this point or else coincides with it, i.e., the surface M extends in one sheet above the whole complex plane and may at some points touch the plane itself; (3) the surface is continuous in the sense that a continuous change in the position of the point z on the complex plane produces a continuous change in the value of $t = |f(z)|$, i.e., in the altitude t of points of the surface. (This was shown in the last subsection.)

The fundamental theorem of algebra consists in proving that the surface M touches the complex plane in one point at least and does not remain everywhere at a positive distance above it.

On the growth of the modulus of a polynomial with increasing distance from the origin. We show that no matter how large a positive number G

---

* See Chapter IX.

is given, we can find a radius R such that for all points z of the complex plane, lying outside of the circle of radius R with center at the origin, the altitude t of points of the surface M above the complex plane is greater than G.

For let us write the polynomial $f(z)$ as

$$a_0 z^n \left[1 + \left(\frac{a_1}{a_0 z} + \frac{a_2}{a_0 z^2} + \cdots + \frac{a_n}{a_0 z^n} \right) \right].$$

The modulus of the expression

$$\left(\frac{a_1}{a_0 z} + \frac{a_2}{a_0 z^2} + \cdots + \frac{a_n}{a_0 z^n} \right)$$

is not greater than the sum of the moduli of the summands

$$\left| \frac{a_1}{a_0 z} \right| + \left| \frac{a_2}{a_0 z^2} \right| + \cdots + \left| \frac{a_n}{a_0 z^n} \right|,$$

and, with an increase in the modulus of z, every one of these summands decreases, so that the sum also decreases. Therefore, for all z whose moduli are greather than some number R', the modulus of this expression in parentheses is smaller, for example, than $\frac{1}{2}$.

But then for all such z, the expression

$$\Omega = \left[1 + \left(\frac{a_1}{a_0 z} + \frac{a_2}{a_0 z^2} + \cdots + \frac{a_n}{a_0 z^n} \right) \right]$$

will have modulus greater than $\frac{1}{2}$. The modulus of the first factor $a_0 z^n$ is equal to $|a_0| \cdot |z|^n$, so that it increases with increasing modulus of z; moreover, it increases beyond all bounds. Therefore, no matter how large a positive number G is given, there exists a positive number R such that for all z, whose moduli are greater than R, $|f(z)| = |a_0| \cdot |z|^n \cdot |\Omega|$ is greater than G.

The existence of minima of the surface M. We will say that at a point α of the complex plane the surface M has a minimum if the value of the altitude t of the point of the surface M at this point α is smaller than or equal to its values at all points of some neighborhood of the point α, i.e., at all points of some circle, however small, with center at the point α.

Let the altitude t of the point of the surface M corresponding to the origin, i.e., to the point $z = 0$ of the complex plane, be equal to g, i.e., $|f(0)| = g$. We take $G > g$. All altitudes t of points of the surface M are nonnegative and continuously change during continuous movement of the point z in the complex plane. The surface M has altitude $t > G$ outside of a circle drawn about the origin with radius R and altitude $t = g < G$ at the center of the circle. D'Alembert regarded it as an obvious consequence that somewhere in the interior of the circle R there is a point where the altitude is a minimum; more precisely, where the value of t is smaller than or equal to its values at all remaining points of the circle R, i.e., the surface M has at least one minimum.

The rigorous proof of the existence of such a minimum is based on the following *axiom of continuity* of the set of real numbers.

If two sequences of real numbers are given: $a_1 \leqslant a_2 \leqslant \cdots \leqslant a_n \cdots$ and $b_1 \geqslant b_2 \geqslant \cdots \geqslant b_n \geqslant \cdots$, such that $b_n > a_n$ for all n and $b_n - a_n \to 0$ as $n \to \infty$, then there exists one and only one real number c, such that $a_n \leqslant c \leqslant b_n$ for all n.

Geometrically, this continuity property means that if on the line sequence of interval $[a_n, b_n]$ (figure 4) is given, such that every successive interval is contained in the preceding interval, and the lengths of the intervals become arbitrarily small, then there exists a point c belonging to all intervals of the sequence. In other words, the intervals "shrink" to a point, and not to "an empty place."

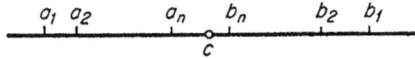

FIG. 4.

Since the length of the segment $[a_n, b_n]$ approaches zero with increasing n, there is only one such point c. From the property of continuity for the set of all points on the number axis immediately follows the property of continuity for complex numbers, i.e., for points of the plane. We give a geometrical formulation of this property.

If in the plane a sequence of rectangles $\Delta_1, \Delta_2, \cdots, \Delta_n, \cdots$ is given, with sides parallel to the coordinate axes, such that every rectangle is contained in the previous one, and such that the length of their diagonals decreases indefinitely, then there exists one and only one point which is contained in all the rectangles of the sequence. This property of continuity of the plane directly follows from the continuity property of the line. For the proof it is sufficient to project the rectangles on the coordinate axes.

Now it is easy to establish the so-called Bolzano-Weierstrass theorem.

If in a rectangle an infinite sequence of points $z_1, z_2, \cdots, z_n, \cdots$ is given, then in the interior or on the boundary of the rectangle there exists a point z_0 such that in any arbitrarily small neighborhood of z_0, i.e., in the interior of an arbitrarily small circle with center at z_0, there are infinitely many points of the sequence $z_1, z_2, \cdots, z_n, \cdots$.

For the proof we denote the given rectangle by Δ_1. We divide it into four equal parts by lines parallel to the coordinate axes. At least one of the parts necessarily contains infinitely many points of the given sequence. This part will be denoted by Δ_2. We again subdivide the rectangle Δ_2 into four equal parts and select among them a Δ_3 which contains infinitely many points of the given sequence, and so on.

We obtain a sequence of imbedded rectangles $\Delta_1, \Delta_2, \Delta_3, \cdots$, whose diagonals decrease indefinitely. By the continuity property we can find a point z_0 contained in all these rectangles. Then this z_0 is the desired point. For, no matter how small a neighborhood of z_0 we take, the rectangles of the sequence $\Delta_1, \Delta_2, \Delta_3, \cdots$, beginning with some one of them will be inside this neighborhood, as soon as their diagonals become smaller than the radius of the neighborhood, and any one of the rectangles contains infinitely many points of the sequence $z_1, z_2, \cdots, z_n, \cdots$. Thus the Bolzano-Weierstrass theorem is proved.

Now it is easy to prove the theorem on the minimum of the modulus $|f(z)|$ of a polynomial. As before, let $|f(0)| = g$, let G be a number greater than g, and let R be such that for $z > R$, we have $|f(z)| > G$.

If $g = 0$, i.e., $f(0) = 0$, then the modulus $|f(z)|$ of the polynomial has a minimum at the point 0, since at all points it is $\geqslant 0$.

If $g > 0$ and $|f(z)| \geqslant g$ for all points z then $|f(z)|$ still has a minimum at the point 0. Let $g > 0$ and let points z exist, in which $|f(z)| < g$; then in the sequence of numbers

$$0, \frac{g}{n}, \frac{2g}{n}, \cdots, \frac{ng}{n} = g \qquad (*)$$

we find the greatest $c_n = (i/n) g$, such that all values $|f(z)| \geqslant c_n$. For the next number $c'_n = [(i + 1)/n] g$ the sequence $(*)$ contains at least one point z_n such that $|f(z_n)| < c'_n$.

Let n increase to infinity. For all n we have $|z_n| \leqslant R$, since if $|z_n| > R$, then $|f(z_n)|$ would be greater than G and consequently greater also than g.

Thus all points z_n lie inside a rectangle with sides $2R$, and with center at the origin. It is possible that some of these points coincide.

By the Bolzano-Weierstrass theorem there exists a point z_0 such that every neighborhood of z_0 contains infinitely many points of the sequence $z_1, z_2, \cdots, z_n, \cdots$.

We establish that the point z_0 furnishes the desired minimum of $|f(z)|$.

For at any point z we have

$$|f(z)| > c_n = c'_n - \frac{g}{n} > |f(z_n)| - \frac{g}{n}$$
$$= |f(z_0)| + [|f(z_n)| - |f(z_0)|] - \frac{g}{n}.$$

This inequality is valid for any n. If we take for n a sequence of values for which z_n indefinitely approaches z_0, then on account of the continuity of $|f(z)|$, the difference $|f(z_n)| - |f(z_0)|$ becomes arbitrarily small in absolute value with g/n.

Consequently, $|f(z)| \geqslant |f(z_0)|$, i.e., $|f(z)|$ actually has a minimum at the point z_0.

D'Alembert's lemma. In view of the fact that all the altitudes t of points on the modulus surface M are nonnegative, it is clear that any root of the polynomial $f(z)$, i.e., any point z of the complex plane where the polynomial $f(z)$ itself (and consequently its modulus $|f(z)|$ also) is equal to zero, corresponds to a minimum of the modulus surface M. However, as d'Alembert showed, the converse is also true: At any minimum the surface M extends down to the complex plane itself, and consequently at that point there is a root of the polynomial $f(z)$. In other words, at any point at which the altitude t is positive and not zero, there is no minimum of the surface M. This follows from the so-called d'Alembert's lemma:

If α is a given complex number such that $f(\alpha) \neq 0$, then a complex number h can always be found with arbitrarily small modulus, such that $|f(\alpha + h)| < |f(\alpha)|$.

Proof. We consider the polynomial

$$f(\alpha + h) = a_0(\alpha + h)^n + a_1(\alpha + h)^{n-1} + \cdots + a_{n-1}(\alpha + h) + a_n$$

in two indeterminates α and h and arrange it in ascending powers of h. In this polynomial there will be a term not containing h at all, namely

$$a_0\alpha^n + a_1\alpha^{n-1} + \cdots + a_{n-1}\alpha + a_n = f(\alpha) \neq 0,$$

since it was assumed that $f(\alpha) \neq 0$. There will also be a term with h^n, namely $a_0 h^n$, since it was assumed that $a_0 \neq 0$. As to the terms with intermediate powers of h, some of them, and in some cases all of them, may be missing. Let the lowest power of h which occurs in this polynomial be m, where $1 \leqslant m \leqslant n$, i.e., this expression will have the form

$$f(\alpha + h) = f(\alpha) + Ah^m + Bh^{m+1} + Ch^{m+2} + \cdots + a_0 h^n.$$

Let us write this as:

$$f(\alpha + h) = f(\alpha) + Ah^m + Ah^m \left(\frac{B}{A} h + \frac{C}{A} h^2 + \cdots + \frac{a_0}{A} h^{n-m} \right),$$

where $A \neq 0$, and B, C, etc., may or may not be equal to zero.

After this preparation the proof of d'Alembert's lemma runs as follows. For h it is sufficient to take a complex number with modulus so small that the length of the vector Ah^m is smaller than the length of the vector

$f(\alpha)$ and with argument such that the direction of the vector Ah^m is opposite to the direction of the vector $f(\alpha)$. Then the vector $f(\alpha) + Ah^m$ will be shorter than the vector $f(\alpha)$. But if the modulus of h is taken sufficiently small, the modulus of the expression

$$\left(\frac{B}{A}h + \frac{C}{A}h^2 + \cdots + \frac{a_0}{A}h^{n-m}\right)$$

can be made arbitrarily small, for example, smaller than one, and consequently, the vector

$$\Delta = Ah^m\left(\frac{B}{A}h + \frac{C}{A}h^2 + \cdots + \frac{a_0}{A}h^{n-m}\right)$$

is shorter than the vector Ah^m and therefore, the vector $f(\alpha + h) = f(\alpha) + Ah^m + \Delta$, as is seen in (figure 5), is also shorter than the vector $f(\alpha)$, even if the direction of the vector Δ is in the opposite direction of the vector Ah^m.

The details of this proof are as follows:

1. Since in multiplication, the arguments of the factors are added, we have to take the argument of h such that

$$\arg A + m \cdot \arg h = \arg f(\alpha) + 180°,$$

i.e., it is necessary to take

$$\arg h = \frac{\arg f(\alpha) - \arg A + 180°}{m}.$$

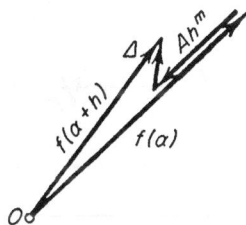

FIG. 5.

2. The modulus of

$$\left(\frac{B}{A}h + \frac{C}{A}h^2 + \cdots + \frac{a_0}{A}h^{n-m}\right)$$

is not greater than the sum of moduli of its summands

$$T = \left|\frac{B}{A}h\right| + \left|\frac{C}{A}h^2\right| + \cdots + \left|\frac{a_0}{A}h^{n-m}\right|;$$

moreover, with decreasing modulus of h, each of the summands of this sum can be arbitrarily decreased and consequently so can the whole sum. Therefore, if h is a complex number with the above given argument, and h_0 is a modulus such that if h has a smaller modulus than h_0 and satisfies the two conditions $|Ah^m| < |f(\alpha)|$ and $T < 1$, then for such h we will have $|f(\alpha + h)| < |f(\alpha)|$, which proves d'Alembert's lemma.

From d'Alembert's lemma it immediately follows that every minimum of the modulus surface M of the polynomial $f(z)$ gives a root of this polynomial. Indeed, if at the point α, $f(\alpha) \neq 0$, then by virtue of d'Alembert's lemma at arbitrarily close points $\alpha + h$ we would have $|f(\alpha + h)| < |f(\alpha)|$, i.e., there would not exist a circle with center at α, at all of whose points the modulus of $f(z)$ is not smaller than the modulus of $f(\alpha)$, and therefore at the point α we would not have a minimum of the modulus of $f(z)$. With this the fundamental theorem of algebra is proved.

The general form of the modulus surface M. The modulus surface M of the polynomial $f(z)$ lies above the complex plane z. It has the form shown in figure 6. It can be shown that at greater altitudes t, the surface M differs very little from the surface obtained by revolving the nth degree parabola $t = |a_0| x^n$ about the Ot-axis.

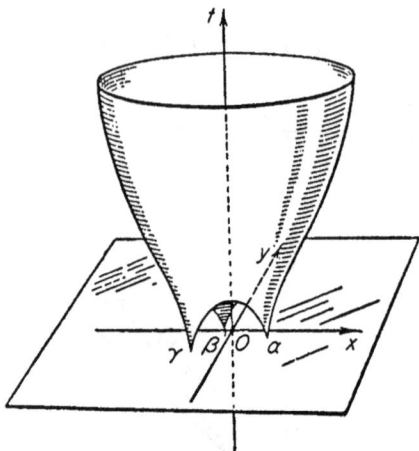

FIG. 6.

But for small t the surface M has minima, whose number is equal to the number of distinct roots of the equation $f(z) = 0$. At all these minima the surface M touches the complex plane z itself.

§4. Investigation of the Distribution of the Roots of a Polynomial on the Complex Plane

A number of problems important in practice are connected with this question: Without solving a given equation, obtain some information about the distribution of its roots on the complex plane. The first such problem, historically, was to determine the number of real roots of an equation. That is, if an equation with real coefficients is given, then by some test depending on its coefficients, to determine, without solving the equation, whether it has real roots and if it does, how many; or how many positive and how many negative roots it has; or how many real roots lying between given limits a and b.

Derivatives of a polynomial. In this section an essential role will be played by the derivative of a polynomial. The definition of the derivative of a function was given in Chapter II.

For the polynomial $a_0x^n + a_1x^{n-1} + \cdots + a_{n-1}x + a_n$ the derivative is given, as is well known, by the polynomial

$$na_0x^{n-1} + (n-1)a_1x^{n-2} + \cdots + a_{n-1}.$$

The concept of derivative in Chapter II was considered only for functions of a real variable. In algebra it is necessary to consider the variable as taking on arbitrary complex values and to introduce polynomials with complex coefficients.

However, the former definition of derivative can be retained, namely as the limit of the ratio of the increment of the function to the increment of the independent variable. The formula for computing the derivative of a polynomial with complex coefficients, and the basic laws for the derivative of sum, product, and power remain the same as before.*

Simple and multiple roots of a polynomial. In §2 of this chapter it was established that if the number a is a root of the polynomial $f(x)$, then $f(x)$ is divisible by $x - a$ without remainder. If $f(x)$ is not divisible by $(x - a)^2$, then the number a is called a simple root of the polynomial $f(x)$. Generally, if the polynomial $f(x)$ is divisible by $(x - a)^k$ but not by $(x - a)^{k+1}$, then the number a is called a root of multiplicity k.

A root a of multiplicity k is often regarded as k different roots. The basis for this is that the factor $(x - a)^k$, present in the factorization of $f(x)$ into linear factors, is the product of k factors, each equal to $(x - a)$.

By virtue of the fact that every polynomial of degree n can be factored into the product of n linear factors, the number of roots of the polynomial is equal to its degree, if we take into account the multiplicity of each root.

The following theorems are true:

1. A simple root of a polynomial is not a root of its derivative.

2. A multiple root of a polynomial is a root of its derivative of multiplicity one less.

For, let $f(x) = (x - a)^k f_1(x)$ and let $f_1(x)$ not be divisible by $(x - a)$, i.e., $f_1(a) \neq 0$. Then

$$f'(x) = k(x - a)^{k-1}f_1(x) + (x - a)^k f_1'(x)$$
$$= (x - a)^{k-1}[kf_1(x) + (x - a)f_1'(x)] = (x - a)^{k-1}F(x).$$

The polynomial $F(x) = kf_1(x) + (x - a)f_1'(x)$ is not divisible by $(x - a)$, since $F(a) = kf_1(a) \neq 0$.

---

* See Chapter IX.

Consequently, $f'(x)$ for $k = 1$ is not divisible by $x - a$, and for $k > 1 f'(x)$ is divisible by $(x - a)^{k-1}$ but not by $(x - a)^k$. With this both theorems are proved.

Rolle's theorem and some of its consequences. According to the well-known theorem of Rolle,[*] if the real numbers a and b are roots of a polynomial with real coefficients, then there exists a number c lying between a and b which is a root of the derivative.

From Rolle's theorem the following interesting theorems follow:

1. If all roots of the polynomial $f(x) = a_0 x^n + \cdots + a_n$ are real, then all roots of its derivative are also real. In addition, between two adjacent roots of $f(x)$ there exists one root of $f'(x)$ and this root is simple. Indeed, let $x_1 < x_2 \cdots < x_k$ be the roots of $f(x)$ with multiplicities $m_1, m_2, \cdots, m_k$, respectively. Clearly, $m_1 + m_2 + \cdots + m_k = n$.

Then the derivative $f'(x)$, by the above theorem on multiple roots, will have roots $x_1, x_2, \cdots, x_k$ with multiplicities $m_1 - 1, m_2 - 1, \cdots, m_k - 1$, and by Rolle's theorem there is at least one root $y_1, y_2, \cdots, y_{k-1}$ in the interior of each of the intervals $(x_1, x_2), (x_2, x_3), \cdots, (x_{k-1}, x_k)$ between two successive roots of $f(x)$. Thus, the number of real roots of $f'(x)$ is equal (with regard to multiplicities) to at least $(m_1 - 1) + (m_2 - 1) + \cdots + (m_k - 1) + k - 1 = n - 1$. But $f'(x)$ as an $(n - 1)$th-degree polynomial has (with regard to multiplicities) $n - 1$ roots. Consequently, all roots of $f'(x)$ are real, $y_1, y_2, \cdots, y_{k-1}$ are simple roots, and roots other than $x_1, x_2, \cdots, x_k$ and $y_1, y_2, \cdots, y_{k-1}$ of the polynomial $f'(x)$ do not exist.

2. If all roots of a polynomial $f(x)$ are real and of these p are positive, then $f'(x)$ has p or $p - 1$ positive roots.

For, let $x_1 < x_2 < \cdots < x_k$ be all positive roots of the polynomial $f(x)$ with multiplicities $m_1, m_2, \cdots, m_k$, respectively. Then $m_1 + m_2 + \cdots + m_k = p$. The derivative $f'(x)$ will have the following positive roots: $x_1, x_2, \cdots, x_k$ with multiplicities $m_1 - 1, m_2 - 1, \cdots, m_k - 1$; simple roots $y_1, y_2, \cdots, y_{k-1}$ lying in the intervals $(x_1, x_2), \cdots, (x_{k-1}, x_k)$; and it can also have a simple root y_0 lying in the interval (x_0, x_1) where x_0 is the largest nonpositive root of $f(x)$. Consequently, the number of positive roots is equal to $(m_1 - 1) + \cdots + (m_k - 1) + k - 1 = p - 1$ or $(m_1 - 1) + \cdots + (m_k - 1) + (k - 1) + 1 = p$ which was required to be proved.

Descartes' law of signs. In his significant book of 1637 "Geometry," in which the first presentation of analytic geometry was given, Descartes,

---

[*] This theorem is the simplest form of the mean value theorem, which was mentioned in Chapter II.

among other things, gave the first significant algebraic theorem concerning the distribution of roots of a polynomial on the complex plane, the so-called "Descartes law of signs." It can be stated as follows:

If the coefficients of an equation are real and all its roots are also known to be real, then the number of its positive roots, with account taken of multiplicities, is equal to the number of changes of sign in the sequence of its coefficients. If it also has complex roots, then this number is equal to or an even number less than the number of these changes in sign.

We first explain what we mean by the number of changes of sign in the sequence of coefficients of the equation. To obtain this number we write down all coefficients of the equation, for example in the order of decreasing powers of the unknown, including the coefficient of x^n and the constant term, but omitting coefficients equal to zero, and consider all pairs of successive numbers of the sequence so obtained. If in such a pair the signs of the numbers are different, then we call this a change of sign. For example, if the given equation is

$$x^7 + 3x^5 - 5x^4 - 8x^2 + 7x + 2 = 0$$

then the sequence of its coefficients is

$$1, 3, -5, -8, 7, 2$$

and there are 2 changes of sign.

Now we pass to the proof of the first part of the theorem.*

Without loss of generality we can assume that the leading coefficient a_0 of the polynomial $f(x) = a_0 x^n + \cdots + a_n$ is positive.

First of all, we establish that if $f(x)$ has only real roots and of these p are positive (counting multiplicities) then $(-1)^p$ is the sign of the last coefficient of $f(x)$ different from zero.

Indeed, let

$$f(x) = a_0 x^n + \cdots + a_k x^{n-k}$$
$$= a_0 x^{n-k}(x - x_1) \cdots (x - x_p)(x - x_{p+1}) \cdots (x - x_{n-k}),$$

where $x_1, \cdots, x_p$ are the positive roots of $f(x)$, $x_{p+1}, \cdots, x_{n-k}$ are the negative roots of $f(x)$, account being taken of the multiplicity of each root. Then $a_k = a_0(-1)^p x_1 \cdots x_p(-x_{p+1}) \cdots (-x_{n+k})$ and, since all the numbers $a_0, x_1, \cdots, x_p, -x_{p+1}, \cdots, -x_{n-k}$ are positive, the sign of a_k is $(-1)^p$.†

The subsequent proof is based on the method of mathematical induction.

---

* We could give another, direct proof, not involving derivatives, but it would be somewhat longer.

†We note that this assertion is also correct for the case when some of the roots of $f(x)$ are complex.

For first-degree polynomials the theorem is trivial. Indeed, a first-degree polynomial $a_0 x + a_1$ has a unique root $-a_1/a_0$, which is positive if and only if a_0 and a_1 have opposite signs.

Let us assume now that the theorem is proved for all polynomials of $(n-1)$th degree with real roots, and with this assumption we will prove it for any polynomial $f(x) = a_0 x^n + \cdots + a_{n-1} x + a_n$ of degree n.

1. $a_n = 0$. We consider the polynomial $f_1(x) = a_0 x^{n-1} + \cdots + a_{n-1}$. The positive roots of the polynomials $f(x)$ and $f_1(x)$ are the same; the number of changes of sign in the sequence of their coefficients is also the same. For the polynomial $f_1(x)$ Descartes' law is valid; consequently it is valid for the polynomial $f(x)$.

2. $a_n \neq 0$. We consider the derivative

$$f'(x) = na_0 x^{n-1} + (n-1) a_1 x^{n-2} + \cdots + a_{n-1}.$$

It is clear that the number of changes of sign in the sequence of coefficients of the polynomial $f(x)$ is equal to the analogous number for the derivative $f'(x)$, if the signs of a_n and the last nonzero coefficient of the derivative coincide, or it is one more, if the signs are opposite.

By what was said above at the beginning of the proof, in the first case the number of positive roots of $f(x)$ and of $f'(x)$ have the same parity (are both even or both odd), and in the second case they have opposite parity. But as we deduced from Rolle's theorem, the number of positive roots of a polynomial, if all its roots are real, can be either equal to the number of positive roots of its derivative, or be one more. Taking this into consideration, we note that in the first case $f(x)$ has the same number of positive roots as $f'(x)$, and in the second case one more. For $f'(x)$ Descartes' law is valid by the induction assumption, i.e., the number of positive roots of $f'(x)$ is equal to the number of changes of sign in the sequence of its coefficients. Consequently, in both cases the number of positive roots of $f(x)$ is equal to the number of changes of sign in the sequence of coefficients, and this is the required proof.

The second part of Descartes' law is not more complicated to establish, and we will omit the proof here.

Remark 1. The first assertion of Descartes' theorem is particularly important, since in many practical problems it is automatically known whether all roots of a given equation are positive. In this case it can be quickly determined, how many roots are positive and how many negative. Also it can be seen at once, how many zero roots the equation has.

Remark 2. If in the given polynomial we set $x = y + a$ where a is an arbitrary given real number, i.e., we form the polynomial $f(y + a)$, then the positive roots y of this polynomial will be those and only those that are obtained from the roots x of the given polynomial $f(x)$ that are greater than a. Therefore the number of roots of the given polynomial $f(x)$, all of whose roots are real, lying between given limits a and b ($b > a$), is equal to the number of changes of sign for the polynomial $f(y + a)$ minus the number of changes of sign for the polynomial $f(z + b)$. If, however, not all roots of $f(x)$ are real, then it can be shown that this number is equal to this difference or some even number less. This is the so-called Budan theorem.

Sturm's theorem. Descartes' law of signs, as well as Budan's theorem do not, however, give an answer to the problem: Does a given equation with real coefficients have at least one real root, how many real roots does it have altogether, and how many real roots does it have lying between given limits a and b? For more than two centuries mathematicians attempted to solve these problems but without result. A long series of efforts in this direction were made by Descartes, Newton, Sylvester, Fourier, and many others, but they did not succeed in solving even the first of these problems, until, finally in 1835 the French mathematician Sturm suggested a method that solved all three problems.

Sturm's method is really not very complicated, but it is of such a character that one might seek it for a long time and not find it. Sturm himself was very happy that he had succeeded in solving this remarkable and exceedingly important pratical problem of algebra. In his lectures, when he came to the presentation of his result, he usually said: "Here is the theorem whose name I bear." But it must be said that Sturm did not solve this problem by mere chance; he pondered for many years on questions related to it.

Let $f(z)$ be a polynomial with real coefficients and $f_1(z)$ be the derivative $f'(z)$. Let us divide the polynomial $f(z)$ by $f_1(z)$ and denote the remainder in this division by $f_2(z)$, taking it with the opposite sign. Then, divide $f_1(z)$ by $f_2(z)$ and denote the remainder, taken with opposite sign, by $f_3(z)$, etc.

It can be shown that the last nonzero polynomial $f_s(z)$ of the constructed sequence will be a constant number c.

Sturm's theorem is as follows: If $a < b$ are two real numbers, which are not roots of the polynomial $f(z)$. then substituting in the polynomials

$$f(z), f_1(z), \cdots, f_{s-1}(z), c$$

$z = a$ and $z = b$, we obtain two sequences of real numbers

$$f(a), f_1(a), f_2(a), \cdots, f_{s-1}(a), c, \tag{I}$$
$$f(b), f_1(b), f_2(b), \cdots, f_{s-1}(b), c, \tag{II}$$

such that the number of changes of sign in sequence (I) is greater than or equal to the number of changes of sign in sequence (II) and the difference between these numbers of changes of sign is exactly equal to the number of real roots of $f(z)$ lying between a and b, or in other words, the number of these roots is equal to the loss of changes of sign in sequence (I) in going from a to b.

The proof of Sturm's theorem is not more difficult than the proof of Descartes' theorem, but we will not give it here.

Sturm's theorem enables us to compute the number of roots of a polynomial with real coefficients on any segment of the real axis. Therefore the application of Sturm's theorem to any given polynomial gives a clear picture of distribution of roots of a polynomial on the real axis, in particular it enables us to *separate* the roots, i.e., to construct segments in each of which only one root of the polynomial is contained.

In many applications, the solution of the analogous problem for the complex roots of a polynomial is equally important. Since complex numbers are represented by points not on the line but in the plane, it is impossible to speak of "segments" in which complex roots are contained; instead of a segment, we have to consider a region, i.e., a part of the plane, chosen in one way or another.

Thus, with respect to complex roots the following problem arises:

Given a polynomial $f(z)$ and a region in the complex plane, it is required to find the number of roots of the polynomial inside this region.

We assume that the region is bounded by a closed contour (figure 7) and that on the contour the polynomial $f(z)$ does not have roots.

Imagine that the point z goes around the contour of the region once in the positive direction. Every value of the polynomial is also represented by points on the plane. With continuous change of z the polynomial $f(z)$ also changes continuously. Therefore, while z goes once around the

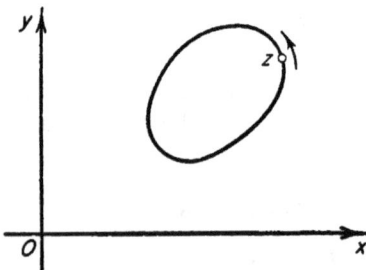

FIG. 7.

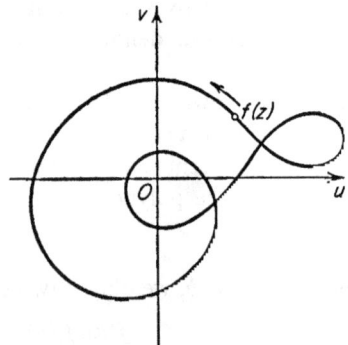

FIG. 8.

contour of the region, $f(z)$ describes some closed curve. This curve will not go through the origin of the coordinate system, since $f(z)$ by assumption does not reduce to zero at any of the points of the contour (figure 8).

The answer to the above mentioned problem is given by the following theorem:

Principle of the argument. The number of roots of the polynomial $f(z)$ inside the region bounded by a closed curve C is equal to the number of times the point $f(z)$ winds around the origin as z goes around the contour C once in the positive direction.

For the proof we decompose $f(z)$ into linear factors

$$f(z) = a_0 z^n + a_1 z^{n-1} + \cdots + a_n = a_0(z - z_1)(z - z_2) \cdots (z - z_n).$$

We know that the argument of the product of several complex numbers is equal to the sum of the arguments of the factors. Consequently,

$$\arg f(z) = \arg a_0 + \arg (z - z_1) + \arg (z - z_2) + \cdots + \arg (z - z_n).$$

Let us denote by $\Delta \arg f(z)$ the increment of the argument of $f(z)$, computed under the assumption that z goes once around the contour C. It is clear that $\Delta \arg f(z)$ is 2π multiplied by the number of times the point $f(z)$ winds around the origin.

Clearly,

$$\Delta \arg f(z) = \Delta \arg a_0 + \Delta \arg (z - z_1)$$
$$+ \Delta \arg (z - z_2) + \cdots + \Delta \arg (z - z_n).$$

It is clear that $\Delta \arg a_0 = 0$ since a_0 is a constant. Then, $z - z_1$ is represented by the vector going from the point z_1 to the point z. Let us assume that z_1 is in the interior of the region. Geometrically it is clear (figure 9) that as the point z goes around the contour C the vector $z - z_1$ makes a complete revolution about its initial point, so that $\Delta \arg (z - z_1) = 2\pi$. We assume now that the point z_2 is in the exterior of the region. In this case the vector "oscillates" to one side and back, and returns to its original position without making a revolution about its initial point, so that $\Delta \arg (z - z_2) = 0$. We can reason the same way about all the roots. Consequently $\Delta \arg f(z)$ is equal to 2π multiplied by the number of roots of $f(z)$ lying in the interior of the region. Hence the number of roots of $f(z)$ inside the region is equal to the number of times the point $f(z)$ winds around the origin, and this is the required proof.

This theorem enables us to solve the problem in every particular case, and to draw the curve traced by the point $f(z)$ with any degree of accuracy. To do this it is necessary to take a sufficiently dense set of points z on the contour C, to compute the corresponding values $f(z)$ and to join them by a continuous curve. However, in some cases we can get by without these tedious computations. We indicate one of the methods with a numerical example.

Example. Let us find the number of roots of the polynomial $f(z) = z^{11} + 5z^2 - 2$ inside a circle of radius 1 with center at the origin.

On the indicated circle $|z| = 1$, one of the three terms which make up the polynomial $f(z)$, namely $5z^2$, dominates the others. Indeed, $|5z^2| = 5$, but $|z^{11} - 2| \leqslant |z|^{11} + 2 = 3$. This property allows us to reason thus. Let us denote $z^{11} + 5z^2 - 2$ by w, $5z^2$ by N_1, and $z^{11} - 2$ by N_2. While the point z goes once around the unit circle, $N_1 = 5z^2$ winds around a circle of radius 5 twice, since $|N_1| = 5$ and $\arg N_1 = 2 \arg z$. The point w is "tethered" to the point N_1 by a vector whose length is $|N_2| \leqslant 3$, i.e., the distance from the point w to the point N_1 is at all times smaller than the distance from N_1 to the origin of the coordinate system.

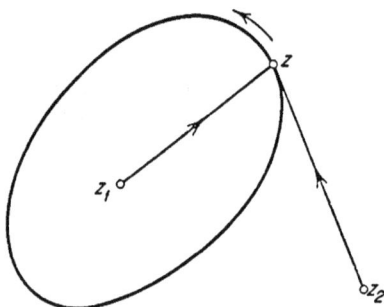
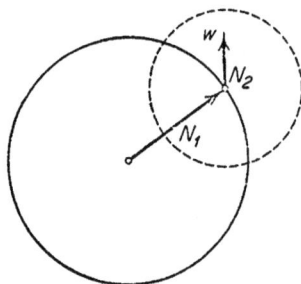

FIG. 9.   FIG. 10.

Consequently, the point w, however it may "wind" around N_1 (figure 10) cannot "independently" go around the origin, and therefore winds around the origin exactly as many times as the point N_1 does, i.e., twice. Consequently, the number of roots of $f(z)$ in the interior of the region in question is equal to two.

Hurwitz's problem. In mechanics, particularly in the theory of oscillations and control, an important role is played by the conditions that permit us to decide whether all the roots of a given polynomial $f(z) = a_0 z^n + a_1 z^{n-1} + \cdots + a_n$ (with real coefficients) have negative real parts, i.e., lie in the half plane left of the imaginary axis.

One of the criteria for solving this problem is easy to obtain from reasons similar to the principle of the argument.

We will assume that $a_0 > 0$.

Let the point z (figure 11) move on the imaginary axis downward from above, i.e., let $z = iy$ as y changes from $+\infty$ to $-\infty$, remaining real. Then $f(z)$ describes a curve with infinite branches. For our investigation the closely related curve described by the function

$$f_1(z) = (i)^{-n} f(z) = a_0 y^n - a_2 y^{n-2} + a_4 y^{n-4} + \cdots - i(a_1 y^{n-1} - a_3 y^{n-3} + \cdots)$$
$$= \phi(y) - i\psi(y),$$

where

$$\phi(y) = a_0 y^n - a_2 y^{n-2} + \cdots,$$
$$\psi(y) = a_1 y^{n-1} - a_3 y^{n-3} + \cdots$$

is more convenient.

Since $\arg i = \pi/2$, therefore $\arg f_1(z) = -n\pi/2 + \arg f(z)$, and consequently the increments of the arguments of $f(z)$ and $f_1(z)$ are the same.

Let us compute the increment of the argument of the point $f_1(z)$ as z moves on the imaginary axis downward.

Let $f(z) = a_0(z - z_1)(z - z_2) \cdots (z - z_n)$. Then

$$\arg f_1(z) = \arg(a_0 i^{-n}) + \arg(z - z_1) + \arg(z - z_2) + \cdots + \arg(z - z_n).$$

It is clear geometrically that the increment of $\arg(z - z_k)$ is equal to π, if z_k lies in the right half plane and to $-\pi$, if z lies in the left half plane (figure 11).

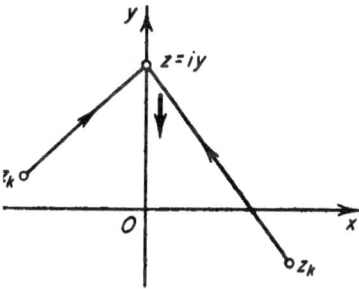

FIG. 11.

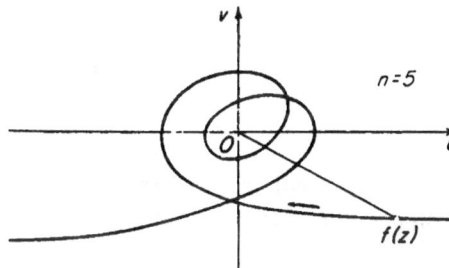

FIG. 12.

Therefore the increment of the argument of $f_1(z)$ is equal to $\pi(N_1 - N_2)$, where N_1 is the number of roots of $f(z)$ in the right half plane and N_2 is the number of roots in the left half plane. For all the roots to lie in the left half plane it is necessary and sufficient that the increment of the argument of the point $f_1(z)$ be equal to $-\pi n$, i.e., that the point $f_1(z)$ make n half revolutions clockwise about the origin (figure 12).

We note that the point $f_1(z) = \phi(y) - i\psi(y)$ intersects the imaginary axis for those values of y that are roots of $\phi(y)$ and the real axis for the roots of $\psi(y)$. Since $\phi(y)$ has no more than n real roots and the number of real roots of $\psi(y)$ is not more than $n-1$, it is easy to see geometrically that $f_1(z)$ can make n complete revolutions in the clockwise direction if and only if the curve comes from the fourth quadrant and intersects in turn the negative half of the imaginary axis, the negative half of the real axis, the positive half of the imaginary axis, the positive half of the real axis, etc., so that the total number of points of intersection with the imaginary axis is equal to n (one for every half revolution), and with the real axis it is equal to $n-1$ (one less than the number of half revolutions). Therefore the coefficient a_1 must be positive, and the roots of the polynomials $\phi(y)$ and $\psi(y)$ must be all real and alternating. This last statement means that if $y_1 > y_2 > \cdots > y_n$ are the roots of $\phi(y)$ arranged in decreasing order, and $\eta_1 > \eta_2 > \cdots > \eta_{n-1}$ are the roots of $\psi(y)$, then $y_1 > \eta_1 > y_2 > \eta_2 > \cdots > y_{n-1} > \eta_{n-1} > y_n$.

Thus, in order that all roots of the polynomial $f(z) > z^n + a_1 z^{n-1} + \cdots + a_n$ with real coefficients and $a_0 > 0$ lie in the left half plane, it is necessary and sufficient that the coefficient a_1 be positive and the roots of the polynomials $\phi(y) = a_0 y^n - a_2 y^{n-2} + a_4 y^{n-4} - \cdots$ and $\psi(y) = a_1 y^{n-1} - a_3 y^{n-3} + \cdots$ be all real and alternating.

This condition is equivalent to the well-known condition of Hurwitz to the effect that all the following determinants are positive:

$$
a_1, \quad
\begin{vmatrix} a_1 & a_0 \\ a_3 & a_2 \end{vmatrix}, \quad
\begin{vmatrix} a_1 & a_0 & a_{-1} \\ a_3 & a_2 & a_1 \\ a_5 & a_4 & a_3 \end{vmatrix}, \quad \cdots, \quad
\begin{vmatrix}
a_1 & a_0 & a_{-1} & \cdots & a_{2-n} \\
a_3 & a_2 & a_1 & \cdots & a_{4-n} \\
\cdot & \cdot & \cdot & \cdots & \cdot \\
\cdot & \cdot & \cdot & \cdots & \cdot \\
\cdot & \cdot & \cdot & \cdots & \cdot \\
a_{2n-1} & a_{2n-2} & a_{2n-3} & \cdots & a_n
\end{vmatrix},
$$

where all a_i with indices less than 0 or greater than n are replaced by zero (on determinants, see Chapter XVI, §3).

§5. Approximate Calculation of Roots

Sturm's method in combination with the lower limit of the difference of two distinct real roots allows us to construct the "separation" of real roots of a polynomial with real coefficients, i.e., allows us to determine for each root limits a and b between which only this one root can be found. It remains to discover a suitable method for finding, in the segment $a < b$, numbers $\alpha_1 < \alpha_2 < \alpha_3 < \cdots$ and $\beta_1 > \beta_2 > \beta_3 > \cdots$, which converge as rapidly as possible to the desired root, the first sequence being an approximation by defect and the second by excess. Each of the two approximations α_k and β_k clearly differs from the desired root x by less than their difference $\beta_k - \alpha_k$, since the root lies between them. Thus we can find upper bounds for the error when we stop at any given approximation.

Graph of a polynomial. Let the given nth-degree polynomial with real coefficients be

$$f(x) = a_0 x^n + a_1 x^{n-1} + \cdots + a_{n-1} x + a_n .$$

Let us consider the curve that represents in rectangular coordinates the equation $y = f(x)$, i.e., the graph of this polynomial. This curve is sometimes called an nth-order parabola. First of all, it is clear that for any real x there is one and only one definite $y = f(x)$; consequently, the graph f ranges arbitrarily far to the right and to the left. In addition, for continuous change of x, $f(x)$ as well as $f'(x)$ change continuously, i.e., without jumps. Therefore, the graph f is a smooth curve. For x large in absolute value the first term $a_0 x^n$ exceeds in absolute value the sum of all remaining terms, since they are all of lower degree. From this it follows that if n is even and $a_0 > 0$, then the graph f on the right and on the left goes to infinity upward (and if $a_0 < 0$, downward); but if n is odd and $a_0 > 0$, then on the right it goes upward and on the left downward (if $a_0 < 0$, then conversely).

The points of intersection of the graph f with the Ox-axis, i.e., those points where $y = f(x) = 0$, correspond to the real roots of the equation $f(x) = 0$; there are no more than n of them. At the maxima and minima of the graph $y = f(x)$, the derivative $f'(x) = 0$; consequently, the number of maxima and minima is not greater than $n - 1$. If on some section $f''(x) > 0$, the first derivative increases there, i.e., the graph is concave upward; if $f''(x) < 0$, then the graph is concave downward. Because some roots of $f'(x) = 0$ may be complex, the number of maxima and minima of the graph f may be smaller than $n - 1$.

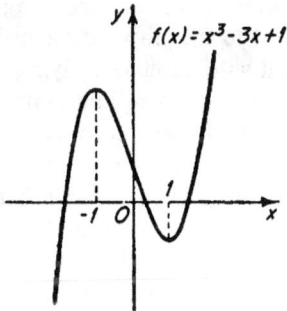

$f(x) = x^3 - 3x + 1$

FIG. 13.

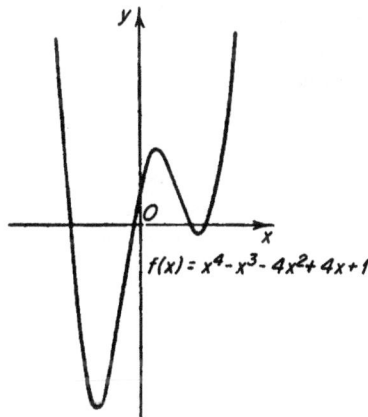

$f(x) = x^4 - x^3 - 4x^2 + 4x + 1$

FIG. 14.

Here are examples of the graphs of polynomials

$$f(x) = x^3 - 3x + 1 \quad \text{(figure 13)},$$
$$f(x) = x^4 - x^3 - 4x^2 + 4x + 1 \quad \text{(figure 14)}.$$

After constructing the graph of a polynomial it is easy to find approximations to its roots. Namely, the roots are the abscissas of the points of intersection of the graph with the Ox-axis.

The method of "undershot" and "overshot." Let us substitute in the polynomial $f(x)$ some integral rational number, for example 3, and then substitute 4, 5, $\cdots$. If in substituting 4, 5, 6 we still obtain the same sign as for 3, but for 7 the opposite sign, then it is clear that between 6 and 7 the polynomial $f(x)$ has at least one root. Now we substitute 6, 6.1, 6.2, $\cdots$ and find two neighbors of this sequence of numbers, for example 6.4 and 6.5 which when substituted give different signs. Accordingly, there will be at least one root between them. Then we substitute 6.4, 6.41, 6.42, 6.43, $\cdots$ and find even closer limits for the root, for example, 6.42 and 6.43, etc. This is the method of "undershooting and overshooting." The method can be considerably simplified by applying at each step of the calculations a supplementary transformation of the polynomial, and then at each step after the first, it will be necessary to substitute only whole numbers and not fractions, and moreover, only the whole numbers 1, 2, $\cdots$, 9. But we will not dwell on this simplification.

The method of tangents and the method of chords. The method of tangents, called Newton's method, and the method of chords, or of linear interpolation, called also the method of false position (*regula falsi*), are used either separately or together to obtain estimates of error. Suppose between a and b we have only one root of the polynomial $f(x)$, so that $f(a)$ and $f(b)$ are of opposite sign, and let us also suppose that the second derivative $f''(x)$ between a and b is of constant sign. In this case the part of the graph of $f(x)$ between a and b has one of four forms (figure 15).

In cases I and II in figure 15, the tangent to the graph at the point with abscissa a intersects the Ox-axis at a point with abscissa α_1 lying between the desired root and a. If we calculate the abscissa α_1 and consider now the tangent to the graph from the point with abscissa α_1, we analogously find a point α_2 lying between the point α_1 and the desired root, and then find a corresponding α_3 and so on. In this way we will obtain better and better approximations with defect. As can be seen from the diagram, these values approach the desired root with great rapidity.

In cases III and IV it is necessary, on the other hand, to start with the abscissa b, and then obtain points $\beta_1, \beta_2, \beta_3, \cdots$, i.e., better and better approximations with excess. Which of the four cases actually occurs, is easy to determine by the signs of $f(a)$, $f(b)$, and $f''(x)$ for $a < x < b$.

Since the equation of the tangent to the curve $y = f(x)$ at its points with abscissa a is

$$y - f(a) = f'(a)(x - a),$$

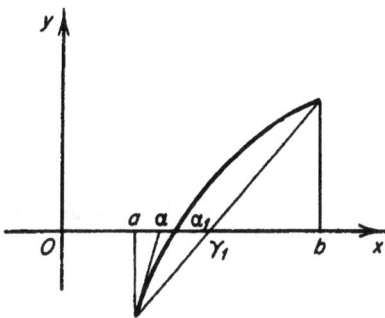

Case I

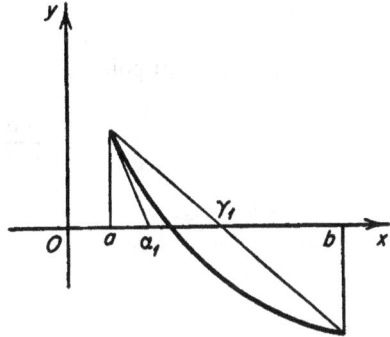

Case II

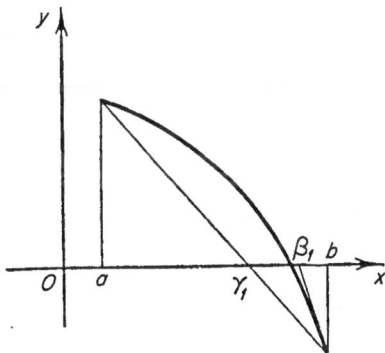

Case III

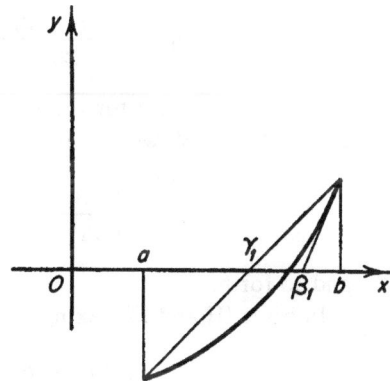

Case IV

FIG. 15.

the abscissa α_1 of the point of its intersection with the Ox-axis is obtained from the equality

$$0 - f(a) = f'(a)(\alpha_1 - a)$$

that is

$$\alpha_1 = a - \frac{f(a)}{f'(a)}.$$

Then

$$\alpha_2 = \alpha_1 - \frac{f(\alpha_1)}{f'(\alpha_1)}, \quad \alpha_3 = \alpha_2 - \frac{f(\alpha_2)}{f'(\alpha_2)}$$

and so on.

Analogously,

$$\beta_1 = b - \frac{f(b)}{f'(b)} \, , \beta_2 = \beta_1 - \frac{f(\beta_1)}{f'(\beta_1)} \, , \beta_3 = \beta_2 - \frac{f(\beta_2)}{f'(\beta_2)}$$

and so on.

This is Newton's method.*

The method of linear interpolation or false position, consists of the following. The equation of a chord, as the equation of a line passing through two given points, has the form

$$\frac{x - a}{b - a} = \frac{y - f(a)}{f(b) - f(a)} \, ,$$

and the abscissa γ_1 of the point of its intersection with the Ox-axis, as obtained from the equation

$$\frac{x - a}{b - a} = \frac{0 - f(a)}{f(b) - f(a)}$$

is equal to

$$\gamma_1 = - \frac{(b - a)f(a)}{f(b) - f(a)} + a = \frac{af(b) - bf(a)}{f(b) - f(a)} \, .$$

Taking this number for the new b in cases I and II and for the new a in cases III and IV, we find in cases I and II

$$\gamma_2 = \frac{af(\gamma_1) - \gamma_1 f(a)}{f(\gamma_1) - f(a)} \, , \quad \gamma_3 = \frac{af(\gamma_2) - \gamma_2 f(a)}{f(\gamma_2) - f(a)}$$

and so forth.

In cases III and IV, taking γ_1 for the new a we find

$$\gamma_2 = \frac{\gamma_1 f(b) - bf(\gamma_1)}{f(b) - f(\gamma_1)} \, , \quad \gamma_3 = \frac{\gamma_2 f(b) - bf(\gamma_2)}{f(b) - f(\gamma_2)}$$

and so forth.

The combination of these two methods is particularly important, since (as may be seen from the diagrams) it allows us, if the approximations from above and below are known, to estimate the error, which is clearly

---

* From these formulas we also obtain a rigorous proof of the two assertions made from a consideration of the diagrams. Namely, the values α_n (or β_n) with increasing n change monotonically, for example in case I they increase and are bounded, i.e., by virtue of the Weierstrass lemma, they approach some limit α. Replacing α_n in these formulas by its limit α, we obtain $\alpha = \alpha - [f(\alpha)/f'(\alpha)]$ from which $f(\alpha) = 0$, i.e., α is a root of f.

not greater than the difference between these approximations, since the desired root is between them.

Remark. It is important to note that the fact that $f(x)$ is a polynomial, and not some other function of x, does not play any role at all either in Newton's method, or in the method of linear interpolation, i.e., both of these methods and their combination can be adapted, under the aforementioned conditions, to transcendental equations.

Lobačevskiĭ's method. One of the most widely used methods of calculation of roots, especially of complex roots, is the method* proposed by N. I. Lobačevskiĭ in his book "Algebra," published in 1834. The basic idea of this method goes back to Bernoulli.

We note, first of all, that if we are given a polynomial whose roots are $x_1, x_2, \cdots, x_n$, then it is easy to write down the polynomial, also of the nth-degree, whose roots are $x_1^2, x_2^2, \cdots, x_n^2$, i.e., the squares of the roots of the given polynomial. Indeed, if $x_1, x_2, \cdots, x_n$ are the roots of the polynomial

$$x^n + a_1 x^{n-1} + a_2 x^{n-2} + \cdots + a_n,$$

then it may be written as

$$(x - x_1)(x - x_2) \cdots (x - x_n),$$

and the polynomial

$$x^n - a_1 x^{n-1} + a_2 x^{n-2} - \cdots \pm a_n,$$

whose roots are the roots of the given polynomial taken with opposite sign, may be written as

$$(x + x_1)(x + x_2) \cdots (x + x_n).$$

The product of these two polynomials is consequently

$$(x^2 - x_1^2)(x^2 - x_2^2) \cdots (x^2 - x_n^2)$$

and therefore contains only even powers of x. Setting $x^2 = y$, we obtain an nth-degree polynomial in y

$$y^n + b_1 y^{n-1} + b_2 y^{n-2} + \cdots + b_n,$$

which may be written as

$$(y - x_1^2)(y - x_2^2) \cdots (y - x_n^2),$$

---

* This method was discovered independently by Dandelin (1826), N. I. Lobačevskiĭ (1834), and Graeffe (1837).

since its roots are $x_1^2, x_2^2, \cdots, x_n^2$. Instead of directly multiplying the polynomial

$$x^n + a_1 x^{n-1} + a_2 x^{n-2} + \cdots + a_n$$

by the polynomial

$$x^n - a_1 x^{n-1} + a_2 x^{n-2} - \cdots \pm a_n ,$$

we can obtain the coefficients b_k according to the following scheme. In the first row above a horizontal line, we write $1, a_1, a_2, \cdots, a_n$ and then below the line, under each of these coefficients a_k, we write first its square a_k^2, then minus twice the product of its neighbors

$$- 2 a_{k-1} a_{k+1} ,$$

then plus twice the product of the coefficients

$$+ 2 a_{k-2} a_{k+2} ,$$

symmetric with respect to a_k, etc., alternating in sign until all further coefficients on one side or the other are equal to zero. The coefficients b_k are then obtained as the sum of the corresponding columns of numbers written under the line.

After obtaining these coefficients $1, b_1, b_2, \cdots, b_n$ of the polynomial whose roots are $1, x_1^2, x_2^2, \cdots, x_n^2$, we next construct the coefficients $1, c_1, c_2, \cdots, c_n$ of the polynomial whose roots are the squares of the roots of the polynomial

$$y^n + b_1 y^{n-1} + b_2 y^{n-2} + \cdots + b_n ,$$

i.e., $x_1^4, x_2^4, \cdots, x_n^4$. Then analogously we obtain the coefficients $1, d_1, d_2, \cdots, d_n$ of the polynomial whose roots are $x_1^8, x_2^8, \cdots, x_n^8$; and then the polynomial whose roots are $x_1^{16}, x_2^{16}, \cdots, x_n^{16}$, and so forth.

Let us consider only the fundamental idea of Lobačevskiĭ's method; moreover, we restrict ourselves for simplicity to the case when all roots of the equation are real and distinct in absolute value. Let

$$| x_1 | > | x_2 | > \cdots > | x_n |,$$

i.e., let x_1 be the root largest in absolute value, x_2 the next largest, and so on. Let N be a sufficiently large number and let the polynomial

$$X^n + A_1 X^{n-1} + A_2 X^{n-2} + \cdots + A_n$$

have roots equal to the Nth power of the roots x_1, x_2, $\cdots$, x_n of the given polynomial, i.e.,

$$- A_1 = x_1^N + x_2^N + \cdots + x_n^N,$$

$$A_2 = x_1^N x_2^N + x_1^N x_3^N + \cdots + x_{n-1}^N x_n^N,$$

$$\cdots\cdots\cdots\cdots\cdots\cdots\cdots\cdots\cdots\cdots\cdots\cdots\cdots\cdots\cdots$$

$$\pm A_n = x_1^N x_2^N \cdots x_n^N.$$

Then in the sequence of numbers $|\, x_1^N\, |,\ |\, x_2^N\, |,\ \cdots,\ |\, x_n^N\, |$ for large N each sucessive number is so much smaller than its predecessor that in these expressions for A_1, A_2, $\cdots$, A_n we may retain only the first summand, the sum of all remaining summands being neglected in comparison with the first. We thus obtain the approximate formulas

$$x_1^N \approx - A_1, \qquad x_1^N x_2^N \approx A_2,$$

$$x_1^N x_2^N x_3^N \approx A_3, \cdots, x_1^N x_2^N x_3^N \cdots x_n^N \approx \pm A_n,$$

or, dividing pairwise and taking the Nth roots, we have the following formulas for x_k:

$$x_1 = \sqrt[N]{-A_1}, \ x_2 = \sqrt[N]{-\frac{A_2}{A_1}}, \ x_3 = \sqrt[N]{-\frac{A_3}{A_2}}, \cdots, x_n = \sqrt[N]{-\frac{A_n}{A_{n-1}}}.$$

It can be shown that it is sufficient to extend the computation up to the polynomial whose coefficients taken with signs $+ - + - \cdots$ will be equal with the necessary degree of exactness, to the squares of the corresponding coefficients of the preceding polynomial.

A detailed exposition of Lobačevskiĭ's method can be found in the well-known book of Academician A. N. Krylov "Lectures on approximate calculations."

Suggested Reading

E. Artin, *Galois theory*, 2nd ed., University of Notre Dame, 1944.

——, *Geometric algebra*, Interscience, New York, 1957.

G. Birkhoff and S. MacLane, *A survey of modern algebra*, 2nd ed., Macmillan, New York, 1953.

F. Klein, *Famous problems of elementary geometry. The duplication of the cube, the trisection of an angle, the quadrature of the circle*, Dover, New York, 1956.

——, *Lectures on the icosahedron and the solution of equations of the fifth degree*, Dover, New York, 1956.

M. Marden, *The geometry of the zeros of a polynomial in a complex variable*, American Mathematical Society, Providence, R. I., 1949.

CHAPTER V

ORDINARY
DIFFERENTIAL EQUATIONS

§1. Introduction

Examples of differential equations. The equations that we have encountered up to now have been for the most part concerned with finding the numerical value of one magnitude or another. When, for example, in the search for maxima and minima of functions, we solved an equation and found those points for which the rate of change of a function vanishes, or when in Chapter IV we considered the problem of finding the roots of polynomials, we were in each case looking for isolated numbers. But in the applications of mathematics there often arise problems of a qualitatively different sort, in which the unknown is itself a function, a law expressing the dependence of certain variables on others. For example, in investigating the process of the cooling of a body, our task is to determine how its temperature will change in the course of time; to describe the motion of a planet or a star we must determine the dependence of their coordinates on time, and so forth.

We can quite often construct an equation for finding the required unknown functions, such equations being called functional equations. The nature of these may, generally speaking, be extremely varied; in fact, it may be said that we have already met the simplest and most primitive functional equations when we were considering implicit functions.

The problem of finding unknown functions will concern us in Chapters V, VI, and VII. In the present chapter, and in the following one, we will consider the most important class of equations serving to determine such functions, namely *differential equations*; that is, equations in which not only the unknown function occurs, but also its derivatives of various orders.

311

The following equations may serve as examples:

$$\frac{dx}{dt} + P(t)\,x = Q(t), \frac{d^2x}{dt^2} + m^2 x = A \sin \omega t, \frac{d^2x}{dt^2} = tx,$$

$$\frac{\partial u}{\partial t} = \frac{\partial^2 u}{\partial x^2}, \frac{\partial^2 u}{\partial t^2} = \frac{\partial^2 u}{\partial x^2}, \frac{\partial^2 u}{\partial x^2} + \frac{\partial^2 u}{\partial y^2} = 0. \qquad (1)$$

In the first three of these, the unknown function is denoted by the letter x and the independent variable by t; in the last three, the unknown function is denoted by the letter u and it depends on two arguments, x and t, or x and y.

The great importance of differential equations in mathematics, and especially in its applications, is due chiefly to the fact that the investigation of many problems in physics and technology may be reduced to the solution of such equations.

Calculations involved in the construction of electrical machinery or of radiotechnical devices, computation of the trajectory of projectiles, investigation of the stability of an aircraft in flight, or of the course of a chemical reaction, all depend on the solution of differential equations.

It often happens that the physical laws governing a phenomenon are written in the form of differential equations, so that the differential equations themselves provide an exact quantitative (numerical) expression of these laws. The reader will see in the following chapters how the laws of conservation of mass and of heat energy are written in the form of differential equations. The laws of mechanics discovered by Newton allow one to investigate the behavior of any mechanical system by means of differential equations.

Let us illustrate by a simple example. Consider a material particle of mass m moving along an axis Ox, and let x denote its coordinate at the instant of time t. The coordinate x will vary with the time, and knowledge of the entire motion of the particle is equivalent to knowledge of the functional dependence of x on the time t. Let us assume that the motion is caused by some force F, the value of which depends on the position of the particle (as defined by the coordinate x), on the velocity of motion $v = dx/dt$ and on the time t, i.e., $F = F(x, dx/dt, t)$. According to the laws of mechanics, the action of the force F on the particle necessarily produces an acceleration $w = d^2x/dt^2$ such that the product of w and the mass m of the particle is equal to the force, and so at every instant of the motion we have the equation

$$m\frac{d^2x}{dt^2} = F\left(x, \frac{dx}{dt}, t\right). \qquad (2)$$

This is the differential equation that must be satisfied by the function $x(t)$ describing the behavior of the moving particle. It is simply a representation of laws of mechanics. Its significance lies in the fact that it enables us to reduce the mechanical problem of determining the motion of a particle to the mathematical problem of the solution of a differential equation.

Later in this chapter, the reader will find other examples showing how the study of various physical processes can be reduced to the investigation of differential equations.

The theory of differential equations began to develop at the end of the 17th century, almost simultaneously with the appearance of the differential and integral calculus. At the present time, differential equations have become a powerful tool in the investigation of natural phenomena. In mechanics, astronomy, physics, and technology they have been the means of immense progress. From his study of the differential equations of the motion of heavenly bodies, Newton deduced the laws of planetary motion discovered empirically by Kepler. In 1846 Leverrier predicted the existence of the planet Neptune and determined its position in the sky on the basis of a numerical analysis of the same equations.

To describe in general terms the problems in the theory of differential equations, we first remark that every differential equation has in general not one but infinitely many solutions; that is, there exists an infinite set of functions that satisfy it. For example, the equation of motion for a particle must be satisfied by any motion induced by the given force $F(x, dx/dt, t)$, independently of the starting point or the initial velocity. To each separate motion of the particle there will correspond a particular dependence of x on time t. Since under a given force F there may be infinitely many motions the differential equation (2) will have an infinite set of solutions.

Every differential equation defines, in general, a whole class of functions that satisfy it. The basic problem of the theory is to investigate the functions that satisfy the differential equation. The theory of these equations must enable us to form a sufficiently broad notion of the properties of all functions satisfying the equation, a requirement which is particularly important in applying these equations to the natural sciences. Moreover, our theory must guarantee the means of finding numerical values of the functions, if these are needed in the course of a computation. We will speak later about how these numerical values may be found.

If the unknown function depends on a single argument, the differential equation is called an *ordinary differential equation*. If the unknown function depends on several arguments and the equation contains derivatives with respect to some or all of these arguments, the differential equation is called a *partial differential equation*. The first three of the equations in (1) are ordinary and the last three are partial.

The theory of partial differential equations has many peculiar features which make them essentially different from ordinary differential equations. The basic ideas involved in such equations will be presented in the next chapter; here we will examine only ordinary differential equations.

Let us consider some examples.

Example 1.   The law of decay of radium says that the rate of decay is proportional to the initial amount of radium present. Suppose we know that a certain time $t = t_0$ we had R_0 grams of radium. We want to know the amount of radium present at any subsequent time t.

Let $R(t)$ be the amount of undecayed radium at time t. The rate of decay is given by the value of $-(dR/dt)$. Since this is proportional to R, we have

$$-\frac{dR}{dt} = kR, \tag{3}$$

where k is a constant

In order to solve our problem, it is necessary to determine a function from the differential equation (3). For this purpose we note that the function inverse to $R(t)$ satisfies the equation

$$-\frac{dt}{dR} = \frac{1}{kR}, \tag{4}$$

since $dt/dR = (1/dR)/dt$. From the integral calculus it is known that equation (4) is satisfied by any function of the form

$$t = -\frac{1}{k}\ln R + C,$$

where C is an arbitrary constant. From this relation we determine R as a function of t. We have

$$R = e^{-kt+kC} = C_1 e^{-kt}. \tag{5}$$

From the whole set of solutions (5) of equation (3) we must select one which for $t = t_0$ has the value R_0. This solution is obtained by setting $C_1 = R_0 e^{kt_0}$.

From the mathematical point of view, equation (3) is the statement of a very simple law for the change with time of the function R; it says that the rate of decrease $-(dR/dt)$ of the function is proportional to the value of the function R itself. Such a law for the rate of change of a function is satisfied not only by the phenomena of radioactive decay but also by many other physical phenomena.

We find exactly the same law for the rate of change of a function, for example, in the study of the cooling of a body, where the rate of decrease in the amount of heat in the body is proportional to the difference between the temperature of the body and the temperature of the surrounding medium, and the same law occurs in many other physical processes. Thus the range of application of equation (3) is vastly wider than the particular problem of the radioactive decay from which we obtained the equation.

Example 2. Let a material point of a mass m be moving along the horizontal axis Ox in a resisting medium, for example in a liquid or a gas, under the influence of the elastic force of two springs, acting under Hooke's law (figure 1), which states that the elastic force acts toward the

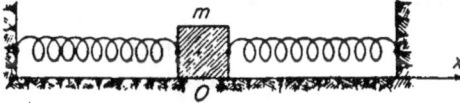

FIG. 1.

position of equilibrium and is proportional to the deviation from the equilibrium position. Let the equilibrium position occur at the point $x = 0$. Then the elastic force is equal to $-bx$ where $b > 0$.

We will assume that the resistance of the medium is proportional to the velocity of motion, i.e., equal to $-a(dx/dt)$, where $a > 0$ and the minus sign indicates that the resisting medium acts against the motion. Such an assumption about the resistance of the medium is confirmed by experiment.

From Newton's basic law that the product of the mass of a material point and its acceleration is equal to the sum of the forces acting on it, we have

$$m\frac{d^2x}{dt^2} = -bx - a\frac{dx}{dt}. \tag{6}$$

Thus the function $x(t)$, which describes the position of the moving point at any instant of time t, satisfies the differential equation (6). We will investigate the solutions of this equation in one of the later sections.

If, in addition to the forces mentioned, the material point is acted upon by still another force, F outside of the system, then the equation of motion (6) takes the form

$$m\frac{d^2x}{dt^2} = -bx - a\frac{dx}{dt} + F \tag{6'}$$

Example 3. A mathematical pendulum is a material point of mass m, suspended on a string whose length will be denoted by l. We will assume that at all stages the pendulum stays in one plane, the plane of the drawing (figure 2). The force tending to restore the pendulum to the vertical position OA is the force of gravity mg, acting on the material point. The position of the pendulum at any time t is given by the angle ϕ by which it differs from the vertical OA. We take the positive direction of ϕ to be counterclockwise. The arc $AA' = l\phi$ is the distance moved by the material point from the position of equilibrium A. The velocity of motion v will be directed along the tangent to the circle and will have the following numerical value:

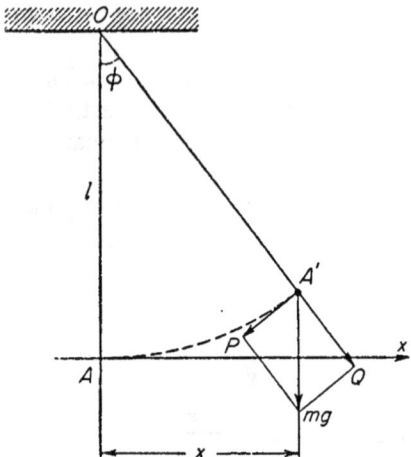

$$v = l\frac{d\phi}{dt}.$$

FIG. 2.

To establish the equation of motion, we decompose the force of gravity mg into two components Q and P, the first of which is directed along the radius OA' and the second along the tangent to the circle. The component Q cannot affect the numerical value of the rate v, since clearly it is balanced by the resistance of the suspension OA'. Only the component P can affect the value of the velocity v. This component always acts toward the equilibrium position A, i.e., toward a decrease in ϕ, if the angle ϕ is positive, and toward an increase in ϕ, if ϕ is negative. The numerical value of P is equal to $-mg \sin \phi$, so that the equation of motion of the pendulum is

$$m\frac{dv}{dt} = -mg \sin \phi$$

or

$$\frac{d^2\phi}{dt^2} = -\frac{g}{l} \sin \phi. \tag{7}$$

It is interesting to note that the solutions of this equation cannot be expressed by a finite combination of elementary functions. The set of elementary functions is too small to give an exact description of even such a simple physical process as the oscillation of a mathematical pendulum. Later we will see that the differential equations that are solvable by elementary functions are not very numerous, so that it very frequently happens that investigation of a differential equation encountered in physics or mechanics leads us to introduce new classes of functions, to subject them to investigation, and thus to widen our arsenal of functions that may be used for the solution of applied problems.

Let us now restrict ourselves to small oscillations of the pendulum for which, with small error, we may assume that the arc AA' is equal to its projection x on the horizontal axis Ox and $\sin \phi$ is equal to ϕ. Then $\phi \approx \sin \phi = x/l$ and the equation of motion of the pendulum will take on the simpler form

$$\frac{d^2x}{dt^2} = -\frac{g}{l} x. \tag{8}$$

Later we will see that this equation is solvable by trigonometric functions and that by using them we may describe with sufficient exactness the "small oscillations" of a pendulum.

Example 4. Helmholtz' acoustic resonator (figure 3) consists of an air-filled vessel V, the volume of which is equal to v, with a cylindrical neck F. Approximately, we may consider the air in the neck of the container as cork of mass

$$m = \rho s l, \tag{9}$$

where ρ is the density of the air, s is the area of the cross section of the neck, and l is its length. If we assume that this mass of air is displaced from a position of equilibrium by an amount x, then the pressure of the air in the container with volume v is changed from the initial value p by some amount which we will call Δp.

FIG. 3.

We will assume that the pressure p and the volume v satisfy the adiabatic law $pv^k = C$. Then, neglecting magnitudes of higher order, we have

$$\Delta p \cdot v^k + pkv^{k-1} \cdot \Delta v = 0$$

and

$$\Delta p = -kp \frac{\Delta v}{v} = -\frac{kps}{v} x. \tag{10}$$

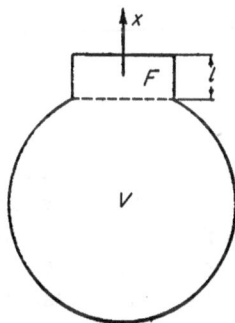

(In our case, $\Delta v = sx$.) The equation of motion of the mass of air in the neck may be written as:

$$m \frac{d^2x}{dt^2} = \Delta p \cdot s. \tag{11}$$

Here $\Delta p \cdot s$ is the force exerted by the gas within the container on the column of air in the neck. From (10) and (11) we get

$$\rho l \frac{d^2x}{dt^2} = -\frac{kps}{v} x, \tag{12}$$

where ρ, p, v, l, k, and s are constants.

Example 5. An equation of the form (6) also arises in the study of electric oscillations in a simple oscillator circuit. The circuit diagram is given in (figure 4). Here on the left we have a condenser of capacity C, in series with a coil of inductance L, and a resistance R. At some instant let the condenser have a voltage across its terminals. In the absence of inductance from the circuit, the current would flow until such time as the terminals of the condenser were at the same potential. The presence of an inductance alters the situation, since the circuit will now generate electric oscillations. To find a law for these oscillations, we denote by $v(t)$, or simply by v, the voltage across the condenser at the instant t, by $I(t)$ the current at the instant t, and by R the resistance. From well-known laws of physics, $I(t)R$ remains constantly equal to the total electromotive force, which is the sum of the voltage across the condenser and the inductance $-L(dI/dt)$. Thus,

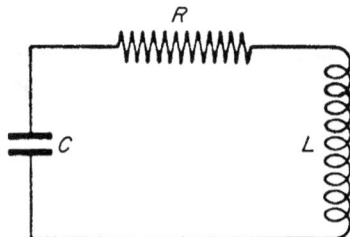

Fig. 4.

$$IR = -v - L\frac{dI}{dt}. \tag{13}$$

We denote by $Q(t)$ the charge on the condenser at time t. Then the current in the circuit will, at each instant, be equal to dQ/dt. The potential difference $v(t)$ across the condenser is equal to $Q(t)/C$. Thus $I = dQ/dt = C(dv/dt)$ and equation (13) may be transformed into

$$LC\frac{d^2v}{dt^2} + RC\frac{dv}{dt} + v = 0. \tag{14}$$

Example 6. The circuit diagram of an electron-tube generator of electromagnetic oscillations is shown in figure 5. The oscillator circuit consisting of a capacitance C, across a resistance R and an inductance L, represents the basic oscillator system. The coil L' and the tube shown in the center of figure 5 form a so-called "feedback." They connect a source of energy, namely the battery B, with the L-R-C circuit; K is the cathode of the tube, A the plate, and S the grid. In such an L-R-C circuit "self-oscillations" will arise. For any actual system in an oscillatory state the energy is transformed into heat or is dissipated in some other form to the surrounding bodies, so that to maintain a stationary state of oscillation it is necessary to have an outside source of energy. Self-oscillations differ from other oscillatory processes in that to maintain a stationary oscillatory state of the system the outside source does not have to be periodic. A self-oscillatory system is constructed in such a way that a constant source of energy, in our case the battery B, will maintain a stationary oscillatory state. Examples of self-oscillatory systems are a clock, an electric bell, a string and bow moved by the hand of the musician, the human voice, and so forth.

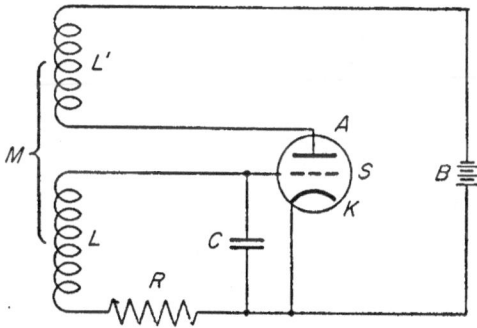

FIG. 5.

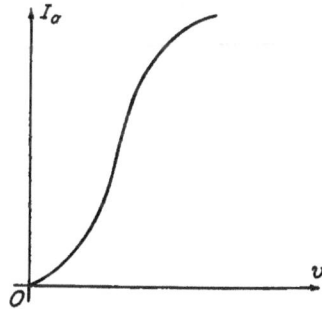

FIG. 6.

The current $I(t)$ in the oscillatory L-R-C circuit satisfies the equation

$$L \frac{dI}{dt} + RI + v = M \frac{dI_a}{dt}. \tag{15}$$

Here $v = v(t)$ is the voltage across the condenser at the instant t, $I_a(t)$ is the plate current through the coil L'; M is the coupling coefficient between the coils L and L'. In comparison with equation (13), equation (15) contains the extra term $M(dI_a/dt)$.

We will assume that the plate current $I_a(t)$ depends only on the voltage between the grid S and the cathode of the tube (i.e., we will neglect the reactance of the anode), so that this voltage is equal to the voltage $v(t)$ across the condenser C. The character of the functional dependence of I_a on v is given in figure 6. The curve as sketched is usually taken to be a cubical parabola and we write an approximate equation for it by:

$$I_a = a_1 v + a_2 v^2 + a_3 v^3.$$

Substituting this into the right side of equation (15), and using the fact that

$$\frac{dv}{dt} = I,$$

we get for v the equation

$$L \frac{d^2v}{dt^2} + [R - M(a_1 + 2a_2 v + 3a_3 v^2)] \frac{dv}{dt} + v = 0. \qquad (16)$$

In the examples considered, the search for certain physical quantities characteristic of a given physical process is reduced to the search for solutions of ordinary differential equations.

Problems in the theory of differential equations. We now give exact definitions. *An ordinary differential equation of order n in one unknown function y is a relation of the form*

$$F[x, y(x), y'(x), y''(x), \cdots, y^{(n)}(x)] = 0 \qquad (17)$$

between the independent variable x and the quantities

$$y(x), y'(x) = \frac{dy}{dx}, y''(x) = \frac{d^2y}{dx^2}, \cdots, y^{(n)}(x) = \frac{d^n v}{dx^n}.$$

The order of a differential equation is the order of the highest derivative of the unknown function appearing in the differential equation. Thus the equation in example 1 is of the first order, and those in examples 2, 3, 4, 5, and 6, are of the second order.

A function $\phi(x)$ is called a *solution of the differential equation* (17) if substitution of $\phi(x)$ for y, $\phi'(x)$ for y', $\cdots, \phi^{(n)}(x)$ for $y^{(n)}$ produces an identity.

Problems in physics and technology often lead to a system of ordinary differential equations with several unknown functions, all depending on the same argument and on their derivatives with respect to that argument.

For greater concreteness, the explanations that follow will deal chiefly with one ordinary differential equation of order not higher than the second and with one unknown function. With this example one may explain the essential properties of all ordinary differential equations and of systems of such equations in which the number of unknown functions is equal to the number of equations.

We have spoken earlier of the fact that, as a rule, every differential equation has not one but an infinite set of solutions. Let us illustrate this first of all by intuitive considerations based on the examples given in equations (2–6). In each of these, the corresponding differential equation is already fully defined by the physical arrangement of the system. But in each of these systems there can be many different motions. For example, it is perfectly clear that the pendulum described by equation (8) may oscillate with many different amplitudes. To each of these different oscillations of the pendulum there corresponds a different solution of equation (8), so that infinitely many such solutions must exist. It may be shown that equation (8) is satisfied by any function of the form

$$x = C_1 \cos \sqrt{\frac{g}{l}}\, t + C_2 \sin \sqrt{\frac{g}{l}}\, t, \tag{18}$$

where C_1 and C_2 are arbitrary constants.

It is also physically clear that the motion of the pendulum will be completely determined only in case we are given, at some instant t_0, the (initial) value x_0 of x (the initial displacement of the material point from the equilibrium position) and the initial rate of motion $x_0' = (dx/dt)\,|_{t=0}$. These intial conditions determine the constants C_1 and C_2 in formula (18).

In exactly the same way, the differential equations we have found in other examples will have infinitely many solutions.

In general, it can be proved, under very broad assumptions concerning the given differential equation (17) of order n in one unknown function that it has infinitely many solutions. More precisely: If for some "initial value" of the argument, we assign an "initial value" to the unknown function and to all of its derivatives through order $n - 1$, then one can find a solution of equation (17) which takes on these preassigned initial values. It may also be shown that such initial conditions completely determine the solution, so that there exists only one solution satisfying the initial conditions given earlier. We will discuss this question later in more detail. For our present aims, it is essential to note that the initial values of the function and the first $n - 1$ derivatives may be given arbitrarily. We have the right to make any choice of n values which define an "initial state" for the desired solution.

If we wish to construct a formula that will if possible include all solutions of a differential equation of order n, then such a formula must contain n

independent arbitrary constants, which will allow us to impose n initial conditions. Such solutions of a differential equation of order n, containing n independent arbitrary constants, are usually called *general solutions* of the equation. For example, a general solution of (8) is given by formula (18) containing two arbitrary constants; a general solution of equation (3) given by formula (5).

We will now try to formulate in very general outline the problems confronting the theory of differential equations. These are many and varied, and we will indicate only the most important ones.

If the differential equation is given together with its initial conditions, then its solution is completely determined. The construction of formulas giving the solution in explicit form is one of the first problems of the theory. Such formulas may be constructed only in simple cases, but if they are found, they are of great help in the computation and investigation of the solution.

The theory should provide a way to obtain some notion of the behavior of a solution: whether it is monotonic or oscillatory, whether it is periodic or approaches a periodic function, and so forth.

Suppose we change the initial values for the unknown function and its derivatives; that is, we change the initial state of the physical system. Then we will also change the solution, since the whole physical process will now run differently. The theory should provide the possibility of judging what this change will be. In particular, for small changes in the initial values will the solution also change by a small amount and will it therefore be stable in this respect, or may it be that small changes in the initial conditions will give rise to large changes in the solution so that the latter will be unstable?

We must also be able to set up a qualitative, and where possible, quantitative picture of the behavior not only of the separate solutions of an equation, but also of all of the solutions taken together.

In machine construction there often arises the question of making a choice of parameters characterizing an apparatus or machine that will guarantee satisfactory operation. The parameters of an apparatus appear in the form of certain magnitudes in the corresponding differential equation. The theory must help us make clear what will happen to the solutions of the equation (to the working of the apparatus) if we change the differential equation (change the parameters of the apparatus).

Finally, when it is necessary to carry out a computation, we will need to find the solution of an equation numerically, and here the theory will be obliged to provide the engineer and the physicist with the most rapid and economical methods for calculating the solutions.

§2. Linear Differential Equations with Constant Coefficients

For certain important classes of ordinary differential equations the general solution may be expressed in terms of simple well-known functions. One of these classes consists of those differential equations with constant coefficients that are linear with respect to the unknown function and its derivatives (in short, linear). The differential equations (3), (6), (8), and (14) are examples of such equations. A linear equation is called homogeneous if it has no term which does not contain the unknown variable, and nonhomogeneous if there is such a term.

Homogeneous linear equations of the second order with constant coefficients. Such equations have the form

$$m \frac{d^2x}{dt^2} + a \frac{dx}{dt} + bx = 0, \tag{6}$$

where m, a, and b are constants. We will assume that m is positive; this does not restrict the generality, since we can always ensure this situation if need be by changing the sign of all coefficients, provided that $m \neq 0$, which we will assume.

We will look for a solution of this equation in the form of an exponential function $e^{\lambda t}$ and ask how the constant λ should be chosen so that the function $x = e^{\lambda t}$ satisfies the equation. Putting $x = e^{\lambda t}$, $dx/dt = \lambda e^{\lambda t}$ and $d^2x/dt^2 = \lambda^2 e^{\lambda t}$ in the left side of equation (6), we get

$$e^{\lambda t}(m\lambda^2 + a\lambda + b).$$

Thus, in order that $x(t) = e^{\lambda t}$ be a solution of equation (6) it is necessary and sufficient that

$$m\lambda^2 + a\lambda + b = 0. \tag{19}$$

If λ_1 and λ_2 are two real roots of equation (19), then it is easy to prove that a solution of equation (6) is given by every function of the form

$$x = C_1 e^{\lambda_1 t} + C_2 e^{\lambda_2 t}, \tag{20}$$

where C_1 and C_2 are arbitrary constants.

Below we will show that formula (20) gives all solutions of equation (6) in the case that equation (19) has distinct real roots.

We note the following important properties of the solution of equation (6):

1. The sum of two solutions is also a solution.

2. A solution multiplied by a constant is also a solution.

In case λ_1 is a multiple root of equation (19), i.e., $m\lambda_1^2 + a\lambda_1 + b = 0$ and $2m\lambda_1 + a = 0$,* then a solution of equation (6) will also be given by the function $te^{\lambda_1 t}$, since if we substitute this function and its derivatives into the left side of equation (6) we get

$$te^{\lambda_1 t}(m\lambda_1^2 + a\lambda_1 + b) + e^{\lambda_1 t}(2m\lambda_1 + a),$$

which is seen from the previous equations to be identically zero.

The general solution of equation (6) in this case has the form

$$x = C_1 e^{\lambda_1 t} + C_2 t e^{\lambda_1 t}. \tag{21}$$

Now let equation (19) have complex roots. These roots will be complex conjugates of each other since m, a, and b are real numbers. Let $\lambda = \alpha \pm i\beta$ The equation

$$m(\alpha \pm i\beta)^2 \pm a(\alpha + i\beta) \pm b = 0$$

is equivalent to the two equations

$$m\alpha^2 - m\beta^2 + a\alpha \pm b = 0 \text{ and } 2m\alpha\beta + a\beta = 0. \tag{22}$$

It is easy to show that in this case the functions $x = e^{\alpha t} \cos \beta t$ and $x = e^{\alpha t} \sin \beta t$ are solutions of equation (6). Thus, for example, putting the function $x(t) = e^{\alpha t} \cos \beta t$ and its derivatives in the left side of equation (6), we get

$$e^{\alpha t} \cos \beta t(m\alpha^2 - m\beta^2 + a\alpha + b) - e^{\alpha t} \sin \beta t(2m\alpha\beta + a\beta).$$

By equation (22) this expression is identically equal to zero.

The general solution of equation (6), if equation (19) has complex roots, has the form

$$x = C_1 e^{\alpha t} \sin \beta t \pm C_2 e^{\alpha t} \cos \beta t, \tag{23}$$

where C_1 and C_2 are arbitrary constants.

In this way, if we know the roots of equation (19), called the *characteristic equation*, we can write down the general solution of equation (6).

We note that the general solution of a linear homogeneous equation of order n with constant coefficients

$$a_n \frac{d^n x}{dt^n} + a_{n-1} \frac{d^{n-1} x}{dt^{n-1}} + \cdots + a_1 \frac{dx}{dt} + a_0 x = 0$$

may be written in a similar manner as a polynomial in exponential and

---

* The sum of the roots λ_1 and λ_2 of the quadratic equation (19) is $\lambda_1 + \lambda_2 = -a/m$, and if the roots are the same, that is $\lambda_1 = \lambda_2$, then the second of the previous equations is true.

trigonometric functions, provided we know the roots of the algebraic equation

$$a_n \lambda^n + a_{n-1} \lambda^{n-1} + \cdots + a_0 = 0,$$

which again is called the *characteristic equation*. Thus, the problem of integrating a linear ordinary differential equation with constant coefficients is reduced to an algebraic problem.

We now show that formulas (20), (21), and (23) give all the solutions of equation (6). We note that C_1 and C_2 in these formulas may always be so chosen that the function $x(t)$ satisfies arbitrary initial conditions $x(t_0) = x_0$, $x'(t_0) = x_0'$. For this C_1 and C_2 need only to be determined from the system of equations

$$x_0 = C_1 e^{\lambda_1 t_0} + C_2 e^{\lambda_2 t_0},$$
$$x_0' = \lambda_1 C_1 e^{\lambda_1 t_0} + \lambda_2 C_2 e^{\lambda_2 t_0}.$$

in the case of formula (20), or by two similar equations in the case of formulas (21) and (23). Clearly, if there existed a solution of equation (6) not contained among the solutions we have constructed, then there would exist two distinct solutions of equation (6) satisfying the same initial conditions. Their difference $x_1(t)$ would not be identically zero and would satisfy the zero initial conditions $x_1(t_0) = 0$, $x_1'(t_0) = 0$. We will show that a solution of equation (6) which satisfies the zero initial conditions can only be $x_1(t) = 0$. Let us first show this under the assumption that $m > 0$, $a > 0$, and $b > 0$. We multiply the two sides of the equation

$$m \frac{d^2 x_1}{dt^2} + a \frac{dx_1}{dt} + bx_1 = 0 \tag{24}$$

by $2(dx_1/dt)$. Since

$$2 \frac{dx_1}{dt} \cdot \frac{d^2 x_1}{dt^2} = \frac{d}{dt} \left(\frac{dx_1}{dt} \right)^2 \quad \text{and} \quad 2x_1(t) \frac{dx_1}{dt} = \frac{d}{dt} (x_1^2),$$

equation (24) may be put in the form

$$\frac{d}{dt} \left[m \left(\frac{dx_1}{dt} \right)^2 \right] + 2a \left(\frac{dx_1}{dt} \right)^2 + b \frac{d}{dt} (x_1^2) = 0.$$

Integrating this identity between t_0 and t, we get

$$m \left(\frac{dx_1}{dt} \right)^2 + 2a \int_{t_0}^{t} \left(\frac{dx_1}{dt} \right)^2 dt + bx_1^2(t) = 0.$$

This equation is possible only if $x_1(t) \equiv 0$. Otherwise, for $t = t_0$, we would clearly have a positive quantity on the left and zero on the right, with a similar situation for $t < t_0$.

In order to establish our proposition for all constant coefficients m, a, and b, we consider the function $y_1(t) = x_1(t)e^{-\alpha t}$ which, as it is easy to show, also satisfies the zero boundary condtions. If the value of $\alpha > 0$ is chosen sufficiently large, then the function $y_1(t)$ will satisfy some equation of the form (6) for $a > 0$, $b > 0$, and $m > 0$. This equation is easily derived by substituting the function $x_1(t) = y_1(t)e^{\alpha t}$ and its derivatives into equation (6). Then, as was shown earlier, we have $y_1(t) \equiv 0$, which means that $x_1(t) = y_1(t)e^{\alpha t} \equiv 0$.

Thus we have shown that formulas (20), (21), and (23) give all the solutions of equation (6).

Let us see what information these formulas give about the character of the solutions of equations (6). To this end we note the formulas

$$\lambda_{1,2} = -\frac{a}{2m} \pm \sqrt{\frac{a^2}{4m^2} - \frac{b}{m}} \tag{25}$$

for the roots of equation (19). In accordance with the physical applications which led us to equation (6), we will assume $m > 0$, $a \geqslant 0$, and $b > 0$.

Case 1. $a^2 > 4bm$. The two roots of the characteristic equation (19) are real, negative, and distinct. In this case the function $x(t)$ given by by formula (20) is a general solution of equation (6). All the functions given by this formula together with their first derivatives tend to zero for $t \to +\infty$, and there is no more than one value of t for which they vanish. It follows that the function $x(t)$ has no more than one maximum or minimum. Physically, this means that the resistance of the medium is sufficiently large to prevent oscillations. The moving point cannot pass through the equilibrium position $x = 0$ more than once. From then on, after attaining a maximum distance from the point $x = 0$, it will begin a slow approach to the point but will never pass through it again.

Case 2. $a^2 = 4bm$. The two roots of equation (19) are equal to each other and the general solution of equation (6) given by formula (21). In this case again all solutions $x(t)$ and their first derivatives tend to zero for $t \to +\infty$. Here $x(t)$ and $x'(t)$ cannot vanish more than once. The character of the motion of the material point with abscissa $x(t)$ is the same as in the first case.

Case 3. $a^2 < 4bm$. The roots of the characteristic equation (19) have nonzero imaginary parts. The general solution of equation (6) is given by formula (23). The point x performs oscillations along the x-axis with a constant period $2\pi/\beta$, which is the same for all solutions of (6), and with amplitude $Ce^{\alpha t}$, where $\alpha = -(a/2m)$.

The oscillations of a physical system which take place without the action of an exterior force are called *characteristic oscillations* (eigenvibrations) of the system. From the previous discussion, it follows that the period of such oscillations for the systems discussed in examples 2, 3, 4 and 5, depends only on the structure of the system and will be the same for all oscillations which could possibly arise in it. In example 2 this period is equal to $2\pi\sqrt{b/m - a^2/4m^2}$; in example 4 to $2\pi\sqrt{kps/v\rho l}$; and example 5 to $2\pi\sqrt{1/LC - R^2/4L^2}$.

If $a = 0$, i.e., if the medium offers no resistance to the motion, then the amplitude of the oscillations is constant: the point oscillates harmonically. But if $a > 0$, i.e., if the medium offers resistance to the motion, although this resistance is small ($a^2 < 4bm$), then the amplitude of the oscillations tends to zero and the oscillations die out.

Finally, the solution $x(t) \equiv 0$ of equation (6) in all cases indicates a state of rest for the point x at the position $x = 0$, which is called the position of equilibrium.

If the real parts of both roots of equation (19) are negative, then it can be seen from formulas (20), (21), and (23), that all the solutions of equation (6), together with their derivatives, tend to zero for $t \to +\infty$; that is, the oscillations die out with the passage of time.

However, if the real part of even one of the roots of equation (19) is positive, then there are solutions of equation (6) not tending to zero for $t \to +\infty$, so that some of the solutions of (6) would not even be bounded for $t \to +\infty$. Such a case can occur only for negative b or negative a, if $m > 0$. Physically, this would correspond to the case in which the elastic force does not attract the point x to the equilibrium position but repels it or else that the resistance of the medium is negative. Such cases cannot be realized in the physical examples considered at the beginning of this chapter, but they are entirely realizable in other physical models.

If the real part of the roots λ_1 and λ_2 of equation (19) is equal to zero, which is possible only if the coefficient a in equation (19) is zero, then for $\alpha = 0$ the point $x(t)$, as can be seen from formula (23), carries out harmonic oscillations with bounded amplitude and bounded velocity.

Nonhomogeneous linear equations with constant coefficients. Let us consider in detail the equation

$$m\frac{d^2x}{dt^2} + a\frac{dx}{dt} + bx = A\cos\omega t. \tag{26}$$

This is the equation of linear oscillations of a material point under the action of an elastic force, of the resistance of a medium and of an external periodic force $A \cos \omega t$ (see equation (6') in §1).

Equation (26) is a nonhomogeneous linear equation and (6) is the corresponding homogeneous equation.

We will now look for the general solution to equation (26).

We note that the sum of a solution of a nonhomogeneous equation and a solution of the corresponding homogeneous equation is also a solution of the nonhomogeneous linear equation. Thus, in order to find a general solution of equation (26), it is sufficient to find any one particular solution. The general solution of equation (26) will then be given in the form of the sum of this particular solution and a general solution of the corresponding homogeneous equation.

It is natural to expect that the motion will follow the rhythm of the external periodic force and to look for a particular solution of equation (26) in the form $x = B \cos (\omega t + \delta)$, where B and δ are as yet undetermined constants. We will attempt to determine B and δ in such a way that the function $x = B \cos (\omega t + \delta)$ will satisfy equation (26). Calculating the derivatives $dx/dt = -B\omega \sin (\omega t + \delta)$ and $d^2x/dt^2 = -B\omega^2 \cos (\omega t + \delta)$ and substituting them into equation (26), we get

$$m[-B\omega^2 \cos (\omega t + \delta)] + a[-B\omega \sin (\omega t + \delta)]$$
$$+ bB \cos (\omega t + \delta) = A \cos \omega t.$$

Applying well-known formulas, we have

$$B[(b - m\omega^2) \cos (\omega t + \delta) - a\omega \sin (\omega t + \delta)]$$
$$= B \sqrt{(b - m\omega^2)^2 + a^2\omega^2} \cos (\omega t + \delta') = A \cos \omega t,$$

where $\delta' = \delta + \gamma$ and $\gamma = \arctan a\omega/(b - m\omega^2)$. Obviously, if we set

$$\delta = -\arctan \frac{a\omega}{b - m\omega^2} \quad \text{and} \quad B = \frac{A}{\sqrt{(b - m\omega^2)^2 + a^2\omega^2}},$$

the function $x = B \cos (\omega t + \delta)$ will satisfy equation (26).

A solution of the form $B \cos (\omega t + \delta)$ will always exist if $(b - m\omega^2)^2 + a^2\omega^2 \neq 0$. In case $(b - m\omega^2)^2 + a^2\omega^2 = 0$, i.e., when $a = 0$ and $b = m\omega^2$, equation (26) has the form

$$m \frac{d^2x}{dt^2} + m\omega^2 x = A \cos \omega t.$$

A particular solution in this case, as is easily established, is $x = (At/2 \sqrt{mb}) \sin \omega t$.

§2. *LINEAR DIFFERENTIAL EQUATIONS* 329

Solutions of the nonhomogeneous equation (26) are called forced oscillations. The multiplier $\phi(\omega) = 1/\sqrt{(b - m\omega^2)^2 + a^2\omega^2}$ characterizes the relation of the amplitude B of the forced oscillation to the amplitude A of the disturbing force. The graph of the function $\phi(\omega)$ is called the resonance curve. The frequency ω for which $\phi(\omega)$ attains its maximum is called the resonant frequency. Let us calculate it. If $\phi(\omega)$ attains the maximum at $\omega_1 \neq 0$, then for this value of ω the derivative $\phi'(\omega)$ vanishes, i.e., $- 4(b - m\omega_1^2) m\omega_1 + 2a^2\omega_1 = 0$, so that $\omega_1 = \sqrt{b/m - a^2/2m^2}$. For this value of ω_1

$$\phi(\omega_1) = \frac{1}{a\sqrt{b/m - a^2/4m^2}}.$$

Hence it can be seen that the amplitude of the forced oscillation for $\omega = \omega_1$ is greater for smaller values of a. For very small a, the frequency ω_1 is very close to the value $\sqrt{b/m}$, i.e., to the frequency the free oscillations. For $a = 0$ and $b = m\omega^2$, as we saw, the forced oscillation has the form

$$x = \frac{At}{2\sqrt{mb}} \sin \omega t,$$

i.e., the amplitude of this oscillation increases beyond all bounds as $t \to + \infty$, a situation which represents the mathematical meaning of resonance. Resonance will occur if the period of the external force is the same as the period of one of the characteristic oscillations of the system. In the practical world, in cases where the period of the external force and the period of the characteristic oscillations are close together, the displacements of the system may become extremly large.

The possibility of large oscillations is often made use of in the construction of various kinds of amplifiers, for example in radio technology. But large oscillations may also lead to the breaking up of structures such as bridges or the framework of machines. Thus it is very important to foresee the possibility of resonance or of oscillations close to it.

From the remarks made earlier, any solution of equation (26) can be written as a sum of the forced oscillation we have found and of one of the solutions of the homogeneous equation given in formulas (20), (21), and (23). For $a > 0$ and $b > 0$ the solution of the homogeneous equation tends to zero for $t \to + \infty$, i.e., any motion eventually approximates the forced oscillations. If $a = 0$ and $b > 0$, the forced oscillation is superposed on a nondecaying characteristic oscillation of the system. For $b = m\omega^2$ and $a = 0$, we have resonance.

If a periodic external force $f(t)$ is imposed on the sytem, the forced oscillations of the system may be found in the following manner. We may represent $f(t)$ with sufficient exactness as a segment of a trigonometric series*

$$\sum_{i=1}^{n} (a_i \cos \omega_i t + b_i \sin \omega_i t). \tag{27}$$

Let us find the forced oscillations corresponding to each term of this sum. Then the oscillation corresponding to the force $f(t)$ will be found by adding together the oscillations corresponding to the various terms of the sum (27). If any of these frequencies is identical with the frequency of a characteristic oscillation of the system, we will have resonance.

§3. Some General Remarks on the Formation and Solution of Differential Equations

There are not many differential equations with the property that all their solutions can be expressed explicitly in terms of simple functions, as is the case for linear equations with constant coefficients. It is possible to give simple examples of differential equations whose general solution cannot be expressed by a finite number of integral of known functions, or as one says, in quadratures.

As Liouville showed in 1841, the solution of the Riccati equation of the form $dy/dx + ay^2 = x^2$, for $a > 0$, cannot be expressed as a finite combination of integrals of elementary functions. So it becomes important to develop methods of approximation to the solutions of differential equations, which will be applicable to wide classes of equations.

The fact that in such cases we find not exact solutions but only approximations should not bother us. First of all, these approximate solutions may be calculated, at least in principle, to any desired degree of accuracy. Second, it must be emphasized that in most cases the differential equations describing a physical process are themselves not altogether exact, as can be seen in all the examples discussed in §1.

An especially good example is provided by the equation (12) for the acoustic resonator. In deriving this equation, we ignored the compressibility of the air in the neck of the container and the motion of the air in the container itself. As a matter of fact, the motion of the air in the neck sets into motion the mass of the air in the vessel, but these two motions have different velocities and displacements. In the neck the displacement of the particles of air is considerably greater than in the container. Thus we ignored the motion of the air in the container, and

---

* Cf. Chapter XII, §7.

took account only of its compression. For the air in the neck, however, we ignored the energy of its compression and took account only of the kinetic energy of its motion.

To derive the differential equation for a physical pendulum, we ignored the mass of the string on which it hangs. To derive equation (14) for electric oscillations in a circuit, we ignored the self-inductance of the wiring and the resistance of the coils. In general, to obtain a differential equation for any physical process, we must always ignore certain factors and idealize others. In view of this, A. A. Andronov drew especial attention to the fact that for physical investigations we are especially interested in those differential equations whose solutions do not change much for arbitrary small changes, in some sense or another, in the equations themselves. Such differential equations are called "intensive." These equations deserve particularly complete study.

It should be stated that in physical investigations not only are the differential equations that describe the laws of change of the physical quantities themselves inexactly defined but even the number of these quantities is defined only approximately. Strictly speaking, there are no such things as rigid bodies. So to study the oscillations of a pendulum, we ought to take into account the deformation of the string from which it hangs and the deformation of the rigid body itself, which we approximated by taking it as a material point. In exactly the same way, to study the oscillations of a load attached to springs, we ought to consider the masses of the separate coils of the springs. But in these examples it is easy to show that the character of the motion of the different particles, which make up the pendulum and its load together with the springs, has little influence on the character of the oscillation. If we wished to take this influence into account, the problem would become so complicated that we would be unable to solve it to any suitable approximation. Our solution would then bear no closer relation to physical reality than the solution given in §1 without consideration of these influences. Intelligent idealization of a problem is always unavoidable. To describe a process, it is necessary to take into account the essential features of the process but by no means to consider every feature without exception. This would not only complicate the problem a great deal but in most cases would result in the impossibility of calculating a solution. The fundamental problem of physics or mechanics, in the investigation of any phenomenon, is to find the smallest number of quantities, which with sufficient exactness describe the state of the phenomenon at any given moment, and then to set up the simplest differential equations that are good descriptions of the laws governing the changes in these quantities. This problem is often very difficult. Which features are the essential ones and which are nonessential is a question that in the final analysis can be decided only by long experience. Only by comparing the answers provided by an idealized argument with the results of experiment can we judge whether the idealization was a valid one.

The mathematical problem of the possibility of decreasing the number of quantities may be formulated in one of the simplest and most characteristic cases, as follows.

Suppose that to begin with we characterize the state of a physical system at time t by the two magnitudes $x_1(t)$ and $x_2(t)$. Let the differential equations expressing their rates of change have the form

$$\frac{dx_1}{dt} = f_1(t, x_1, x_2),$$

$$\epsilon \frac{dx_2}{dt} = f_2(t, x_1, x_2),$$

$$(28)$$

In the second equation the coefficient of the derivative is a small constant parameter ϵ. If we put $\epsilon = 0$, the second of equations (28) will cease to be a differential equation. It then takes the form

$$f_2(t, x_1, x_2) = 0.$$

From this equation, we define x_2 as a function of t and x_1 and we substitute it into the first of the equations (28). We then have the differential equation

$$\frac{dx_1}{dt} = F(t, x_1)$$

for the single variable x_1. In this way the number of parameters entering into the situation is reduced to one. We now ask, under what conditions will the error introduced by taking $\epsilon = 0$ be small. Of course, it may happen that as $\epsilon \to 0$ the value dx_2/dt grows beyond all bounds, so that the right side of the second of equations (28) does not tend to zero as $\epsilon \to 0$.

§4. Geometric Interpretation of the Problem of Integrating Differential Equations; Generalization of the Problem

For simplicity we will consider initially only one differential equation of the first order with one unknown function

$$\frac{dy}{dx} = f(x, y),$$

$$(29)$$

where the function $f(x, y)$ is defined on some domain G in the (x, y) plane.

This equation determines at each point of the domain the slope of the tangent to the graph of a solution of equation (29) at that point. If at each point (x, y) of the domain G we indicate by means of a line segment the the direction of the tangent (either of the two directions may be used) as determined by the value of $f(x, y)$ at this point, we obtain a field of directions. Then the problem of finding a solution of the differential equation (29) for the initial conditon $y(x_0) = y_0$ may be formulated thus: In the domain G we have to find a curve $y = \phi(x)$, passing through the point $M_0(x_0, y_0)$, which at each of its points has a tangent whose slope is given by equation (29), or briefly, which has at each of its points a preassigned direction.

From the geometric point of view this statement of the problem has two unnatural features:

1. By requiring that the slope of the tangent at any given point (x, y) of the domain G be equal to $f(x, y)$, we automatically exclude tangents parallel to Oy, since we generally consider only finite magnitudes; in particular, it is assumed that the function $f(x, y)$ on the right side of equation (29) assumes only finite values.

2. By considering only curves which are graphs of functions of x, we also exclude those curves which are intersected more than once by a line perpendicular to the axis Ox, since we consider only single-valued functions; in particular, every solution of a differential equation is assumed to be a single-valued function of x.

So let us generalize to some extent the preceding statement of the problem of finding a solution to the differential equation (29). Namely, we will now allow the tangent at some points to be parallel to the axis Oy. At these points, where the slope of the tangent with respect to the axis Ox has no meaning, we will take the slope with respect to the axis Oy. In other words, we consider, together with the differential equation (29), the equation

$$\frac{dx}{dy} = f_1(x, y), \qquad (29')$$

where $f_1(x, y) = 1/f(x, y)$, if $f(x, y) \neq 0$, using the second equation when the first is meaningless. The problem of integrating the differential equations (29) and (29') then becomes: In the domain G to find all curves having at each point the tangent defined by these equations. These curves will be called integral curves (integral lines) of the equations (29) and (29') or of the tangent field given by these equations. In place of the plural "equations (29), (29')", we will often use the singular "equation (29), (29')". It is clear that the graph of any solution of equation (29) will also be an integral curve of equation (29), (29'). But not every integral curve of equation (29), (29′) will be the graph of a solution of equation (29). This case will occur, for example, if some perpendicular to the axis *Ox* intersects this curve at more than one point.

In what follows, if it can be clearly shown that

$$f(x, y) = \frac{M(x, y)}{N(x, y)},$$

then we will write only the equation

$$\frac{dy}{dx} = \frac{M(x, y)}{N(x, y)},$$

and omit writing

$$\frac{dx}{dy} = \frac{N(x, y)}{M(x, y)}.$$

Sometimes in place of these equations we introduce a parameter *t*, and write the system of equations

$$\frac{dx}{dt} = N(x, y), \frac{dy}{dt} = M(x, y),$$

where *x* and *y* are considered as functions of *t*.

Example 1. The equation

$$\frac{dy}{dx} = \frac{y}{x} \tag{30}$$

defines a tangent field everywhere except at the origin. This tangent field is sketched in figure 7. All the tangents given by equation (30) pass through the origin.

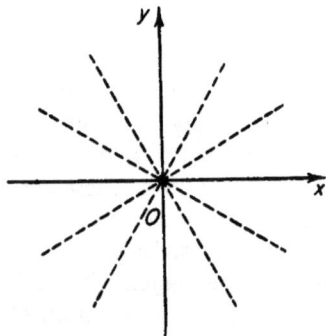

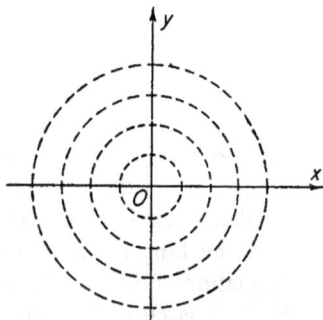

FIG. 7.   FIG. 8.

It is clear that for every k the function

$$y = kx \tag{31}$$

is a solution of equation (30). The collection of all integral curves of this equation is then defined by the relation

$$ax = by = 0, \tag{32}$$

where a and b are arbitrary constants, not both zero. The axis Oy is an integral curve of equation (30), but it is not the graph of a solution of it.

Since equation (30) does not define a tangent field at the origin, the curves (31) and (32) are, strictly speaking, integral curves everywhere except at the origin. Thus it is more correct to say that the integral curves of equation (30) are not straight lines passing through the origin but half lines issuing from it.

Example 2. The equation

$$\frac{dy}{dx} = -\frac{x}{y} \tag{33}$$

defines a field of tangents everywhere except at the origin, as sketched in figure 8. The tangents defined at a given point (x, y) by equations (30) and (33) are perpendicular to each other. It is clear that all circles centered at the origin will be integral curves of equation (33). However the solutions of this equation will be the functions

$$y = + \sqrt{R^2 - x^2}, y = - \sqrt{R^2 - x^2}, -R \leqslant x \leqslant R.$$

For brevity in what follows we will sometimes say "a solution passes through the point (x, y)" in place of the more exact statement "the graph of a solution passes through the point (x, y)."

§5. Existence and Uniqueness of the Solution of a Differential Equation; Approximate Solution of Equations

The question of existence and uniqueness of the solution of a differential equation. We return to the differential equation (17) of arbitrary order n. Generally, it has infinitely many solutions and in order that we may pick from all the possible solutions some one specific one, it is necessary to attach to the equation some supplementary conditions, the number of which should be equal to the order n of the equation. Such conditions may be of extremely varied character, depending on the physical, mechanical, or other significance of the original problem. For example, if we have to investigate the motion of a mechanical system beginning with some specific initial state, the supplementary conditions will refer to a specific (initial) value of the independent variable and will be called initial conditions of the problem. But if we want to define the curve of a cable in a suspension bridge, or of a loaded beam resting on supports at each end, we encounter conditions corresponding to different values of the independent variable, at the ends of the cable or at the points of support of the beam. We could give many other examples showing the variety of conditions to be fulfilled in connection with differential equations.

We will assume that the supplementary conditions have been defined and that we are required to find a solution of equation (17) that satisfies them. The first question we must consider is whether any such solution exists at all. It often happens that we cannot be sure of this in advance. Assume, say, that equation (17) is a description of the operation of some physical apparatus and suppose we want to determine whether periodic motion occurs in this apparatus. The supplementary conditions will then be conditions for the periodic repetition of the initial state in the apparatus, and we cannot say ahead of time whether or not there will exist a solution which satisfies them.

In any case the investigation of problems of existence and uniqueness of a solution makes clear just which conditions can be fulfilled for a given differential equation and which of these conditions will define the solution in a unique manner. But the determination of such conditions and the proof of existence and uniqueness of the solution for a differential equation corresponding to some physical problem also has great value for the physical theory itself. It shows that the assumptions adopted in setting up the mathematical description of the physical event are on the one hand mutually consistent and on the other constitute a complete description of the event.

The methods of investigating the existence problem are manifold, but among them an especially important role is played by what are called direct methods. The proof of the existence of the required solution is provided by the construction of approximate solutions, which are proved to converge to the exact solution of the problem. These methods not only establish the existence of an exact solution, but also provide a way, in fact the principal one, of approximating it to any desired degree of accuracy.

For the rest of this section we will consider, for the sake of definiteness, a problem with initial data, for which we will illustrate the ideas of Euler's method and the method of successive approximations.

Euler's method of broken lines. Consider in some domain G of the (x, y) plane the differential equation

$$\frac{dy}{dx} = f(x, y). \tag{34}$$

As we have already noted, equation (34) defines in G a field of tangents. We choose any point (x_0, y_0) of G. Through it there will pass a straight line L_0 with slope $f(x_0, y_0)$. On the straight line L_0 we choose a point (x_1, y_1), sufficiently close to (x_0, y_0); in figure 9 this point is indicated by the number 1. We draw the straight line L_1 through the point (x_1, y_1) with slope $f(x_1, y_1)$ and on it mark the point (x_2, y_2); in the figure this point is denoted by the number 2. Then on the straight line L_2 corresponding to the point (x_2, y_2) we mark the point (x_3, y_3), and continue in the same manner with $x_0 < x_1 < x_2 < x_3 < \cdots$. It is assumed, of course, that all the points (x_0, y_0), (x_1, y_1), (x_2, y_2), $\cdots$ are in the domain G. The broken line joining these points is called an Euler broken line. One may also construct an Euler broken line in the direction of decreasing x; the corresponding vertices on our figure are denoted by -1, -2, -3.

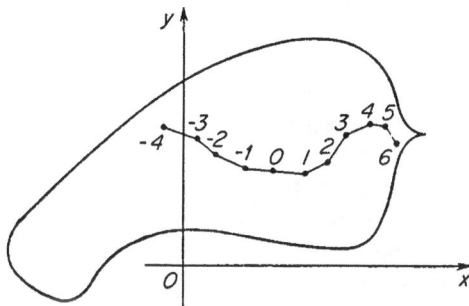

FIG. 9.

It is reasonable to expect that every Euler broken line through the point (x_0, y_0) with sufficiently short segments gives a representation of an integral curve l passing through the point (x_0, y_0), and that with decrease in the length of the links, i.e., when the length of the longest link tends to zero, the Euler broken line will approximate this integral curve.

Here, of course, it is assumed that the integral curve exists. In fact it is not hard to prove that if the function $f(x, y)$ is continuous in the domain G, one may find an infinite sequence of Euler broken lines, the length of the largest links tending to zero, which converges to an integral curve l. However, one usually cannot prove uniqueness: there may exist different sequences of Euler broken lines that converge to different integral curves passing through one and the same point (x_0, y_0). M. A. Lavrent'ev has constructed an example of a differential equation of the form (29) with a continuous function $f(x, y)$, such that in any neighborhood of any point P of the domain G there passes not one but at least two integral curves.

In order that through every point of the domain G there pass only one integral curve, it is necessary to impose on the function $f(x, y)$ certain conditions beyond that of continuity. It is sufficient, for example, to assume that the function $f(x, y)$ is continuous and has a bounded derivative with respect to y on the whole domain G. In this case it may be proved that through each point of G there passes one and only one integral curve and that every sequence of Euler broken lines passing through the point (x_0, y_0) converges uniformly to this unique integral curve, as the length of the longest link of the broken lines tends to zero. Thus for sufficiently small links the Euler broken line may be taken as an approximation to the integral curve of equation (34).

From the preceding it can be seen that the Euler broken lines are so constituted that small pieces of the integral curves are replaced by line segments tangent to these integral curves. In practice, many approximations to integral curves of the differential equation (34) consist not of straight-line segments tangent to the integral curves, but of parabolic segments that have a higher order of tangency with the integral curve. In this way it is possible to find an approximate solution with the same degree of accuracy in a smaller number of steps (with a smaller number of links in the approximating curve). The coefficients of the equation for the (higher order) parabola

$$y = a_0 + a_1(x - x_k) + a_2(x - x_k)^2 + \cdots + a_n(x - x_k)^n, \qquad (35)$$

which at the point (x_k, y_k) has nth-order tangency with the integral curves of equation (34) through this point, are given by the following formulas:

$$a_0 = y_k, \qquad (36)$$

$$a_1 = \left(\frac{dy}{dx}\right)_{x=x_k} = f(x_k, y_k), \qquad (36')$$

$$2a_2 = \left(\frac{d^2y}{dx^2}\right)_{x=x_k} = \left[\frac{df(x, y)}{dx}\right]_{x=x_k} = f_x'(x_k, y_k) + f_y'(x_k, y_k)\left(\frac{dy}{dx}\right)_{x=x_k}$$

$$= f_x'(x_k, y_k) + f_y'(x_k, y_k)f(x_k, y_k), \qquad (36'')$$

$$6a_3 = \left(\frac{d^3y}{dx^3}\right)_{x=x_k} = \left\{\frac{d}{dx}[f_x'(x, y(x)) + f_y'(x, y(x))f(x, y(x))]\right\}_{x=x_k}$$

$$= f_{xx}''(x_k, y_k) + 2f_{xy}''(x_k, y_k)f(x_k, y_k)$$

$$+ f_{yy}''(x_k, y_k)f^2(x_k, y_k) + f_y'^2(x_k, y_k)f(x_k, y_k)$$

$$+ f_y'(x_k, y_k)f_x'(x_k, y_k). \qquad (36''')$$

The polynomial (35) is needed only in order to compute its value for $x = x_{k+1}$. The actual values of the coefficients $a_0, a_1, a_2, \cdots, a_n$ themselves are not needed. There are many ways of computing the value for $x = x_{k+1}$ of the polynomial (35) whose coefficients are given by formula (36), without computing the coefficients $a_0, a_1, \cdots, a_n$ themselves.

Other approximation methods exist for finding the solution of the differential equation (34), which are based on other ideas. One convenient method was developed by A. N. Krylov (1863–1945).

The method of successive approximations. We now describe another method of successive approximation, which is as widely used as the method of the Euler broken lines. We assume again that we are required to find a solution $y(x)$ of the differential equation (34) satisfying the initial condition

$$y(x_0) = y_0.$$

For the initial approximation to the function $y(x)$, we take an arbitrary function $y_0(x)$. For simplicity we will assume that it also satisfies the initial condition, although this is not necessary. We substitute it into the right side $f(x, y)$ of the equation for the unknown function y and construct a first approximation y_1 to the solution y from the following requirements:

$$\frac{dy_1}{dx} = f[x, y_0(x)], y_1(x_0) = y_0.$$

Since there is a known function on the right side of the first of these equations the function $y_1(x)$ may be found by integration:

$$y_1(x) = y_0 + \int_{x_0}^{x} f[t, y_0(t)] \, dt.$$

It may be expected that $y_1(x)$ will differ from the solution $y(x)$ by less than $y_0(x)$ does, since in the construction of $y_1(x)$ we made use of the differential equation itself, which should probably introduce a correction into the original approximation. One would also think that if we improve the first approximation $y_1(x)$ in the same way, then the second approximation

$$y_2(x) = y_0 + \int_{x_0}^{x} f[t, y_1(t)] \, dt$$

will be still closer to the desired solution.

Let us assume that this process of improvement has been continued indefinitely and that we have constructed the sequence of approximations

$$y_0(x), y_1(x), \cdots, y_n(x), \cdots.$$

Will this sequence converge to the solution $y(x)$?

More detailed investigations show that if $f(x, y)$ is continuous and f_y' is bounded in the domain G, the functions $y_n(x)$ will in fact converge to the exact solution $y(x)$ at least for all x sufficiently close to x_0 and that if we break off the computation after a sufficient number of steps, we will be able to find the solution $y(x)$ to any desired degree of accuracy.

Exactly in the same way as for the integral curves of equation (34), we may also find approximations to integral curves of a system of two or more differential equations of the first order. Essentially the necessary condition here is to be able to solve these equations for the derivatives of the unknown functions. For example, suppose we are given the system

$$\frac{dy}{dx} = f_1(x, y, z), \frac{dz}{dx} = f_2(x, y, z). \tag{37}$$

Asuming that the right sides of these equations are continuous and have bounded derivatives with respect to y and z in some domain G in space, it may be shown under these conditions that through each point (x_0, y_0, z_0) of the domain G, in which the right sides of the equations in (37) are defined, there passes one and only one integral curve

$$y = \phi(x), \quad z = \psi(x)$$

of the system (37). The functions $f_1(x, y, z)$ and $f_2(x, y, z)$ give the direction numbers at the point (x, y, z), of the tangent to the integral curve passing through this point. To find the functions $\phi(x)$ and $\psi(x)$ approximately, we may apply the Euler broken line method or other methods similar to the ones applied to the equation (34).

The process of approximate computation of the solution of ordinary differential equations with initial conditions may be carried out on computing machines. There are electronic machines that work so rapidly that if, for example, the machine is programmed to compute the trajectory of a projectile, this trajectory can be found in a shorter space time than it takes for the projectile to hit its target (cf. Chapter XIV).

The connection between differential equations of various orders and a system of a large number of equations of first order. A system of ordinary differential equations, when solved for the derivative of highest order of each of the unknown functions, may in general be reduced, by the introduction of new unknown functions, to a system of equations of the first order, which is solved for all the derivatives. For example, consider the differential equation

$$\frac{d^2y}{dx^2} = f\left(x, y, \frac{dy}{dx}\right). \tag{38}$$

We set

$$\frac{dy}{dx} = z. \tag{39}$$

Then equation (38) may be written in the form

$$\frac{dz}{dx} = f(x, y, z). \tag{40}$$

Hence, to every solution of equation (38) there corresponds a solution of the system consisting of equations (39) and (40). It is easy to show that to every solution of the system of equations (39) and (40) there corresponds a solution of equation (38).

Equations not explicitly containing the independent variable. The problems of the pendulum, of the Helmholtz acoustic resonator, of a simple electric circuit, or of an electron-tube generator considered in §1 lead to differential equations in which the independent variable (time) does not explicitly appear. We mention equations of this type here, because the corresponding differential equations of the second order may be reduced in each case to a single differential equation of the first order rather than to a system of first-order equations as in the paragraph above for the general equation of the second order. This reduction greatly simplifies their study.

Let us then consider a differential equation of the second order, not containing the argument t in explicit form

$$F\left(x, \frac{dx}{dt}, \frac{d^2x}{dt^2}\right) = 0. \tag{41}$$

We set

$$\frac{dx}{dt} = y \tag{42}$$

and consider y as a function of x, so that

$$\frac{d^2x}{dt^2} = \frac{d}{dt}\left(\frac{dx}{dt}\right) = \frac{dy}{dt} = \frac{dy}{dx} \cdot \frac{dx}{dt} = y\frac{dy}{dx}.$$

Then equation (41) may be rewritten in the form

$$F\left(x, y, y\frac{dy}{dx}\right) = 0. \tag{43}$$

In this manner, to every solution of equation (41) there corresponds a unique solution of equation (43). Also to each of the solutions $y = \phi(x)$

of equation (43) there correspond infinitely many solutions of equation (41). These solutions may be found by integrating the equation

$$\frac{dx}{dt} = \phi(x), \qquad (44)$$

where x is considered as a function of t.

It is clear that if this equation is satisfied by a function $x = x(t)$, then it will also be satisfied by any function of the form $x(t + t_0)$, where t_0 is an arbitrary constant.

It may happen that not every integral curve of equation (43) is the graph of a single function of x. This will happen, for example, if the curve is closed. In this case the integral curve of equation (43) must be split up into a number of pieces, each of which is the graph of a function of x. For every one of these pieces, we have to find an integral of equation (44).

The values of x and dx/dt which at each instant characterize the state of the physical system corresponding to equation (41) are called the *phases* of the system, and the (x, y) plane is correspondingly called the *phase plane* for equation (41). To every solution $x = x(t)$ of this equation there corresponds the curve

$$x = x(t), \quad y = x'(t)$$

in the (x, y) plane; t here is considered as a parameter. Conversely, to every integral curve $y = \phi(x)$ of equation (43) in the (x, y) plane there corresponds an infinite set of solutions of the form $x = x(t + t_0)$ for equation (41); here t_0 is an arbitrary constant. Information about the behavior of the integral curves of equation (43) in the plane is easily transformed into information about the character of the possible solutions of equation (41). Every closed integral curve of equation (43) corresponds, for example, to a periodic solution of equation (41).

If we subject equation (6) to the transformation (42), we obtain

$$\frac{dy}{dx} = \frac{-ay - bx}{my}. \qquad (45)$$

Setting $v = x$ and $dv/dt = y$ in equation (16), in like manner we get

$$L\frac{dy}{dx} = \frac{-[R - M(a_1 + 2a_2x + 3a_3x^2)]\,y - x}{y}. \qquad (46)$$

Just as the state at every instant of the physical system corresponding to the second-order equation (41) is characterized by the two magnitudes*

---

* The values of d^2x/dt^2, d^3x/dt^3, $\cdots$ at the same instant of time are defined by the values of x and dx/dt from equation (41) and from the equations obtained from (45) by differentiation (cf. formula (36)).

(phases) x and $y = dx/dt$, the state of a physical system described by equations of higher order or by a system of differential equations is characterized by a larger number of magnitudes (phases). Instead of a phase plane, we then speak of a phase space.

§6. Singular Points

Let the point $P(x, y)$ be in the interior of the domain G in which we consider the differential equation

$$\frac{dy}{dx} = \frac{M(x, y)}{N(x, y)}. \tag{47}$$

If there exists a neighborhood R of the point P through each point of which passes one and only one integral curve (47), then the point P is called an *ordinary point* of equation (47). But if such a neighborhood does not exist, then the point P is called a *singular point* of this equation. The study of singular points is very important in the qualitative theory of differential equations, which we will consider in the next section.

Particularly important are the so-called *isolated singular points*, i.e., singular points in some neighborhood of each of which there are no other singular points. In applications one often encounters them in investigating equations of the form (47), where $M(x, y)$ and $N(x, y)$ are functions with continuous derivatives of high orders with respect to x and y. For such equations, all the interior points of the domain at which $M(x, y) \neq 0$ or $N(x, y) \neq 0$ are ordinary points. Let us now consider any interior point (x_0, y_0) where $M(x, y) = N(x, y) = 0$. To simplify the notation we will assume that $x_0 = 0$ and $y_0 = 0$. This can always be arranged by translating the original origin of coordinates to the point (x_0, y_0). Expanding $M(x, y)$ and $N(x, y)$ by Taylor's formula into powers of x and y and restricting ourselves to terms of the first order, we have, in a neighborhood of the point $(0, 0)$,

$$\frac{dy}{dx} = \frac{M'_x(0, 0) x + M'_y(0, 0) y + \phi_1(x, y)}{N'_x(0, 0) x + N'_y(0, 0) y + \phi_2(x, y)}, \tag{48}$$

where $\phi_1(x, y)$ and $\phi_2(x, y)$ are functions of x and y for which

$$\lim_{\substack{x \to 0 \\ y \to 0}} \frac{\phi_1(x, y)}{\sqrt{x^2 + y^2}} = 0 \quad \text{and} \quad \lim_{\substack{x \to 0 \\ y \to 0}} \frac{\phi_2(x, y)}{\sqrt{x^2 + y^2}} = 0.$$

Equations (45) and (46) are of this form. Equation (45) does not define either dy/dx or dx/dy for $x = 0$ and $y = 0$. If the determinant

$$\begin{vmatrix} M'_x(0,0) & M'_y(0,0) \\ N'_x(0,0) & N'_y(0,0) \end{vmatrix} \neq 0,$$

then, whatever value we assign to dy/dx at the origin, the origin will be a point of discontinuity for the values dy/dx and dx/dy, since they tend to different limits depending on the manner of approach to the origin. The origin is a singular point for our differential equation.

It has been shown that the character of the behavior of the integral curves near an isolated singular point (here the origin) is not influenced by the behavior of the terms $\phi_1(x, y)$ and $\phi_2(x, y)$ in the numerator and denominator, provided only that the real part of both roots of the equation

$$\begin{vmatrix} \lambda - M'_y(0,0) & -M'_x(0,0) \\ -N'_y(0,0) & \lambda - N'_x(0,0) \end{vmatrix} = 0 \qquad (49)$$

is different from zero. Thus, in order to form some idea of this behavior, we study the behavior near the origin of the integral curves of the equation

$$\frac{dy}{dx} = \frac{ax + by}{cx + dy} \qquad (50)$$

for which the determinant

$$\begin{vmatrix} a & b \\ c & d \end{vmatrix} \neq 0.$$

We note that the arrangement of the integral curves in the neighborhood of a singular point of a differential equation has great interest for many problems of mechanics, for example in the investigation of the trajectories of motions near the equilibrium position.

It has been shown that everywhere in the plane it is possible to choose coordinates ξ, η, connected with x, y by the equations

$$\begin{aligned} x &= k_{11}\xi + k_{12}\eta, \\ y &= k_{12}\xi + k_{22}\eta, \end{aligned} \qquad (51)$$

where the k_{ij} are real numbers such that equation (50) is tranformed into one of the following three types:

$$1)\ \frac{d\eta}{d\xi} = k\frac{\eta}{\xi}, \quad \text{where} \quad k = \frac{\lambda_2}{\lambda_1}. \tag{52}$$

$$2)\ \frac{d\eta}{d\xi} = \frac{\xi + \eta}{\xi}. \tag{53}$$

$$3)\ \frac{d\eta}{d\xi} = \frac{\beta\xi + \alpha\eta}{\alpha\xi - \beta\eta}. \tag{54}$$

Here λ_1 and λ_2 are the roots of the equation

$$\begin{vmatrix} c - \lambda & d \\ a & b - \lambda \end{vmatrix} = 0. \tag{55}$$

If these roots are real and different, then equation (50) is transformed into the form (52). If these roots are equal, then equation (50) is transformed either into the form (52) or into the form (53), depending on whether $a^2 + d^2 = 0$ or $a^2 + d^2 \neq 0$. If the roots of equation (55) are complex, $\lambda = \alpha \pm \beta i$, then equation (51) is transformed into the form (54).

We will consider each of the equations (52), (53), (54). To begin with, we note the following.

Even though the axes Ox and Oy were mutually perpendicular, the axes $O\xi$ and $O\eta$ need not, in general, be so. But to simplify the diagrams, we will assume they are perpendicular. Further, in the transformation (51) the scales on the $O\xi$ and $O\eta$ axes may be changed; they may not be the same as the ones originally chosen on the axes Ox and Oy. But again, for the sake of simplicity, we assume that the scales are not changed. Thus, for example, in place of the concentric circles, as in figure 8, there could in general occur a family of similar and similarly placed ellipses with common center at the origin.

All integral curves of equation (52) are given by a relation of the form

$$a\eta + b\,|\,\xi\,|^k = 0,$$

where a and b are arbitrary constants.

The integral curves of equation (52) are graphed in figure 10; here we we have assumed that $k > 1$. In this case all integral curves except one, the axis $O\eta$, are tangent at the origin to the axis $O\xi$. The case $0 < k < 1$ is the same as the case $k > 1$ with interchange of ξ and η, i.e., we have only to interchange the roles of the axes ξ and η. For $k = 1$, equation (52) becomes equation (30), whose integral curves were illustrated in figure 7.

An illustration of the integral curves of equation (52) for $k < 0$ is given in figure 11. In this case we have only two integral curves that pass through the point O: these are the axis $O\xi$ and the axis $O\eta$. All other integral

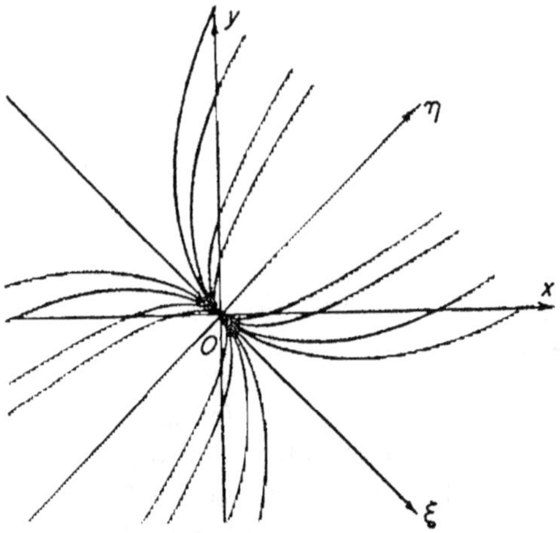

FIG. 10. 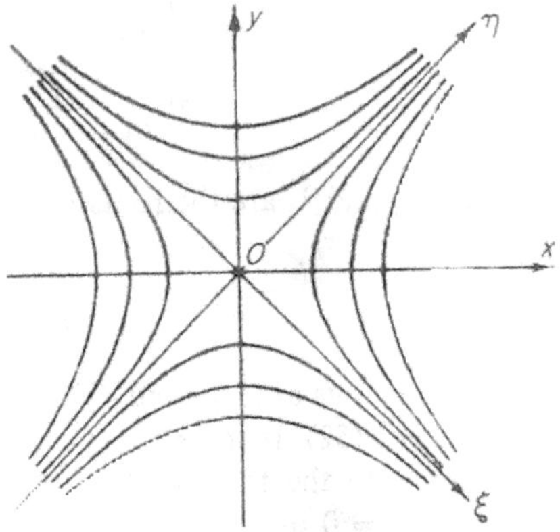   FIG. 11.

curves, after approaching the origin no closer than to some minimal distance, recede again from the origin. In this case we say that the point O is a *saddle point* because the integral curves are similar to the contours on a map representing the summit of a mountain pass (saddle).

All integral curves of equation (53) are given by the equation

$$b\eta = \xi(a + b \ln | \xi |),$$

where a and b are arbitrary constants. These are illustrated schematically in figure 12; all of them are tangent to the axis $O\eta$ at the origin.

If every integral curve entering some neighborhood of the singular point O passes through this point and has a definite direction there, i.e., has a definite tangent at the origin, as is illustrated in figures 10 and 12, then we say that the point O is a *node*.

Equation (54) is most easily integrated, if we change to polar coordinates ρ and ϕ, putting

$$\xi = \rho \cos \phi, \quad \eta = \rho \sin \phi.$$

Then this equation changes into the equation

$$\frac{d\rho}{d\phi} = k\rho, \quad \text{where} \quad k = \frac{\alpha}{\beta},$$

and hence,

$$\rho = Ce^{k\phi}. \qquad (56)$$

If $k > 0$ then all the integral curves approach the point O, winding infinitely often around this point as $\phi \to -\infty$ (figure 13). If $k < 0$,

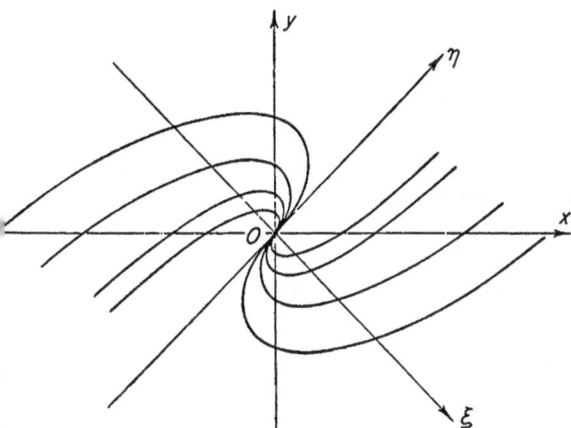

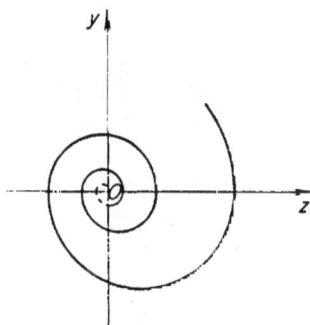

Fig. 12.   Fig. 13.

then this happens for $\phi \to +\infty$. In these cases, the point O is called a *focus*. If, however, $k = 0$, then the collection of integral curves of (56) consists of curves with center at the point O. Generally, if some neighborhood of the point O is completely filled by closed integral curves, surrounding the point O itself, then such a point is called a *center*.

A center may easily be transformed into a focus, if in the numerator and the denominator of the right side of equation (54) we add a term of arbitrarily high order; consequently, in this case the behavior of integral curves near a singular point is not given by terms of the first order.

Equation (55), corresponding to equation (45), is identical with the characteristic equation (19). Thus figures 10 and 12 schematically represent the behavior in the phase plane (x, y) of the curves

$$x = x(t), \quad y = x'(t),$$

corresponding to the solutions of equation (6) for real λ_1 and λ_2 of the same sign; Figure 11 corresponds to real λ_1 and λ_2 of opposite signs, and figures 13 and 8 (the case of a center) correspond to complex λ_1 and λ_2. If the real parts of λ_1 and λ_2 are negative, then the point $(x(t), y(t))$ approaches 0 for $t \to +\infty$; in this case the point $x = 0, y = 0$ corresponds to stable equilibrium. If, however, the real part of either of the numbers

λ_1 and λ_2 is positive, then at the point $x = 0$, $y = 0$, there is no stable equilibrium.

§7. Qualitative Theory of Ordinary Differential Equations

An important part of the general theory of ordinary differential equations is the qualitative theory of differential equations. It arose at the end of the last century from the requirements of mechanics and astronomy.

In many practical problems, it is necessary to establish the character of the solution of a differential equation describing some physical process and to describe the properties of its solutions as the independent variable ranges over a finite or infinite interval. For example, in celestial mechanics, which studies the motion of heavenly bodies, it is important to have information about the behavior of the solutions of differential equations describing the motion of the planets or other heavenly bodies for unbounded periods of time.

As we said earlier, for only a few particularly simple equations can a general solution be expressed in terms of integrals of known functions. So there arose the problem of investigating the properties of the solutions of a differential equation from the equation itself. Since the solution of a differential equation is given in the form of a curve in a plane or in space, the problem consisted of investigating the properties of integral curves, their distribution and their behavior in the neighborhood of singular points. For example, do they lie in a bounded part of the plane or do they have branches tending to infinity, are some of them closed curves, and so forth? The investigation of such questions constitutes the qualitative theory of differential equations.

The founders of the qualitative theory of differential equations are the Russian mathematician A M. Ljapunov and the French mathematician H. Poincaré.

In the preceding section, we considered in detail one of the important questions of the qualitative theory, namely the distribution of integral curves in a neighborhood of a singular point. We turn now to some other basic questions in qualitative theory.

Stability.   In the examples considered at the beginning of the chapter, the question of stability or instability of the equilibrium of a system was easily answered from physical considerations, without investigating the differential equations. Thus in example 3 it is obvious that if the pendulum, in its equilibrium position OA, is moved by some external force to a nearby position OA', i.e., if a small change is made in the initial conditions, then the subsequent motion of the pendulum cannot carry it very far from the equilibrium position, and this deviation will be smaller for smaller original deviations OA', i.e., in this case the equilibrium position will be stable.

For other more complicated cases, the question of stability of the equilibrium position is considerably more complicated and can be dealt with only by investigating the corresponding differential equations. The problem of the stability of equilibrium is closely connected with the question of the stability of motion. Fundamental results in this field were established by A. M. Ljapunov.

Let some physical process be described by the system of equations

$$\frac{dx}{dt} = f_1(x, y, t),$$

$$\frac{dy}{dt} = f_2(x, y, t). \tag{57}$$

For simplicity, we consider only a system of two differential equations, although our conclusions remain valid for a system with a larger number of equations. Each particular solution of the system (57), consisting of two functions $x(t)$ and $y(t)$, will sometimes be called a motion, following the usage of Ljapunov. We will assume that $f_1(x, y, t)$ and $f_2(x, y, t)$ have continuous partial derivatives. It has been shown that, in this case, the solution of the system of differential equations (57) is uniquely defined if at any instant of time $t = t_0$ the initial values $x(t_0) = x_0$ and $y(t_0) = y_0$ are given.

We will denote by $x(t, x_0, y_0)$ and $y(t, x_0, y_0)$ the solution of the system of equations (57) satisfying the initial conditions

$$x = x_0 \text{ and } y = y_0 \text{ for } t = t_0.$$

A solution $x(t, x_0, y_0), y(t, x_0, y_0)$ is called *stable in the sense of Ljapunov* if for all $t > t_0$ the functions $x(t, x_0, y_0)$ and $y(t, x_0, y_0)$ have arbitrarily small changes for sufficiently small changes in the initial values x_0 and y_0.

More exactly, for a solution to be stable in the sense of Ljapunov, the differences

$$|\, x(t, x_0 + \delta_1, y_0 + \delta_2) - x(t, x_0, y_0)\,|,$$

$$|\, y(t, x_0 + \delta_1, y_0 + \delta_2) - y(t, x_0, y_0)\,| \tag{58}$$

may be made less than any previously given number ϵ for all $t > t_0$, if the numbers δ_1 and δ_2 are taken sufficiently small in absolute value.

Every motion that is not stable in the sense of Ljapunov is called *unstable*.

In his investigation, the motion $x(t, x_0, y_0)$ and $y(t, x_0, y_0)$ was called by Ljapunov unperturbed, and the motion $x(t, x_0 + \delta_1, y_0 + d_2)$, $y(t, x_0 + \delta_1, y_0 + \delta_2)$ with nearby initial conditions was called perturbed. In this way stability in the sense of Ljapunov for an unperturbed motion means that for all $t > t_0$ the perturbed motion must differ only a little from the unperturbed.

The stability of equilibrium is a special case of stability of motion, corresponding to the case in which the unperturbed motion is

$$x(t, x_0, y_0) \equiv 0 \text{ and } y(t, x_0, y_0) \equiv 0.$$

Conversely, the question of the stability of any motion $x = \phi_1(t)$ and $y = \phi_2(t)$ of the system (57) may be reduced to the question of the stability of equilibrium for some system of differential equations. To this end we replace the unknown functions $x(t)$ and $y(t)$ in the system (57) by the new unknown functions

$$\xi = x - \phi_1(t) \text{ and } \eta = y - \phi_2(t). \tag{59}$$

In the system (57) transformed in this way, the motion $x = \phi_1(t)$ and $y = \phi_2(t)$ will correspond to the motion $\xi \equiv 0$ and $\eta \equiv 0$, i.e., the position of equilibrium. In what follows we will everywhere assume that the transformation (59) has been made, so that we may consider stability in the sense of Ljapunov only for the solution $x = 0$, $y = 0$.

The condition of stability in the sense of Ljapunov now means that, for δ_1 and δ_2 sufficiently small and $t > t_0$, the trajectory in the (x, y) plane of a perturbed motion does not pass outside of the square with sides of length 2 parallel to the coordinate axes and with center at the point $x = 0, y = 0$.

We will be interested in those cases in which, without knowing an integral of the system (57), we can nevertheless arrive at conclusions about the stability or instability of a motion. Stability is a very important practical question in the motion of projectiles, or of aircraft; and the stability of orbits is important in celestial mechanics, where the motion of planets and other heavenly bodies leads to this kind of investigation.

We assume that the functions $f_1(x, y, t)$ and $f_2(x, y, t)$ may be represented in the form

$$f_1(x, y, t) = a_{11}x + a_{12}y + R_1(x, y, t),$$

$$f_2(x, y, t) = a_{21}x + a_{22}y + R_2(x, y, t),$$

$$\tag{60}$$

where the a_{ij} are constants, and $R_1(x, y, t)$ and $R_2(x, y, t)$ are functions of x, y, and t such that

$$| R_1(x, y, t) | \leqslant M(x^2 + y^2) \text{ and } | R_2(x, y, t) | \leqslant M(x^2 + y^2), \quad (61)$$

where M is a positive constant.

If in the system (57) we substitute equations (60), neglecting $R_1(x, y, t)$ and $R_2(x, y, t)$, we get a system of differential equations with constant coefficients

$$\frac{dx}{dt} = a_{11}x + a_{12}y,$$

$$\frac{dy}{dt} = a_{21}x + a_{22}y, \quad (62)$$

which is called the *system of first approximation to the nonlinear system* (57).

Before the time of Ljapunov, researchers confined themselves to investigating stability of the first approximation, believing that the results obtained would carry over to the question of stability for the basic non-linear system (57). Ljapunov was the first to show that in the general case this conclusion is false. On the other hand, he gave a series of very wide conditions under which the question of stability for the nonlinear system is completely solved by the first approximation. One of these conditions is the following. If the real parts of both the roots of the equation

$$\begin{vmatrix} a_{11} - \lambda & a_{12} \\ a_{21} & a_{22} - \lambda \end{vmatrix} = 0$$

are negative and the functions $R_1(x, y, t)$ and $R_2(x, y, t)$ fulfill condition (61), then the solution $x(t) \equiv 0$, $y(t) \equiv 0$ is stable in the sense of Ljapunov. If the real part of either of the roots is positive, then the solution $x(t) \equiv 0$, $y(t) \equiv 0$ of an equation satisfying the conditions (61) is unstable. Ljapunov also gave a series of other sufficient conditions for stability and instability of a motion.*

If the right sides of equations (57) do not depend on t, then dividing the first equation of the system (57) by the second we get

$$\frac{dy}{dx} = \frac{f_1(x, y)}{f_2(x, y)}. \quad (63)$$

The origin will be a singular point for this equation. In the case of stability of equilibrium, this point may be a focus, a node, or a center, but cannot be a saddle point.

---

*A. M. Ljapunov, *The general problem of stability of motion.*

Thus the character of a singular point may be determined from the stability or instability of the equilibrium position.

The behavior of integral curves in the large. It is sometimes important to construct a schematized representation of the behavior of the integral curves "in the large"; that is, in the entire domain of the given system of differential equations, without attempting to preserve the scale. We will consider a space in which this system defines a field of directions as the phase space of some physical process. Then the general scheme of the integral curves, corresponding to the system of differential equations, will give us an idea of the character of all processes (motions) which can possibly occur in this system. In figures 10–13 we have constructed approximate schematized representations of the behavior of the integral curves in the neighborhood of an isolated singular point.

One of the most fundamental problems in the theory of differential equations is the problem of finding as simple a method as possible for constructing such a scheme for the behavior of the family of integral curves of a given system of differential equations in the entire domain of definition, in order to study the behavior of the integral curves of this system of differential equations "in the large." This problem remains almost untouched for spaces of dimension higher than 2. It is still very far from being solved for the single equation of the form

$$\frac{dy}{dx} = \frac{M(x, y)}{N(x, y)} \tag{64}$$

even when $M(x, y)$ and $N(x, y)$ are polynomials.

In what follows, we will assume that the functions $M(x, y)$ and $N(x, y)$ have continuous partial derivatives of the first order.

If all the points of a simply connected domain G, in which the right side of the differential equation (64) is defined, are ordinary points, then the family of integral curves may be represented schematically as a family of segments of parallel straight lines; since in this case one integral curve will pass through each point, and no two integral curves can intersect. For an equation (64) of more general form, which may have singular points, the structure of the integral curves may be much more complicated. The case in which equation (64) has an infinite set of singular points (i.e., points where the numerator and the denominator both vanish) may be excluded, at least when $M(x, y)$ and $N(x, y)$ are polynomials. Thus we restrict our consideration to those cases in which equation (64) has a finite number of isolated singular points. The behavior of the integral curves that are near to one of these singular points forms the essential element in setting up a schematized representation of the behavior of all the integral curves of the equation.

A very typical element in such a scheme for the behavior of all the integral curves of equation (64) is formed by the so-called *limit cycles*. Let us consider the equation

$$\frac{d\rho}{d\phi} = \rho - 1, \tag{65}$$

where ρ and ϕ are polar coordinates in the (x, y) plane.

The collection of all integral curves of equation (65) is given by the formula

$$\rho = 1 + Ce^\phi, \tag{66}$$

where C is an arbitrary constant, different for different integral curves. In order that ρ be nonnegative, it is necessary that ϕ have values no larger than $-\ln |C|$, $C < 0$. The family of integral curves will consist of

1. the circle $\rho = 1$ ($C = 0$);

2. the spirals issuing from the origin, which approach this circle from the inside as $\phi \to -\infty$ ($C < 0$);

3. the spirals, which approach the circle $\rho = 1$ from the outside as $\phi \to -\infty$ ($C > 0$) (figure 14).

The circle $\rho = 1$ is called a limit cycle for equation (65). In general a closed integral curve l is called a *limit cycle*, if it can be enclosed in a disc all points of which are ordinary for equation (64) and which is entirely filled by nonclosed integral curves.

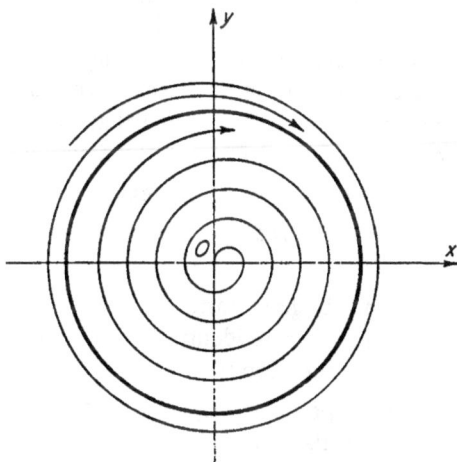

FIG. 14.

From equation (65) it can be seen that all points of the circle are ordinary. This means that a small piece of a limit cycle is not different from a small piece of any other integral curve.

Every closed integral curve in the (x, y) plane gives a periodic solution $[x(t), y(t)]$ of the system

$$\frac{dx}{dt} = N(x, y), \frac{dy}{dt} = M(x, y), \tag{67}$$

describing the law of change of some physical system. Those integral curves in the phase plane that as $t \to +\infty$ approximate a limit cycle are motions that as $t \to +\infty$ approximate periodic motions.

Let us suppose that for every point (x_0, y_0) sufficiently close to a limit cycle l, we have the following situation: If (x_0, y_0) is taken as initial point (i.e., for $t = t_0$) for the solution of the system (67), then the corresponding integral curve traced out by the point $[x(t), y(t)]$, as $t \to +\infty$ approximates the limit cycle l in the (x, y) plane. (This means that the motion in question is approximately periodic.) In this case the corresponding limit cycle is called *stable*. Oscillations that act in this way with respect to a limit cycle correspond physically to self-oscillations. In some self-oscillatory systems, there may exist several stable oscillatory processes with different amplitudes, one or another of which will be established by the initial conditions. In the phase plane for such "self-oscillatory systems," there will exist corresponding limit cycles if the processes occuring in these systems are described by an equation of the form (67).

The problem of finding, even if only approximately, the limit cycles of a given differential equation has not yet been satisfactorily solved. The most widely used method for solving this problem is the one suggested by Poincaré of constructing "cycles without contact." It is based on the following theorem. We assume that on the (x, y) plane we can find two closed curves L_1 and L_2 (cycles) which have the following properties:

1. The curve L_2 lies in the region enclosed by L_1.

2. In the annulus Ω, between L_1 and L_2, there are no singular points of equation (64).

3. L_1 and L_2 have tangents everywhere, and the directions of these tangents are nowhere identical with the direction of the field of directions for the given equation (64).

4. For all points of L_1 and L_2 the cosine of the angle between the interior normals to the boundary of the domain Ω and the vector with components $[N(x, y), M(x, y)]$ never changes sign.

Then between L_1 and L_2, there is at least one limit cycle of equation (64).

Poincaré called the curves L_1 and L_2 *cycles without contact*.

The proof of this theorem is based on the following rather obvious fact. We assume that for decreasing t (or for increasing t) all the integral curves

$$x = x(t), \quad y = y(t)$$

of equation (64) (or, what amounts to the same thing, of equations (67), where t is a parameter), which intersect L_1 or L_2, enter the annulus Ω

between L_1 and L_2. Then they must necessarily tend to some closed curve l lying between L_1 and L_2, since none of the integral curves lying in the annulus can leave it, and there are no singular points there.

But the problem of finding cycles without contact is also a complicated one and no general methods are known for solving it. For particular examples it has been possible to find cycles without contact, thereby proving the existence of limit cycles.

In radio technology it is important to find limit cycles (self-oscillatory processes) for equation (16) for the electron-tube generator. For equations of the type of (16), N. M. Krylov and N. N. Bogoljubov gave a method, about twenty years ago, for approximate computation of a certain limit cycle that exists for this equation. At about the same time the Soviet physicists L. I. Mandel'stam, N. D. Papaleksi, and A. A. Andronov gave a proof of the possibility of applying what is called the method of the small parameter, a method that to some extent had been used earlier in practice, though without any rigorous justification. Andronov was also the first to make systematic practical use, in the analysis of self-oscillatory systems, of the theoretical methods already developed by Ljapunov and Poincaré. In this manner he obtained a whole series of important results.

As was mentioned earlier, an important role is played in physics by "insensitive" systems (cf. §3). Andronov, together with L. S. Pontrjagin, set up a catalogue of the elements from which one could construct a complete chart of the behavior of the integral curves in the (x, y) plane for an insensitive differential equation of the form (64). It had been long known, for example, that a center near a singular point is easily destroyed by small changes in the equations (64). Thus in the construction of a chart of the behavior of the integral curves of equation (64), we cannot have a center, i.e., a family of closed integral curves surrounding a singular point, if the equation is "insensitive."

The question of the behavior of the integral curves in the large is still far from its final solution. We note that the analogous and probably simpler question of the form of real algebraic curves in the plane, i.e., curves defined by the equation

$$P(x, y) = 0,$$

where $P(x, y)$ is a polynomial of degree n, is also far from a complete solution. The form of these curves is completely known only for $n < 6$.

The solutions of the system (64) define motions in the plane. If we replace each point (x_0, y_0) in the plane by the corresponding point $[x(t, x_0, y_0), y(t, x_0, y_0)]$, where $x(t, x_0, y_0)$ and $y(t, x_0, y_0)$ are the solution of the system (64) with initial conditions $x = x_0$ and $y = y_0$ for $t = t_0$, we obtain a transformation of the points of the plane depending on the parameter t. Similar transformations depending on a parameter, together with the motions they generate, may be considered on a sphere, a torus, or other manifolds. The properties of these motions are studied in the theory of dynamical systems. In a neighborhood of every point these motions are the solutions of some system of differential equations. In the past decade the theory of dynamical systems has been developed on a broad basis in the works of V. V. Stepanov, A. Ja. Hinčin, N. N. Bogoljubov, N. M. Krylov, A. A. Markov, V. V. Nemyckiĭ and others, and also in the works of G. D. Birkhoff and other mathematicians.

In this chapter we have given a brief outline of the present state of the theory of ordinary differential equations and have attempted to describe the problems that are considered in this theory. Our study in no sense pretends to be complete. We have had to omit consideration of many branches of the theory that arise in the study of more special problems or that require broader mathematical knowledge than the reader of this book is assumed to possess. For example, we have nowhere touched upon the general and important area in which the theory of differential equations with complex arguments is considered. We have had no opportunity to examine the theory of boundary-value problems and in particular, of eigenfunctions, which is of great importance in the applications.

We have also been able to pay very little attention to approximative methods for the numerical or analytical solution of differential equations. For these questions, we recommend that the reader consult the specialized literature.

Suggested Reading

R. P. Agnew, *Differential equations*, 2nd ed., McGraw-Hill, New York, 1960.

E. A. Coddington, *An introduction to ordinary differential equations*, Prentice-Hall, Englewood Cliffs, N. J., 1961.

E. A. Coddington and N. Levinson, *Theory of ordinary differential equations*, McGraw-Hill, New York, 1955.

W. Hurewicz, *Lectures on ordinary differential equations*, Technology Press and Wiley, New York, 1958.

S. Lefschetz, *Differential equations: geometric theory*, Interscience, New York, 1957.

INDEX

This revised, enlarged index was prepared
through the generous efforts of Stanley Gerr.

357

CONTENTS OF THE SERIES

VOLUME TWO

PART 3

CONTENTS

CONTENTS OF THE SERIES

VOLUME THREE

PART 5

PART 6

CONTENTS

www.ingramcontent.com/pod-product-compliance
Lightning Source LLC
Chambersburg PA
CBHW060753220326
41598CB00022B/2423

* 9 780262 510059 *